Energy Savings Calculations for Commercial Building Energy Efficiency Upgrades

Code-compliant building materials and equipment will typically have a lower initial cost; however, the lifetime energy savings of the high efficiency equipment will often justify the upfront cost premium and result in a more cost-effective solution. *Energy Savings Calculations for Commercial Building Energy Efficiency Upgrades* assists energy professionals, contractors, building owners, and managers in developing energy savings estimates that can facilitate a quick assessment of the potential energy savings that might be realized when replacing existing building components with the highest efficiency equipment. It also provides algorithms to estimate greenhouse gas emission reductions that may be achieved by building energy efficiency upgrades and the impact these upgrades can have on building electrification-decarbonization projects. This book:

- Focuses on the development of energy savings estimates based upon a whole building's energy consumption and the energy consumption associated with building end-uses such as space heating, space cooling, ventilation, lighting, and so forth.
- Includes over 70 illustrative examples using algorithms to demonstrate how energy savings and greenhouse gas emission reductions may be estimated utilizing different strategies and equipment.

Anthony J. Buonicore is Chairman and Director of Engineering at Sustainable Real Estate Solutions, Inc. (SRS). He previously served as chief executive officer of Buonicore Partners, Inc. (BP), an energy risk management consulting firm and publisher of the daily energy news service, *Building Energy Performance Assessment News*, prior to the firm being acquired by SRS. He has been active in the energy and environmental industry for more than 50 years. His broad career experience has involved policy planning and technical assignments in government and private industry. He has been a leader in the energy-environmental industry since the early 1970s, serving as general chairman of the American Institute of Chemical Engineers' First National Conference on Energy and the Environment in 1973 and as founder and first chairman of the Air & Waste Management Association's Energy-Environmental Interactions Technical Committee in 1974. He has also been a pioneer in the technical development of commercial PACE programs in the U.S., including technical review of thousands of PACE projects. He has more than 100 publications and holds two patents. He is a senior member of the Association of Energy Engineers, a member of ASHRAE, past president and Fellow Member of the Air & Waste Management Association, a diplomate in the American Academy of Environmental Engineers, and a licensed professional engineer. He is chairman of ASTM's Building Energy Performance Assessment (BEPA) Task Group, which developed the E2797 and E3224 BEPA standards, and teaches ASTM's Building Energy Performance training course. He holds both bachelor's and master's degrees in chemical engineering from Manhattan College.

Energy Savings Calculations for Commercial Building Energy Efficiency Upgrades

Anthony J. Buonicore

CRC Press
Taylor & Francis Group
Boca Raton London New York

CRC Press is an imprint of the
Taylor & Francis Group, an **informa** business

Designed cover image: Shutterstock.

First edition published 2024
by CRC Press
2385 NW Executive Center Drive, Suite 320, Boca Raton FL 33431

and by CRC Press
4 Park Square, Milton Park, Abingdon, Oxon, OX14 4RN

CRC Press is an imprint of Taylor & Francis Group, LLC

ISBN: 978-1-032-69273-9 (hbk)
ISBN: 978-1-032-69279-1 (pbk)
ISBN: 978-1-032-69277-7 (ebk)

DOI: 10.1201/9781032692777

Typeset in Times
by Apex CoVantage, LLC

This book is dedicated to my college professor, "Big Lou," who pushed me to be a better engineer, presented me with countless opportunities to learn and grow, and supported me as a trusted advisor throughout my career in engineering and business.

Contents

Figures

Tables

Preface

This book was prepared with the energy professional and contractor in mind who is replacing existing energy-consuming equipment such as heating, cooling, ventilation, and lighting systems with new high efficiency equipment in commercial and multifamily buildings. In the U.S., there are approximately 5.9 million non-residential commercial buildings[1] and 1.8 million multifamily buildings (defined as having five or more units).[2] More than 95% of these buildings are less than 50,000 square feet in size and are categorized as small to medium-sized buildings (SMBs). Assuming the estimated useful life of most energy-consuming equipment in these buildings ranges between 15 to 25 years, this would mean at least 400,000 buildings each year with equipment at the end of its useful life needing to be replaced. And this does not include buildings with equipment having a remaining useful life but voluntarily being replaced by building owners. This represents a significant market particularly for energy professionals and contractors.

While institutional clients with buildings typically greater than 50,000–100,000 square feet (which is less than 5% of the U.S. building population) may be able to afford upfront, relatively expensive building simulation modeling to evaluate energy savings, most building owners cannot. The (non-institutional) building owners who need to replace equipment that is near, has reached, or exceeded its useful life, usually solicit bids from local contractors to obtain the best pricing. For competitive reasons, contractor bids typically propose the lowest price equipment complying with code. Moreover, due to a lack of expertise in calculating energy cost savings, rarely are energy cost implications even addressed. As such, making an investment case for higher efficiency equipment which usually comes at a cost premium is rarely attempted.

The principal goal of this book is to assist energy professionals, contractors, building owners and managers in cost-effectively developing energy consumption and cost savings estimates. These estimates, for example, can facilitate a quick assessment of potential savings that might be realized when replacing code-compliant equipment with higher efficiency, above code equipment. Code-compliant equipment will typically have the lower first cost, but the lifetime energy cost savings of the higher efficiency equipment can often justify the upfront cost premium and result in a more cost-effective solution. However, to justify the added expense of the higher efficiency equipment necessitates being able to provide the building owner with comparative energy savings. This book is designed to assist in this important effort. It will also assist energy professionals seeking preliminary energy savings estimates when first evaluating a project in the development phase.

Over the years, in responding to thousands of clients seeking to know the energy cost savings that might be associated with a planned energy efficiency equipment upgrade project to their existing building, I have often faced the dilemma of either using expensive, time-consuming, and complicated building simulation modeling (or complicated, data intensive spreadsheets) or finding alternative more cost-effective and easier to use approaches that can quickly provide reasonable results. Moreover, it can be foolhardy to waste significant time and resources to nail things down to the last decimal place when you may not even be awarded the project, or the answer may result in something that is commercially unattractive to the building owner and cause the project to be canceled.

The problem is that while building simulation modeling may be the preferred approach, and certainly it is for buildings that are being newly constructed, the expense of using it to support the vast majority of existing building equipment replacements (particularly in the SMB market) often is not economically justifiable. Moreover, building owners usually request proposals from multiple energy professionals and contractors and are unwilling to pay at this "proposal stage" for the expensive cost of modeling. Even if building simulation modeling is approved by a building owner, it still requires a significant amount of data input and is often accompanied by much guesswork, such as predicting future weather at the building location. As such, this book does not focus on building energy simulation modeling or data-intensive spreadsheets, but rather on algorithms that can quickly provide reasonable energy savings estimates.

As part of the technical review for thousands of Property Assessed Clean Energy (PACE) projects across the country, I have used the algorithms in this book to provide quickly and cost-effectively an estimate of the energy savings associated with high efficiency equipment upgrades in existing commercial and multifamily buildings. In using these algorithms, I have found that they typically are able to estimate energy savings within ±20%. While this may not be as good as building simulation modeling, I have found it to be acceptable to most building owner clients.

The book is divided into 17 chapters and begins with a discussion of the energy savings conundrum; that is, that energy savings can mean different things to different people. Chapter 2 focuses on development of a whole building's energy consumption and the energy consumption associated with building end-uses such as space heating, space cooling, ventilation, and lighting. Chapter 3 discusses how to establish the baseline energy consumption against which the energy savings will be estimated. Chapters 4 to 10 focus on the most common energy-consuming equipment replacements or upgrades that I have encountered, and include a description of the equipment, the energy efficiency metrics, and the energy savings algorithms. I have included Chapters 11 and 12 on cogeneration and solar photovoltaic systems because I have found these more and more often to be considered and included with the building energy efficiency upgrade project. Since many water-saving building fixtures, such as today's faucets and showers, can also reduce energy particularly in multifamily buildings, Chapter 13 addresses these savings.

While the book focuses on the most common equipment replacements and upgrades, there are many additional energy efficiency upgrades that can exist. Some of the more common I have encountered within the SMB market are presented in Chapter 14. Since new equipment upgrades have the potential to interact with each other, Chapter 15 considers the interactivity implications and their impact on energy savings. The chapter also focuses on the sequencing of equipment installation to avoid double counting, on methodologies to deal with mixed-use buildings, and on the most common mistakes that I have encountered in project technical review.

While estimating energy consumption savings is important, it is even more important to be able to translate these savings into the language that commercial real estate owners and managers can understand, and that is their impact on cash flow. Chapter 16 focuses on financial metrics, including from personal experience typical installed costs for energy efficiency upgrades.

Finally, as climate change and greenhouse gas emissions become the political mantra, Chapter 17 provides algorithms to estimate greenhouse gas emission reductions that may be achieved by building energy efficiency upgrades and the impact these upgrades can have on building electrification-decarbonization projects.

Throughout the book I have included dozens of illustrative examples on use of the algorithms presented. It is hoped that these examples will demonstrate the value of the algorithms and how they can best be used in energy savings evaluations for clients.

If your preference is to use a software tool to do many of the calculations presented in this book, I refer you to Sustainable Real Estate Solutions' Energy Performance Improvement Calculator (EPIC). Additional information on this software tool may be found at www.srsworx.com.

If you are interested in taking a two-day in-person or four-day virtual course on estimating energy savings as presented in this book, contact me directly at my email address.

Anthony J. Buonicore
P.E., BCEE, QEP
ajb@paceworx.com

NOTES

1. Commercial Building Energy Consumption Survey (CBECS), 2018.
2. Residential Energy Consumption Survey (RECS), 2020.

About the Author

Anthony J. Buonicore, P.E., BCEE, QEP, is Chairman and Director of Engineering at Sustainable Real Estate Solutions, Inc. (SRS). He previously served as chief executive officer of Buonicore Partners, Inc. (BP), an energy risk management consulting firm and publisher of the daily energy news service, *Building Energy Performance Assessment News*, prior to the firm being acquired by SRS. He has been active in the energy and environmental industry for more than 50 years. His broad career experience has involved policy planning and technical assignments in government and private industry. He has been a leader in the energy-environmental industry since the early 1970s, serving as general chairman of the American Institute of Chemical Engineers' First National Conference on Energy and the Environment in 1973 and as founder and first chairman of the Air & Waste Management Association's Energy-Environmental Interactions Technical Committee in 1974. He has also been a pioneer in the technical development of commercial PACE programs in the U.S., including technical review of thousands of PACE projects. He has more than 100 publications and holds two patents.

Mr. Buonicore is a Senior Member of the Association of Energy Engineers, past president and Fellow Member of the Air & Waste Management Association, a diplomate in the American Academy of Environmental Engineers, a Fellow Member of ASTM, a member of ASHRAE, and a licensed professional engineer. He is Chairman of ASTM's Building Energy Performance Assessment (BEPA) Task Group, which developed the E2797 and E3224 national BEPA standards. He was the 2021 recipient of ASTM'S prestigious Walter C. Voss award presented to an engineer or scientist who has made outstanding contributions to the field of building technology. He holds both bachelor's and master's degrees in chemical engineering from Manhattan College.

Mr. Buonicore has been conducting two-day courses throughout the country for energy professionals and contractors on estimating energy savings in commercial buildings.

1 The Energy Savings Conundrum

1.1 DEFINITION OF ENERGY SAVINGS

Today, owners or managers evaluating a purchasing decision for new energy efficient equipment in their buildings often request the "energy savings" to support their investment decision. While this is often asked, it is just as often misunderstood because energy savings can mean different things to different people. For example, a building owner may only be interested in the estimated utility cost savings and the impact a project may have on cash flow. The energy consumption savings associated with the project would not mean that much. On the other hand, a utility energy efficiency program manager evaluating potential incentives or rebates may only be interested in the equipment's energy consumption savings.

To answer the question "what are the energy savings" properly, more information is needed.

1. Does energy savings refer to energy *cost savings* or energy *consumption savings* or both?
2. Does energy savings include both *consumption savings* and *demand savings*?
3. Does energy savings refer to the savings in any of the following?
    - *Whole building energy consumption.*
    - *Primary energy source consumption* (e.g., whole building *electricity consumption*, and/or whole building *fuel consumption*).
    - *Impacted building end-use energy consumption* (e.g., savings in cooling energy consumption, savings in space heating energy consumption, savings in ventilation energy consumption, savings in lighting energy consumption).
    - *New high efficiency equipment energy consumption* compared to that of the *existing equipment.*
    - Combinations of the above.
4. Are the energy savings estimated using *actual weather conditions* (in a particular year) or weather conditions in an average or *typical meteorological year*?
5. Are the energy savings *first year savings*, or are they *lifetime savings* over the equipment's useful lifetime?

Providing an answer to "what are the energy savings" without qualifying the basis for your answer is meaningless.

DOI: 10.1201/9781032692777-1

1.2 GENERAL APPROACH TO ESTIMATING ENERGY SAVINGS

In my experience with conducting technical due diligence for thousands of energy efficiency projects seeking third-party financing approval, the best practice to estimate energy savings associated with replacement of existing inefficient equipment with new high efficiency equipment includes the following key steps:

1. Determine annual whole building energy consumption (refer to Chapter 2, Section 2.2). This may be done, for example, by collecting at least a year's worth of utility bill data; by referring to published energy use data for the building type, size, and location; or by building simulation modeling. Since predictions of energy savings are based upon average or typical meteorological data (to remove the influence of unusual weather conditions), energy data is weather-normalized (refer to Chapter 2, Section 2.2.3).
2. Estimate the energy consumption of each major building end-use, such as space heating, cooling, domestic hot water (DHW) heating, lighting, and ventilation (refer to Chapter 2, Section 2.3). Building end-use data should be specific to the building type, its location, and be weather-normalized.
3. Identify the energy efficiency metric(s) of the existing equipment being replaced (refer to Chapter 3). This becomes the baseline against which the energy savings are evaluated. If existing equipment data is not readily available from equipment specifications in the owner's manual or nameplate data, or internet research or an inquiry to the manufacturer, it may be necessary to assume that when it was installed it complied with the local building energy code in effect at the time (refer to Chapter 3, Section 3.1.1). If knowledge of the local building energy code applicable at the time is not available, the ASHRAE 90.1 Energy Standard for Buildings[(1)] or the International Energy Conservation Code (IECC)[(2)] that existed at the time the equipment was installed may be helpful in estimating energy efficiency metrics for the equipment (refer to Chapter 3, Sections 3.1.2 and 3.1.3). It may also be necessary to consider potential performance degradation that may have taken place from the time the equipment was installed to the present (refer to Chapter 3, Section 3.3).
4. Identify the energy efficiency metric(s) of the new high efficiency equipment that will be replacing the existing equipment. These metrics should be readily available from the manufacturer's literature.
5. Knowledge of items (1) to (4) will enable an estimation of the energy consumption savings associated with replacing existing equipment with new high efficiency equipment (refer to Chapters 4 to 14).
6. Knowledge of item (5) and the current energy cost (which may be obtained from a building's utility bills), will enable an estimate of the first year's energy cost savings.
7. Knowledge of the estimated useful life of the new equipment (refer to Appendix IX), how the estimated first year energy cost savings might escalate on average over the equipment's estimated useful life, and whether the new equipment may experience a slight performance degradation each year will enable a life cycle cost savings to be calculated (refer to Chapter 16).

8. Knowing the total installed cost of the new equipment, how it will be financed, and knowledge of item (7) will enable a cash flow statement to be generated (refer to Chapter 16). The project's estimated impact on greenhouse gas emissions can also be made (refer to Chapter 17).

The language of building owners and property managers is cash flow. If a contractor desires to install new more efficient equipment, providing a proposal limited to the equipment's installed cost, or even including a statement that there will be energy savings, will likely be insufficient for the owner or manager to make an informed purchase decision. Including such key financial analytics showing the upfront investment, annual energy cost savings, and cash flow impacts will go a long way to accelerate the owner's decision-making process. Moreover, such cash flow analysis may help transform a capital expense into a compelling investment opportunity.

1.3 UTILITY ENERGY SAVINGS

Utilities providing incentives for the installation of new high efficiency equipment to replace existing inefficient equipment often view the energy savings calculations a little differently. The key determinant for utility managers is when the replacement equipment is being installed (i.e., whether at the end of its useful life or before the end of its useful life). If a piece of equipment is being replaced at the end of its useful life, utilities know that any new replacement equipment must comply with the current building energy code (i.e., be "code-compliant") which presumably would require a higher energy efficiency than that required when the equipment was first installed. As such, the energy savings credit towards incentives is not applied to the new equipment unless its energy efficiency is "above code."

In view of this distinction, utilities look at energy savings in two ways:

1. "Early retirement savings."
2. "Lost opportunity savings."

Early retirement savings refers to equipment installed prior to the end of its useful life. At this stage, the energy savings are calculated based upon the equipment's existing energy consumption compared to the energy consumption of the new high efficiency replacement equipment. However, these savings are only realized up to the estimated useful life (EUL) of the equipment. Once the estimated useful life of the equipment is reached, there would only be a lost opportunity savings, defined as the difference in energy consumption of the high efficiency replacement equipment compared to the energy consumption of equipment complying with the local energy code. If only code-compliant equipment is installed, there would be no lost opportunity savings. However, if the replacement equipment's energy efficiency is above code, then the credited savings would be the difference between the energy performance at code compared to that above code. This is demonstrated in Figure 1.1, where the baseline is code-compliant equipment.

Utilities also use the term "deemed (stipulated) savings" when calculating the energy savings associated with a project. Deemed savings define savings values for

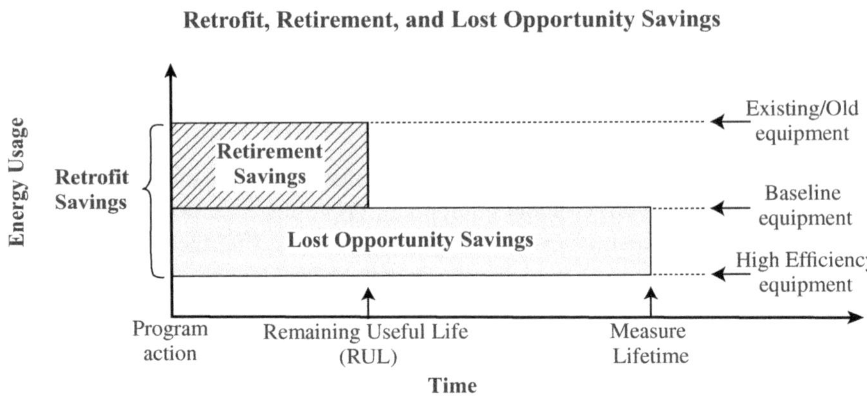

FIGURE 1.1 Utility energy savings definitions.

projects with well-known and documented savings values in defined applications. The use of deemed values in a savings calculation for the utility is an agreement with the utility to accept a predetermined savings value derived from an acceptable calculation approach toward a utility incentive. Many of these energy savings algorithms are employed throughout this book. Deemed values are documented in the technical reference (or resource) manual (sometimes referred to as a deemed savings or program savings manual) of the local utility. They are available in many of the states, including those listed below and may be accessed online.

- Arkansas
- California
- Colorado
- Connecticut
- Delaware
- Hawaii
- Illinois
- Indiana
- Maine
- Massachusetts
- Mid-Atlantic
- Minnesota
- Missouri
- New Jersey
- New York
- Ohio
- Pennsylvania
- Rhode Island
- Tennessee Valley Authority
- Texas
- Wisconsin
- Vermont.

Throughout this book, when energy savings are estimated, they are based on the difference between the energy performance of existing equipment and the energy performance of the new equipment replacing the existing equipment, regardless of when the existing equipment is being replaced. This approach enables a building owner to better understand, compare, and put into perspective energy cost savings since they already have a good sense of what they have been paying for utilities. As such, lost opportunity savings are not considered.

1.4 GRAPHICAL REPRESENTATION OF ENERGY SAVINGS

Figure 1.2 is a graphical representation of energy savings. You can see the baseline period (the area on the left in the figure) where building energy consumption is shown prior to the installation of any energy saving measures. This baseline energy consumption can then be used to project future building energy consumption under typical weather conditions without the energy saving measures installed (the top line on the right). When an energy efficiency project is developed, the lower line on the right would represent the projected energy consumption after the energy efficiency measures (EEMs) are installed, again assuming typical weather conditions. The difference between the two lines would be the projected energy savings.

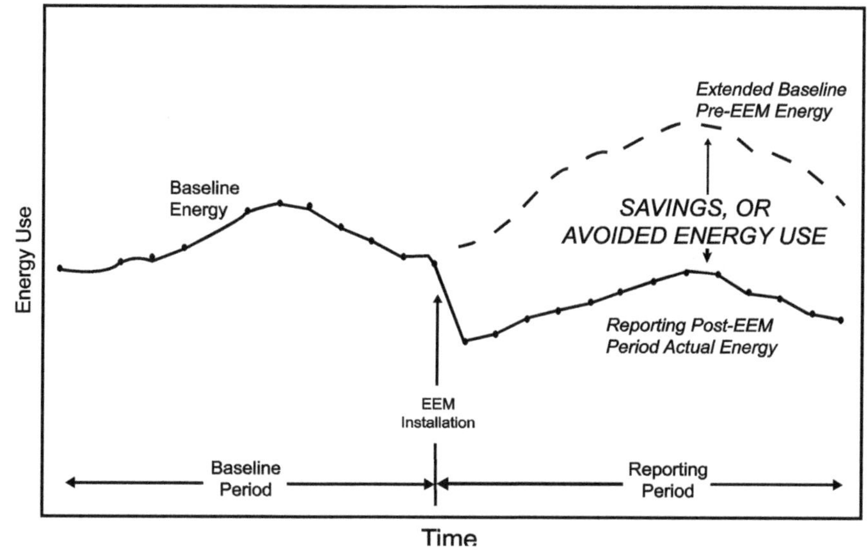

FIGURE 1.2 Graphical representation of energy savings.

Example Problem 1–1

An existing air conditioning system that is 10 years old has an estimated 5 years of remaining useful life (RUL). It is going to be replaced with a new air conditioning system with an above-code energy efficiency, having an estimated useful life (EUL) of 15 years. How would the energy savings be calculated to determine the utility incentive?

How would the energy savings be calculated if the new air conditioning system was just code-compliant?

Suppose the existing air conditioning system was replaced at the end of its useful life for both of the previous cases. How would the energy savings be calculated to receive the utility incentive?

Case 1: Above-code new equipment replacement scenario, with RUL.

Energy consumption savings = [(existing energy consumption for 5 years of remaining useful life) – (energy consumption of the new above-code air conditioning system for 5 years)] + [(energy consumption of the new above-code air conditioning system for 10 years) – energy consumption of a code-compliant air conditioning system for 10 years)]

Case 2: Code-compliant new equipment replacement scenario, with RUL.

Energy consumption savings = [(existing energy consumption for 5 years of remaining useful life) – (energy consumption of the new code-compliant air conditioning system for 5 years)]

Case 3: Above-code new equipment replacement scenario, with no RUL.

Energy consumption savings = [(energy consumption of the new above-code air conditioning system for 15 years) – energy consumption of a code-compliant air conditioning system for 15 years)]

Case 4: Code-compliant new equipment replacement scenario, with no RUL.

There would be no energy consumption savings for the utility to award an incentive.

Example Problem 1–2

You replaced an air conditioning unit at an office building with a high efficiency unit. After reviewing the first year electricity bills, the building owner came back to you and said the energy cost savings you predicted were not being achieved. How might you respond?

The first thing you should do is check the utility bills to confirm what the building owner said. At the same time, you should check the electric rates to see if they may have increased or decreased since you last reviewed them. You should then confirm that there have been no major building renovations over the past year. Next you should review building operation to confirm that the current conditions have not changed. For example, have the operating hours changed or the temperature set-points of the HVAC system been modified, or has the occupancy of the building changed? Finally, you should review the weather over the past year to see how it compares to the typical meteorological year weather (long-term average weather that is typical for the area). For example, if the summer was warmer than expected, air conditioning would be greater than expected. These are the major areas that should be investigated. It is highly likely that the reason for the prediction being off is related to one or more of these issues.

REFERENCES

1. American Society of Heating, Refrigerating and Air-Conditioning Engineers (ASHRAE), www.ashrae.org/technical-resources/standards-and-guidelines/read-only-versions-of-ashrae-standards
2. International Energy Conservation Code (IECC), published by the International Code Council, https://shop.iccsafe.org/international-codes/iecc-references.html

2 Building Energy Consumption

Energy consumption associated with a commercial building will vary with the type of building, the local weather, the building's characteristics (e.g., size, shape and orientation, height, construction, operational characteristics, occupancy), and the specific equipment installed including its heating, ventilation, and air conditioning (HVAC) systems, lighting, plug and process loads, and energy management system.

To develop an energy savings estimate for an equipment replacement, it is first necessary to determine the baseline energy consumption of the whole building and the impacted building end-uses such as space heating, cooling, lighting, ventilation, and so forth. The energy consumption associated with a building energy efficiency upgrade (i.e., installation of an energy efficiency measure [EEM] such as a replacement high efficiency packaged roof-top air conditioning unit) is then evaluated against the energy consumption of the existing system being replaced (baseline) to determine the energy consumption savings.

This chapter focuses on developing the energy consumption of a whole building, often referred to as the building's energy use intensity (EUI) in units of kBtu/ft^2, and the building's major energy consuming end-uses. Whole building energy use may be obtained directly from utility bill data. However, if utility bill data is not available, it may be necessary to use other methods to estimate the building's energy consumption. When the cost and time to use building simulation modeling is not an impediment, such modeling is the preferred approach. However, while this may not be an impediment to owners of institutional buildings (typically greater than 50,000–100,000 ft^2), it often can be for owners of small to medium-sized buildings (typically less than 50,000 ft^2). It is worth noting that more than 95% of commercial buildings in the U.S. are not institutional buildings.(1) As such, there is a need for energy savings estimation methodologies that can more cost-effectively and efficiently focus on small to medium-sized buildings.

2.1 BUILDING TYPES

As indicated previously, building energy consumption will vary depending upon the building type (i.e., its use). Building types have generally been defined for the U.S. Energy Information Agency (EIA) Commercial Building Energy Consumption Survey, or CBECS. Since 1979, EIA has conducted surveys designed to collect information on the stock of U.S. commercial buildings, including their energy-related building characteristics and energy consumption data. There have been 11 CBECS surveys conducted since 1979, with the most recent in 2018. The types of commercial buildings surveyed are listed in Appendix II.(2) These surveys include all buildings in which at least half of the floorspace (ft^2) is used for a purpose that is not

DOI: 10.1201/9781032692777-2

residential, industrial, or agricultural. However, CBECS does include building types that might not traditionally be considered commercial, such as schools, hospitals, correctional facilities, and churches. Since industrial buildings have considerable variability in energy consumption depending on the specific operations undertaken and work schedule, these were not included in CBECS.

Multifamily (apartment) buildings are usually part of the commercial buildings category, but they are not included in CBECS. They are defined as consisting of five or more units and are included in EIA's Residential Energy Consumption Survey (RECS).[3,4] There have been 15 RECS surveys conducted since 1978, the most recent in 2020. Additional information on CBECS and RECS may be found in Section 2.2.2.

The energy consumption (kBtu/yr) of a commercial building in a specific year (or 12-month period) is normalized by the building's gross floor area (GFA) in square feet and is referred to as the EUI, expressed in kBtu/ft^2-yr. GFA is the building's total square footage as measured between the exterior walls of the building. Floor area includes common areas, tenant areas, restrooms, elevator shafts, stairwells, basements, storage rooms, and mechanical rooms. It does not include parking or exterior spaces such as balconies, patios, and driveways, although their energy consumption may be included in the energy required to operate the building. GFA should also be distinguished from rentable or leasable floor area, which is the occupied area on all floor levels for which a tenant is charged for occupancy under a lease. Rentable or leasable floor area typically is less than the GFA. If rentable area is used instead of the GFA, the EUI calculated will be higher and adversely impact comparisons with similar buildings that normalize with GFA (i.e., benchmarking). It is always preferred to have the building's EUI calculated directly from utility bill data.

If utility bill data is not available, building EUI may be estimated from CBECS or RECS data or other data sources (see Section 2.2.2). In virtually all these sources, EUIs are typically presented as average or median data for a specific building type in a specific year. The median is different from the average in that extremes in the sample do not influence the median as much. For example, an estimate of the average square feet per building would be affected by the inclusion of some very large buildings. In contrast, an estimate based on median square feet would not be so affected. Table 2.1 presents the median energy consumption of commercial buildings in the U.S. by census region for 2018 (except for multifamily data in the table that was collected in 2020). Table 2.2 presents the average and median EUI for an expanded list of commercial buildings from the 2018 CBECS. The 25th and 75th percentile EUIs are also provided to give an indication of the breath of variability. In this table, the EUIs are for all buildings in 2018 throughout the U.S. in each building type (use) category.

2.2 WHOLE BUILDING ENERGY CONSUMPTION

Energy savings is determined by comparing the energy consumption of the new equipment to be installed in a building to the energy consumption of the existing equipment being replaced. The energy consumption of the existing equipment is defined as the baseline energy consumption. The energy savings of the new equipment is the energy savings associated with the equipment's impact on the impacted building end-use(s).

TABLE 2.1
Median Energy Consumption of Commercial Buildings by Census Region*

Building Type	Energy Use Intensity (kBtu/SF-yr)			
	Northeast	Mid-West	South	West
Education	73.2	65.8	60.7	50.2
Food Sales	N/A	284.2	212.1	N/A
Food Service	204.4	318.4	273.0	231.8
Healthcare	137.4	160.2	151.3	128.5
Inpatient	190.0	193.4	197.8	192.2
Outpatient	90.5	95.3	78.3	72.6
Lodging	92.7	91.6	77.5	86.4
Mercantile	88.2	93.3	83.9	91.7
Retail (other than mall)	57.0	68.3	64.5	60.9
Enclosed/strip malls	112.1	125.7	102.1	113.1
Multifamily (five units or more)	41.6	42.8	33.8	34.3
Office	74.3	64.9	66.3	58.9
Public Assembly	86.2	98.8	74.6	66.7
Public Order and Safety	N/A	N/A	89.8	73.8
Religious Worship	54.9	38.8	30.6	30.5
Service	52.2	43.5	60.0	50.9
Warehouse and Storage	35.2	38.3	28.4	22.5

*Multifamily data from 2020 RECS; all other data from 2018 CBECS.

TABLE 2.2
Various EUI for Building Types from 2018 CBECS

Building Type	EUI by Percentile, kBtu/SF-yr			
	Average EUI*	25th	50th (Median)	75th
Education	62.7	31.1	48.6	74.4
College or University	85.8	32.7	48.6	78.4
K–12	59.5	31.9	48.6	69.5
Elementary	53.5	29.5	45.2	59.8
Middle/Junior High School	61.7	51.3	67.9	77.5
High School	65.8	31.6	52.2	75
Preschool or Daycare	57.7	26.8	52.4	74.4
Other	54.4	11.2	34.3	72.4
Food Sales	232	150.5	215.2	311.2
Convenience Store	224.7	167.9	231.9	313.9
Grocery Store/Food Market	243.3	150.5	242.2	273.2
Food Service	263.3	141	232.9	381.5
Fast Food	326.8	243.1	324.9	479.6
Restaurant	265	132.2	225.8	365.8
Other	145.9	80.6	142.8	188.2
Healthcare	144.5	40.7	61.1	98.8
Inpatient	193.3	140.4	208.6	252.9
Outpatient	82	39.2	60.6	89

TABLE 2.2 (*Continued*)
Various EUI for Building Types from 2018 CBECS

		EUI by Percentile, kBtu/SF-yr		
Building Type	**Average EUI***	**25th**	**50th (Median)**	**75th**
Office (Diagnostic)	63	38.1	51.2	67.6
Clinic	92.8	45.1	69.5	102.8
Lodging	85.7	53.3	86.1	120.3
Hotel	77.7	37.7	72.8	98.7
Motel	80	57.6	75.3	108.1
Dormitory	76.7	46.8	103	123.3
Nursing Home/Assisted Living	115.6	62.6	122.2	185
Other	72.2	55.7	75.7	115
Mercantile	88.4	33.1	62.8	108.1
Retail (other than mall)	64.1	25.8	40.8	63.5
Enclosed and Strip Malls	111	78	113.5	140.1
Strip Center	119.3	78	113.5	140.1
Enclosed Mall	51.4	26.8	47.9	68.6
Office	65.6	36.3	53.5	77.2
Professional	63.2	33.5	52.2	72
Financial/Bank	70.6	46.3	63	92.1
Government	64.7	36.3	52.9	68.5
Medical (non-diagnostic)	64.2	34.1	54.5	85.3
Other	54.5	40.4	51.9	77.2
Public Assembly	81.1	26.8	53.2	90
Library	98.6	63.3	75.5	97.7
Entertainment	82.6	8.5	38.5	53.9
Recreation	89.5	30.4	53.2	121.8
Social	64	21.2	54.7	82.1
Other	100.3	61	64.8	112
Public Order and Safety	86.3	54.6	75	99.9
Fire or Police Station	76.1	47.1	75.5	99.9
Courthouse	65.3	33.9	58.6	69.8
Other	117.3	59	77.6	112.5
Religious Worship	35.3	17.8	32.9	48.5
Service	50.9	19.9	37.3	66.7
Post Office	72.1	24.8	38.6	96.1
Vehicle Service or Repair	44.4	18	37.5	53.5
Vehicle Storage	36.8	16.1	29.4	50.4
Other	89.6	36.7	58.7	94.5
Warehouse and Storage	30.2	2.2	14	35.2
Non-Refrigerated	29.8	2.2	14	35.2
Warehouse	26.1	5.5	18.8	40.1
Distribution or Shipping	34.4	15.1	27.6	52.6
Self-Storage	16.2	0	2.1	5.5
Refrigerated	107	105.8	113.5	147.7
All U.S. Commercial Buildings	70.4	22.6	45	80.6

* Average EUI determined by dividing total Btu of all buildings of that type by the total floor space of all buildings in that category.

This section focuses on the various ways to estimate a whole building's energy consumption. Section 2.3 focuses on building end-use energy consumption (e.g., end-use energy consumptions of the cooling, space heating, ventilation, and lighting systems).

Whole building energy consumption can be estimated for different types of commercial buildings in different geographic areas either by directly collecting the building's utility bill data or, if utility bill data is not available, by using similar building benchmarking data. The general process is as follows:

1. If a building's utility bill data is available, the building's energy consumption is calculated directly for the period covered by the utility bills:
   - Calculate the building's electricity and fuel consumption for the year (or 12-month period) associated with the utility bill data collected (see Section 2.2.1).
   - Weather-normalize the energy consumption data to typical (or average) weather conditions (see Section 2.2.3). Weather-normalized EUI data is also used to compare (or benchmark) the building's energy consumption against similar buildings in the same geographic area or climate zone.
2. If no data is available on the building's actual energy consumption but you still need to get an estimate of its energy consumption, there are sources that can be accessed to obtain an estimate:
   - CBECS EUI data in a specific year for different building types in various climate zones (see Section 2.2.2.1).
   - RECS EUI data in a specific year for multifamily property (five or more units) in various locations (see Section 2.2.2.2).
   - EUI data in DOE's Building Performance Database (see Section 2.2.2.3).
   - EPA ENERGY STAR Portfolio Manager EUI data (see Section 2.2.2.4).
   - CEUS EUI data for different building types in CA covering all CA climate zones (see Section 2.2.2.5).
   - EUI data that might be publicly available from other sources, such as EUI data disclosed for buildings in a specific year in municipalities with building energy performance disclosure legislation.

2.2.1 Utility Bill Data

Utility bill data for commercial buildings can be collected for electricity consumption and, if applicable, for fuel consumption such as natural gas. Bills are typically broken down into two major components: (1) generation (or consumption) in kWh for electricity, therms or CCF for natural gas, and gallons for oil or propane; and (2) demand/delivery in kW for electricity and therms or CCF for natural gas. The EUI for a specific year can be calculated knowing the building's total energy consumption (electricity plus fuel) and GFA (in ft^2).

Building owners are much more interested in energy cost savings than in energy consumption savings. As such, once energy consumption savings have been estimated, they must then be converted to energy cost savings. To obtain energy cost savings, it is necessary to understand utility rates for both energy consumption and demand.

2.2.1.1 Energy Consumption Data

To obtain whole building energy consumption, energy use from the utility bills must be analyzed. This is accomplished as follows:

- Collect fuel and electricity utility bills for a minimum of one year, noting the following:
  - Since electricity data (kWh used and kW demand) is rarely on a calendar month basis (i.e., calendarized), it is likely you will need to collect 13 months of electric bills to calendarize the data to monthly use.
  - Natural gas bills (CCF or therms) typically are already provided on a calendar month basis.
  - Fuel oil (or propane) bills in gallons delivered will always need to be calendarized.
- The time period over which the utility bills are collected to establish the baseline whole building energy consumption must be representative (e.g., during the baseline period), so there should be:
  - No major renovations in the building (such as the addition of a data center or building expansion). A major renovation is defined as a renovation that involves expansion (or reduction) of the building's gross floor area by 10% or more or impacts whole building energy consumption by more than 10%.(5)
  - No major changes in vacancy (vacancy should be relatively stable in the period being analyzed).
  - No major changes to building operation (such as the addition of a second shift).
- Energy consumption data should be determined by month for a minimum of at least a full year.
- Fuel and electricity data should be collected for the same 12-month period.

Once you have the energy consumption data for the whole building and assuming you want to compare it to previous years or benchmark it against similar buildings, the energy consumption should be normalized by building size (ft^2) and weather-normalized to remove the influence of unusual weather conditions (e.g., a warmer than normal winter or a colder than normal summer; see Section 2.2.3). This will result in the whole building's EUI in units of kBtu/ft^2-yr.

2.2.1.2 Energy Demand Data

The delivery portion of the electric bill covers the cost to deliver electricity across the utility's infrastructure equipment. The delivery portion of the bill may include several components, such as a charge for basic service (fixed monthly fee), a transmission fee, a distribution fee, a public benefits charge, and a competitive transition assessment charge (which allows the utility to recover past investments in generation assets). These charges pay for the equipment that delivers electricity to the building, social policy costs, and services and programs that benefit all utility customers. This portion of the bill is not impacted even if you decide to change electricity suppliers.

Demand charges on electric bills are easily recognizable since they are charged on a kW basis ($/kW). If you are a large commercial or industrial utility customer, chances are that a sizeable portion of the utility bill is made up of demand charges. Every piece of equipment using electricity to operate requires the utility to provide a certain amount of power. A building's demand charge is typically based upon the maximum amount of power (peak) required typically over a 15-minute or 30-minute period in a given billing period, which is typically a month. The higher the peak usage in kW, the higher the demand charge. Today, demand charges can amount to anywhere from 30% to 70% of the total bill. As such, energy saving measures that can also reduce demand get a double benefit, that is, a reduction in energy consumption charges (fewer kWh used) and a reduction in demand charges (fewer kW used).

Natural gas bills may also include a demand component. Demand charges do not exist for fuels delivered to the building by truck, such as fuel oil or propane.

2.2.1.3 Energy Cost Data

Utility bills are built around tariffs that define the billing structure, rate schedule, and other charges. The rate schedule applicable to a commercial building will impact the energy cost savings resulting from energy efficiency improvements that can reduce energy consumption and demand. Failure to understand the specifics of the applicable tariff can result in an energy efficiency upgrade project being misrepresented by poorly calculated energy cost savings.

Commercial electricity bills include most (or all) of the following components:

- Customer charge (a fixed cost each month).
- Energy generation/consumption charge.
- Energy demand/delivery charge.
- Possibly a fuel adjustment charge.
- Possible surcharges such as for environmental or regulatory reasons.
- Sales tax (if applicable).

2.2.1.3.1 Electricity Bills

Electricity generation/consumption charges typically rely on flat, tiered, or time-of-use billing. For flat billing there is only one rate for all electricity consumption ($/kWh). With tiered billing, there are two or more rates, with different rates typically associated with different consumption amounts. For example, the first several hundred kWh may be billed at one rate and the next several hundred kWh billed at another rate, and so on and so forth. There also may be seasonal variations for these tiers. With time-of-use billing, there are two or more rates depending on the time of day the electricity is consumed. A utility will typically define on-peak and off-peak times based on how much demand may occur during those times. Customers are charged a higher rate during on-peak times.

Demand charge billing ($/kW) typically results in the customer paying for the highest (or peak) power load (kW) reached typically in a 15-minute interval (or 30-minute interval, depending on the utility) in the billing period. However, the existence of ratchet provisions in the rate schedule can impact charges billed (billed demand). Actual peak demand (or read demand) is defined as the highest average load during a peak demand interval in each billing cycle. It is possible to reduce peak demand when new high efficiency equipment is installed; however, under ratchet provisions, billed demand may not be impacted because it may be based, for example, on the highest peak demand of the prior 12 months. As such, a demand cost savings (billed demand) would not immediately be reflected in the utility bill even if actual peak demand (read demand) has been reduced.

Many utility tariffs often contain a ratchet clause that imposes a minimum demand charge throughout the year based on the highest peak demand reached over the prior year. Ratcheted demand charges are typically determined as follows. The utility records when you establish a "new high" peak demand in any 15-minute period (although some utilities may use 30-minute or 60-minute periods). Assuming the ratchet is 100% (although it may be lower), then going forward for the next 12 months the utility sets the minimum billed demand equal to 100% of the maximum recorded peak kW for the prior 12 months. Even if you make energy efficiency improvements such as the addition of LED lighting that reduces both the kWh energy consumption and kW demand, for the next 12 months you will continue to be billed (billed demand) at 100% of your previously recorded (pre-LED) peak demand. Your utility bill will typically identify read demand and billed demand, but your charge will be based on billed demand (ratchet). Hence, if you projected a demand cost savings in the year following the LED upgrade, unless the utility tariff indicates otherwise, this savings would not be realized by the building owner until after the ratchet expires.

Since electricity bills are rarely provided on a calendar month basis, they must be converted to a monthly basis. Calendarization can be accomplished as follows:

1. Identify the total kWh used in the first billing period (partial month to partial month).
2. Identify the total days in the first billing period.
3. Calculate the average kWh used per day in the first billing period.
4. Split the first billing period's total days into days from the start date to the end of the month (referred to as "first half" days) and days from the beginning of the following month to the meter read end date of that month (referred to as "second half" days).
5. Calculate the total kWh used in the "first half" days.
6. Calculate the total kWh used in the "second half" days.
7. Go to the second billing period.
8. Repeat steps 1 to 6.

9. Add kWh in "second half" days of the preceding month to the kWh of the "first half" days of the second billing period. This is kWh used in the first calendarized month.
10. Repeat this process for all billing periods until you have 12 calendarized months of data. It will take 13 months of utility bills to get a full 12 months of calendarized electricity use data.

Example Problem 2–1

An energy audit is being conducted on a 25,000 GFA retail building. The electricity bill is not provided on a calendar month basis and needs to be converted to a calendar month basis (calendarized). The electric bills are shown below. Calendarize the data.

Using the year May 2018–April 2019:

Calendarization Months	Start date	End Date	Total No. days	kWh used	kWh/day	No. Days Left from "Start Date" to End of Month	kWh in Days left from Start Date	No. Days in Following Month	kWh in Following Month
	13-04-2018	13-05-2018	31	21,440	692	18	12,449	13	8,991
May	14-05-2018	12-06-2018	30	5,290	176	18	3,174	12	2,116
June	13-06-2018	12-07-2018	30	6,270	209	18	3,762	12	2,508
July	13-07-2018	13-08-2018	32	9,840	308	19	5,843	13	3,998
August	14-08-2018	12-09-2018	30	5,850	195	18	3,510	12	2,340
September	13-09-2018	11-10-2018	29	3,310	114	18	2,054	11	1,256
October	12-10-2018	08-11-2018	28	18,000	643	20	12,857	8	5,143
November	09-11-2018	11-12-2018	33	44,900	1,361	22	29,933	11	14,967
December	12-12-2018	13-01-2019	33	53,370	1,617	20	32,345	13	21,025
January	14-01-2019	11-02-2019	29	50,590	1,744	18	31,401	11	19,189
February	12-02-2019	12-03-2019	29	46,220	1,594	17	27,094	12	19,126
March	13-03-2019	10-04-2019	29	42,620	1,470	19	27,923	10	14,697
April	11-04-2019	12-05-2019	33	36,370	1,102	20	22,042	13	14,328

Full Month	kWh/Month	kBtu/month	
May 2018	12,165	41,519	May 2018 = 8,991 + 3,174 = 12,165 kWh/month
June	5,878	20,061	
July	8,351	28,502	
August	7,508	25,625	
September	4,394	14,997	
October	14,113	48,167	
November	35,076	119,713	
December	47,312	161,474	
January 2019	52,425	178,925	
February	46,284	157,966	
March	47,049	160,577	
April	36,739	125,389	
Total	**317,293**	**1,082,915**	

The building used 1,082,915 kBtu in the 12-month period being analyzed. The whole building electricity EUI for the period being analyzed is 43.3 kBtu/ft^2-yr.

2.2.1.3.2 *Natural Gas Bills*

Commercial natural gas bills include most (or all) of the following components:

- Customer charge (a fixed cost each month).
- Consumption (also called generation) charge (in therms or CCF).
- Demand/delivery charge (in therms or CCF).
- Surcharges, such as for energy conservation funding.
- Sales tax (if applicable).

What holds true for utility electricity bills generally also holds true for utility gas bills. There are natural gas consumption charges and natural gas demand/delivery charges. Similar to electricity rates, if ratchet provisions exist there can be a difference in read demand and billed demand on the utility bill.

2.2.1.3.3 *Fuel Oil Bills*

Oil bills are typically based on the gallons of fuel oil received from a third-party supplier. The difference between the current bill and previous bill is the amount of oil used in the period between the bills.

To convert fuel oil to a monthly basis, the following methodology can be used:

1. Determine total fuel oil delivered over the full year.
2. Determine the heating degree days (HDD) by month for the full year (see Section 2.2.3).
3. Allocate the fuel oil data based upon the HDD in each month.

This is accomplished as follows:

$$G_i = (HDD_i / HDD) \times G_T \qquad (2\text{–}1)$$

where G_i = gallons of fuel oil for month i
G_T = total gallons of fuel oil delivered in the 12-month period
HDD_i = HDD for month i
HDD = total HDDs for the 12-month period

Example Problem 2–2

A 48,000 GFA office building located in Stratford, Connecticut, is heated by fuel oil. For the most recent 12-month period through October 2020, 10,700 gallons of fuel oil were used. What is the estimated monthly fuel oil consumption? Monthly heating degree days for Stratford were obtained from www.degreedays.net and are provided below.

Month Ending	HDD	%	Gal Oil	MM Btu	kBtu/ft²
01–11–2019	674	13.60%	1,455	203	4.23
01–12–2019	881	17.77%	1,902	265	5.53
01–01–2020	872	17.59%	1,882	263	5.47
01–02–2020	797	16.08%	1,720	240	5
01–03–2020	639	12.89%	1,379	193	4.01
01–04–2020	514	10.37%	1,110	155	3.23
01–05–2020	250	5.04%	540	75	1.57
01–06–2020	27	0.54%	58	8	0.17
01–07–2020	0	0.00%	0	0	0
01–08–2020	2	0.04%	4	1	0.01
01–09–2020	59	1.19%	127	18	0.37
01–10–2020	242	4.88%	522	73	1.52
Total	**4,957**	**100.00%**	**10,700**	**1,494**	**31.12**

The estimated monthly energy consumption is shown in the next to last column using 139,600 Btu/gal fuel oil and the monthly energy intensity in the last column.

2.2.1.3.4 Using Utility Bills in Energy Cost Savings Analysis

Assume an energy efficiency project is proposed that will save both electricity and natural gas. Once the energy consumption (kWh) and peak demand (kW) savings are calculated, the expected energy cost savings needs to be assessed. Remember, building owner clients typically are more interested in the energy cost savings associated with a proposed energy efficiency improvement project—and this calculation can often be misrepresented by poor application of the utility rate to the projected energy consumption and demand savings.

Example Problem 2–3

Assume a building's electricity bill and natural gas bill in a monthly billing period are as provided below. What are the electric bill's electricity and natural gas generation charges and demand charges for use in energy savings calculations?

Data provided in the utility's electric bill:

Total Charge		\$3,598.20
Total kWh		17,854
Billed Demand		94.6 kW
Customer Service Charge		\$350.00
Generation Charge	17,854 kWh × \$0.0799	\$1,426.53
Demand Charge	94.6 kW × \$9.27	\$876.94
Distribution Demand	94.6 kW × \$7.15	\$676.39
FMCC Delivery Charge	17,854 kWh × \$0.00771	\$137.65
Public Benefits Charge	17,854 kWh × \$0.00732	\$130.69

The kWh charge for use in energy savings calculations would be all charges on a kWh basis or \$1,694.87 / 17,854 kWh = \$0.095 / kWh. The kW demand charge would be all charges on a kW basis or \$1,553.33 / 94.6 kW = \$16.42 / kW.

Data provided in the gas supplier's bill and the utility's natural gas bill:

Third Party Energy Supplier Charge	3,315 CCF × \$0.8692	\$2,881.40
Utility Charges:		
Total Utility Charge		\$1,821.68
Total CCF		3,315
Billed Demand		355 CCF
Customer Service Charge		\$256.52
Delivery Charge	3,315 CCF × \$0.043	\$142.55
Demand Charge	355 CCF × \$2.0952	\$743.80
SER Mechanism	355 CCF × \$0.0251	\$8.91
Conservation Adjustment	3,315 CCF × \$0.046	\$152.49
Transportation Services	3,315 CCF × \$0.0867	\$287.41
Transportation Svc. Demand Charge	355 CCF × \$0.6479	\$230.00
Total Natural Gas Charges (Supplier and Utility)		\$4,703.08

The CCF charge for use in energy savings calculations would be all charges based on the amount of gas supplied (3,315 CCF): the supplier charge of \$2,881.40 plus the delivery charge of \$142.55, plus the conservation adjustment of \$152.49 and transport services charge of \$287.41, for a total of \$3,463.85 / 3,315 CCF or \$1.045 / CCF delivered. Note that this is 20% higher than the supplier charges alone. The total CCF demand charge (i.e., that based on 355 CCF demand) would be \$982.71 / 355 CCF = \$2.77 / demand CCF.

One of the more common errors that exists in estimating energy cost savings involves using "blended rates" for both electricity and natural gas. The blended electricity rate entails dividing the total electricity bill by the total kWh of electricity consumed in that billing period. The blended natural gas rate entails dividing the total natural gas bill by the total CCF of gas consumed. The blended $/kWh rate is then multiplied by the projected kWh saved to estimate the electricity cost savings, and the blended $/CCF is then multiplied by the projected CCF saved to estimate the natural gas cost savings. *Unfortunately, this approach can provide a grossly inaccurate estimate of the energy cost savings, especially for improvements that do not result in any peak demand savings.*

Using the sample electricity bill in Example Problem 2–3, the monthly blended electricity rate would be $3,598.20 / 17,854 kWh or $0.2015/kWh. The correct electricity rate, however, should be based on only those items in the bill that are determined using kWh. This results in an electricity rate of $0.095/kWh. Hence, the blended rate error would result in an electricity cost savings estimate more than twice what it should be.

Similarly, using the total natural gas charges in Example Problem 2–3, the monthly blended rate would be $4,703.08 for 3,315 CCF or $1.4187/CCF. However, the correct natural gas rate to determine the value of gas consumption savings should be $3,463.85 / 3,315 CCF or $1.045/CCF. Hence the blended rate error would result in a natural gas cost savings estimate 36% higher than what it should be.

2.2.1.3.5 Bottom Line on Utility Rate Schedules

1. It is critical to understand utility rate schedules prior to calculating energy cost savings. Carefully review recent utility bills for the owner's building where energy efficiency improvements are being proposed (for at least a 12-month period or a 13-month period assuming the data needs to be calendarized) and review published details about the applicable utility tariff, especially the existence of ratchet provisions and any exceptions to the ratchet (such as may exist for installations involving renewable energy systems).
2. Do not use blended rates to assess the energy cost savings.
3. To determine the $/kWh and $/CCF generation/consumption rates for use in assessing energy cost savings, review the most recent utility electricity and fuel bills and add up all the charges that are specifically based on generation/consumption kWh and CCF. These charges should then be divided by the respective generation/consumption kWh and CCF to determine the appropriate generation/consumption rates. To determine demand charges, add up all the costs based on kW for electricity demand ($/kW) and all the costs based upon demand CCF for natural gas ($/demand CCF). If you notice that billed demand does not change from month to month even though read demand does, this is a good indicator that there is a ratcheting provision which should be factored into the projected cost savings (e.g., there may be no demand cost savings in the first year).

2.2.2 Benchmarking

Energy benchmarking is a process for describing the energy performance of a building at a point in time and for comparing that performance with similar buildings. As this definition implies, there are two key elements in benchmarking: (1) the description of energy performance and (2) the comparison with similar buildings. The description of energy performance is typically accomplished through calculation of a performance metric such as the EUI. Many types of comparison are possible, such as those shown in Table 2.3.

2.2.2.1 Approaches to Energy Benchmarking

Energy benchmarking may be internal or external, and quantitative or qualitative. Internal benchmarking compares data within a building owner's portfolio of buildings, while external benchmarking compares against a broader population of similar buildings. A quantitative approach compares numerical measures of performance to see how building performance changes over time or ranks against that of similar buildings. The qualitative approach analyzes management and operational practices across the entire building portfolio to identify best practices and areas for improvement.

A combination of qualitative and quantitative measures can be a powerful tool for detecting poor performance and identifying best practices that can be harnessed for improvements. For example, a benchmarking exercise might calculate the EUI for a portfolio of ten similar office buildings. If three of the buildings show twice the energy use per square foot as the best performing building, then it is natural to begin to look for an explanation. This can be done, for example, by comparing the systems and operational practices of the poorly performing buildings to those of the better performing buildings.

EUI data may be characterized as either site EUI or source EUI. Site EUI is what you may be familiar with from your utility bills. Site EUI contains a mixture of what are referred to as primary energy (e.g., a raw fuel like natural gas) and secondary energy (e.g., a converted product like electricity or district steam). A unit

TABLE 2.3
Common Comparisons Made When Benchmarking

Comparison	Definition
Best in Class	Compare the building to the best performing building in a population of buildings with similar characteristics.
Average	Compare the building to the average performance of buildings in a population with similar characteristics.
Median	Compare the building to the median performance of buildings in a population with similar characteristics.
Baseline	Compare the building's performance to its historical performance.
Performance Standard	Compare the building to a clearly defined performance standard, such as those that might be established in building energy codes or local building energy performance legislation.

of primary and a unit of secondary energy consumed at the site are not directly comparable because one represents raw fuel while the other represents a converted fuel. Therefore, to assess the relative efficiencies of buildings with varying proportions of primary and secondary energy consumption, it is necessary to convert these two types of energy into equivalent units of raw fuel consumed to generate that one unit of energy consumed on-site. To achieve this equivalency, EPA uses source EUI. Source energy represents the total amount of raw fuel that is required to operate the building, but also includes all transmission, delivery, and production losses (e.g., generating electricity at the utility plant and delivering it to the building). EPA prefers using source energy because it provides an equitable way to combine primary and secondary energy types into a single common unit, ensuring that no building receives either a credit or a penalty based on its energy source or utility. Benchmarking can be conducted on either a site EUI or source EUI basis. However, for comparative energy savings calculations, this book relies on site EUI (e.g., building utility bill data).

When calculating whole building energy consumption on a site EUI basis, EUI has units of kBtu/ft^2-yr. Building electricity consumption in kWh/yr can be converted to kBtu/yr using the conversion factor 3.412 kBtu/kWh. The most common fuel in buildings is natural gas, and fuel consumption is typically in CCF/yr or therms/yr. To convert CCF/yr to kBtu/yr, use the conversion factor 102.9 kBtu/CCF. To convert therms/yr to kBtu/yr, use the conversion factor 100 kBtu/therm. To convert therms to CCF, use the conversion factor 0.972 CCF/therm. Useful conversion factors are included in Appendix I. The sum of the electricity and fuel consumption in kBtu/yr are divided by the building gross floor area in ft^2 to obtain the EUI during the 12-month period over which the utility bills were collected.

Benchmarking is typically performed using the weather-normalized site EUI. As such, the specific 12-month period over which electricity and fuel consumption data was collected is important to note so that weather-normalizing adjustments can be made (see Section 2.2.3).

2.2.2.2 CBECS Data

The Commercial Building Energy Consumption Survey (CBECS) is a national sample survey conducted by the U.S. Energy Information Administration (EIA) that collects information on U.S. commercial buildings, including their energy-related building characteristics and energy consumption data. The definition of commercial buildings in CBECS includes "all buildings [greater than 1,000 SF] in which at least half of the floorspace is used for a purpose that is not residential, industrial, or agricultural." The specific building types covered in CBECS are shown in Tables 2.1 and 2.2. Appendix II includes a more detailed description of each building type.

The first CBECS was conducted in 1979. The most recent CBECS (11th iteration) collected data for calendar year 2018. CBECS 2018 energy consumption data is presented in Appendix V, along with a comparison with the 2003 CBECS data. The CBECS database is accessible for each year that a survey was undertaken. It may be accessed at: www.eia.gov/consumption/commercial/about.php

When actual building energy consumption data is not available, many energy auditors rely on CBECS data in some capacity to assist in the assessment of building energy consumption without a firm understanding of the data and its limitations. Unfortunately, this lack of understanding can lead to erroneous results when projecting energy cost savings. For example, when relying on CBECS site EUI data or building end-use energy consumption data without adjusting for location-specific typical weather conditions can lead to flawed estimates of energy consumption savings, which in turn can lead to inaccurate energy cost savings. This has been a contributor to some building owners having low confidence in the projected energy savings identified in energy audits relying solely on CBECS data without any adjustment. Remember, CBECS data is specific to the year (and weather in that year) when the survey was conducted.

The detailed CBECS tables for the calendar year in which the survey was conducted include:

1. Building characteristics tables, such as the number of buildings of each property type, the amount of floor space, and structural and energy-related characteristics.
2. Energy consumption and expenditures tables, such as whole building total energy consumption data by major energy sources (electricity and fuel) and building end-use energy consumption estimates.

The energy end-use consumption tables provide estimates of the amount of electricity, natural gas, fuel oil, and district heat used for ten major building energy end-uses:

- Space heating
- Cooling
- Ventilation
- DHW heating
- Lighting
- Cooking
- Refrigeration
- Computers (including servers)
- Office equipment
- Other.

Unfortunately, CBECS EUI data does not differentiate building energy consumption by building configuration, occupancy, age of mechanical systems, mechanical system design and operation, maintenance practices, operating schedules, and building energy management systems. As such, average EUI data for a building type can vary widely. The CBECS database does provide insight into the EUI variability by providing 25th percentile, 50th percentile (median), and 75th percentile data, in addition to the average (see Table 2.2). For example, in the 2018 CBECS, the average EUI for office buildings is 65.6 kBtu/ft^2-yr, while at the 25th percentile the EUI is 36.3 kBtu/ft^2-yr, at the 50th percentile (median) the EUI is 53.5 kBtu/ft^2-yr, and at the 75th percentile it is 77.2 kBtu/ft^2-yr. Notwithstanding, CBECS can serve as an excellent resource to obtain insight into the energy consumption associated with major building end-uses as a function of property type if utility bill data for the building is not available.

CBECS data cover four census regions and nine census divisions, as follows:

- Northeast
  - New England (CT, ME, MA, NH, RI, VT)
  - Middle Atlantic (NJ, NY, PA)
- Mid-West
  - East North Central (IL, IN, MI, OH, WI)
  - West North Central (IA, KS, MN, MO, NE, ND, SD)
- South
  - South Atlantic (DE, DC, FL, GA, MD, NC, SC, VA, WV)
  - East South Central (AL, KY, MS, TN)
  - West South Central (AR, LA, OK, TX)
- West
  - Mountain (AZ, CO, ID, MT, NV, NM, UT, WY)
  - Pacific (AK, CA, HI, OR, WA).

In addition to census regions and divisions, it is also important to understand the climate regions used. CBECS classifies buildings by their location into five broad climate regions (see Figure 2.1):

Hot-Humid

A hot-humid climate is generally defined as a region that receives more than 20 in. of annual precipitation and where one or both of the following occur:

- A 67°F or higher wet bulb temperature for 3,000 or more hours during the warmest six consecutive months of the year.
- A 73°F or higher wet bulb temperature for 1,500 or more hours during the warmest six consecutive months of the year.

Mixed-Humid

A mixed-humid climate is generally defined as a region that:

- Receives more than 20 in. of annual precipitation.
- Has approximately 5,400 heating degree days (HDDs, 65°F basis) or fewer.
- Is located where the average monthly outdoor temperature drops below 45°F during the winter months.

Hot-Dry/Mixed-Dry

A hot-dry climate is generally defined as a region that receives less than 20 in. of annual precipitation and is located where the monthly average outdoor temperature remains above 45°F throughout the year. A mixed-dry climate is generally defined as a region that receives less than 20 in. of annual precipitation, has approximately 5,400 HDDs or less, and is located where the average monthly outdoor temperature drops below 45°F during the winter months.

Cold/Very Cold

A cold climate is generally defined as a region with approximately 5,400 HDDs or more and fewer than approximately 9,000 HDDs. A very cold climate is generally defined

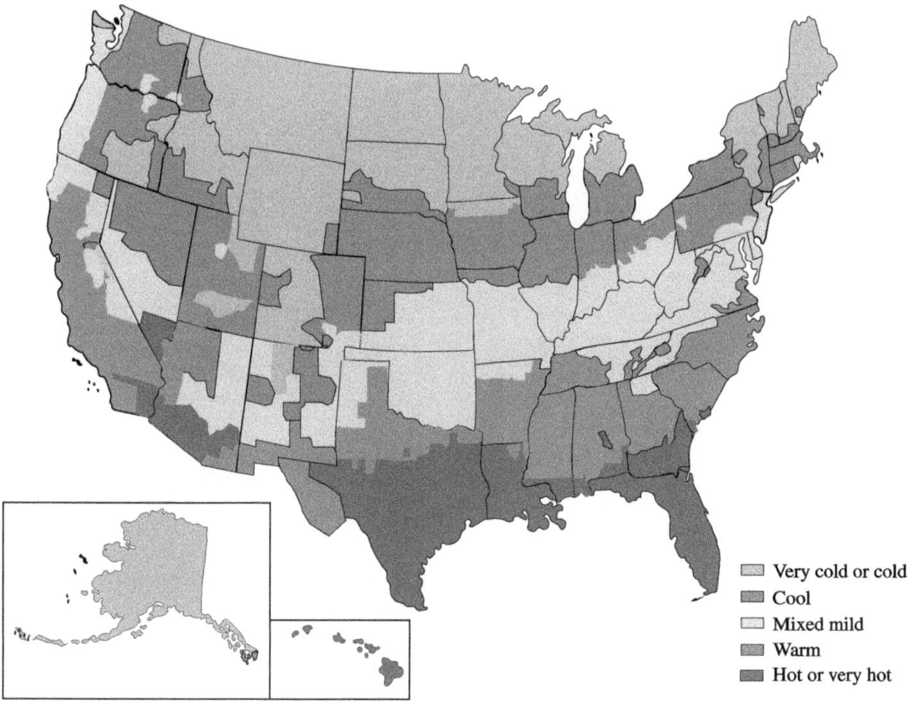

FIGURE 2.1 Climate zones in 2018 CBECS.

as a region with approximately 9,000 HDDs or more and fewer than approximately 12,600 HDDs.

Marine

A marine climate is generally defined as a region that meets the following criteria:

- A mean temperature of the coldest month between 27°F and 65°F.
- A warmest month mean temperature of less than 72°F.
- At least four months with mean temperatures more than 50°F.
- A dry summer season. The month with the heaviest precipitation in the cold season has at least three times as much precipitation as the month with the least precipitation in the rest of the year. The cold season is October through March in the Northern Hemisphere.

Under these broad climate definitions there can be considerable weather variability within any climate zone. It is also unfortunate that energy consumption is not presented by more defined climate zones, such as the 17 climate zones specified in ASHRAE 90.1, Energy Standard for Buildings Except Low Rise Residential Buildings, and the International Energy Conservation Code (IECC). These climate zones are shown in Figure 2.2.

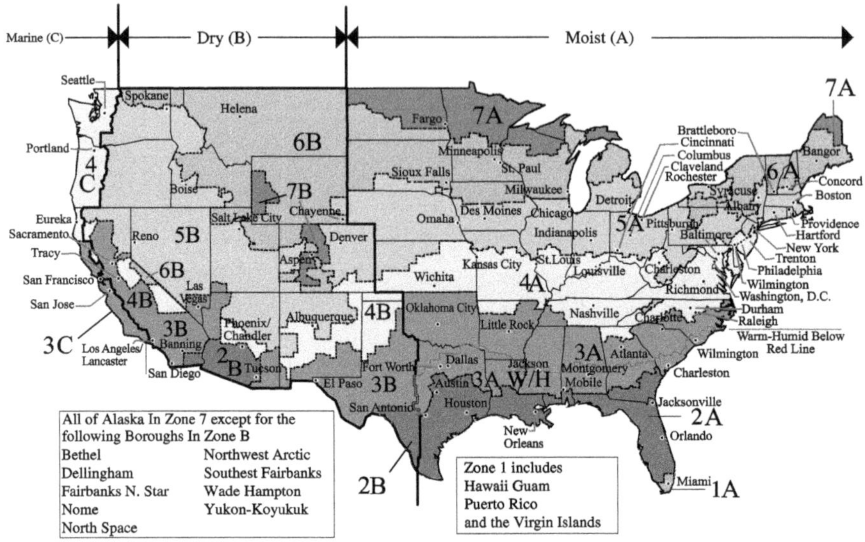

FIGURE 2.2 ASHRAE 90.1 and International Energy Conservation Code (IECC) climate regions.

Interestingly, the 2003 CBECS divided the U.S. into five climatically homogeneous zones (see Figure 2.3), as defined by the National Oceanic and Atmospheric Administration (NOAA). Each NOAA climate division is based on its 30-year average HDD and cooling degree days (CDD) for the period 1971 through 2000.

The 2012 CBECS did not use historical HDD- and CDD-based climate zones and instead relied upon the five major climate regions created by the Building America Program (see Figure 2.3).

U.S. DOE's Pacific Northwest National Laboratory prepared an excellent guide to U.S. climate regions.[6] The guide is available online at: www.energy.gov/sites/prod/files/2015/10/f27/ba_climate_region_guide_7.3.pdf

Climate regions and their associated HDD and CDD data become especially important when adjusting CBECS energy consumption data for a building generally located in a climate region to the building's more specific location (see Section 2.3.2).

2.2.2.2.1 CBECS Data Limitation

Whole building energy consumption data are available in CBECS by general property type and for five general climate zones, intending to represent the 17 much more specific U.S. climate zones. Building end-use energy consumption data are available by property type but are not provided by climate zone. The fact that no climate zone data or only very general climate zone data are available for various property types and end-uses presents an issue that can introduce error in building energy consumption analysis. This is a serious shortcoming in using CBECS data. There can be no question that even state general climates are often insufficient. For example, the energy consumption of a Florida office building located in

2003 CBECS Climate Zones

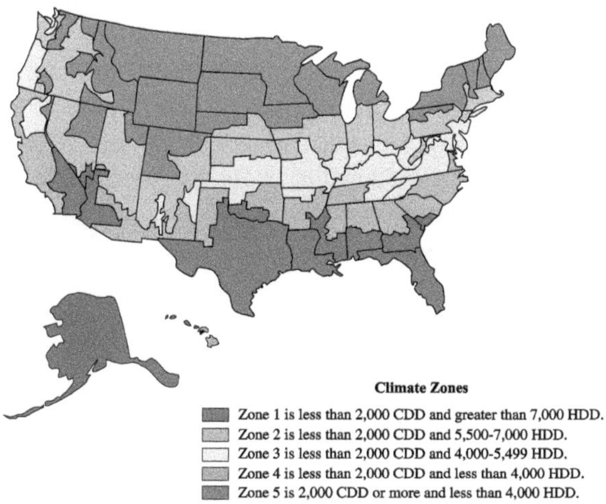

2012 CBECS Climate Zones

The 2012 CBECS no longer used historical HDD- and CDD-based climate zones and instead relied upon climate regions created by the building America program.

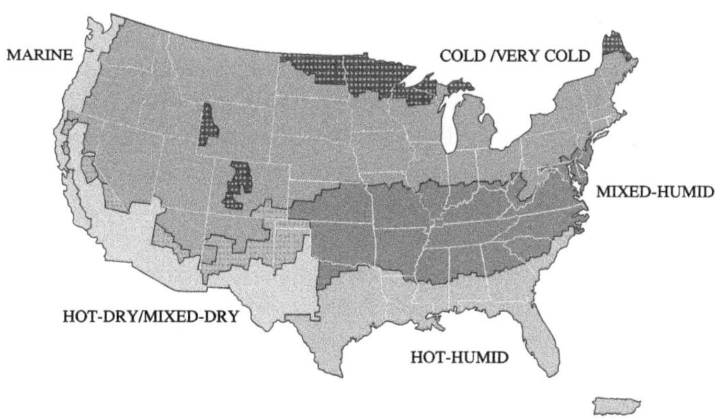

FIGURE 2.3 Climate zones in 2003 and 2012 CBECS.

Miami will be considerably different than the same type and size office building located in Jacksonville. As such, it is strongly recommended that CBECS data be adjusted for weather conditions that may exist at a specific building's location (refer to Section 2.3.2).

2.2.2.3 RECS Data

The U.S. DOE Residential Energy Consumption Survey (RECS) is a national sample survey that collects information on the stock of U.S. residential buildings, including their energy-related building characteristics and energy consumption data. Residential buildings include attached and detached single family homes, apartments with fewer than five units and apartments with five or more units and mobile homes. Only multi-family buildings with five or more units are assumed to be for commercial use. Since this book is solely directed at energy savings in commercial buildings, the only RECS data of interest is that associated with multifamily buildings having five or more units.

The first RECS was conducted for calendar year 1978 and the most recent (15th iteration of the survey) for calendar year 2020. Appendix VI presents multifamily energy consumption data from the 2020 RECS. The appendix also includes energy consumption end-use data by state for all residential properties, including apartment buildings with five or more units.

The RECS database may be accessed at: www.eia.gov/consumption/residential/

2.2.2.4 DOE Building Performance Database

The DOE's Building Performance Database (BPD) is a decision-support platform comprising a database and data analysis tools that enable statistical analysis of building energy performance, and operational and physical characteristic data. The database contains measured data for both existing commercial and residential buildings. The feature that distinguishes the BPD from other similar databases is the ability to analyze narrowly defined peer groups—such as market regions and/or specific building or equipment types—at a localized or factor-specific level appropriate for decision-makers such as building owners.

The goal of the BPD is to leverage the law of large numbers to achieve validity at the local level, not to achieve a representative national sample (which CBECS does). Thus, the BPD is a complementary effort to CBECS, and CBECS is one of the data sources for the BPD. The BPD contains data from over 70 source data sets, totaling almost 1 million residential and commercial buildings submitted voluntarily by both public and private data contributors. The dataset includes more than 230,000 commercial buildings. Primary data contributors include federal government agencies such as the Environmental Protection Agency (EPA) and the Department of Housing and Urban Development (HUD), state and local governments, large portfolio owners, energy efficiency programs, and private companies. The BPD also contains data collected by municipalities under building energy performance disclosure ordinances (e.g., Austin, Boston, Cambridge, Chicago, Minneapolis, New York City, Philadelphia, San Francisco, Seattle, and Washington, D.C.).

All building records contain basic building characteristics such as location, building type, and energy use. These fields constitute the BPD's minimum data requirements. These data are useful for analyses such as comparing a building's energy performance to similar buildings in the BPD or identifying high- or low-performing buildings in a peer group. The BPD database may be accessed at: https://bpd.lbl.gov

To obtain building EUI data after accessing the site, follow the steps below:

1. Log in (or register an account if you do not have one).
2. Click on Building Classification (e.g., Commercial).
3. Deselect All building types and only select on the building type of interest (e.g., office).
4. Click on Location and Data Source and select where the building is located (e.g., by city, by state, by zip code, or by climate zone [15 climate zones]). Clicking on Data Source will provide data from specific databases, such as in municipalities with benchmarking ordinances.
5. On the accompanying Histogram, select Site EUI for the X-axis and see both the median site EUI (on the X-axis) and the number of buildings (on the Y-axis) in the dataset.
6. Go down to the Summary Table and note the 25th, 50th (median) and 75th percentile site EUI for this category of buildings in the location selected.

EUI data in the BPD database is not weather-normalized, but rather represents data as it was input to the BPD from the data source in the year it was collected.

2.2.2.5 ENERGY STAR Portfolio Manager Database

EPA's ENERGY STAR Portfolio Manager is an online tool that can be used to measure and track energy and water consumption, as well as greenhouse gas emissions. It can provide a building's normalized (for building activity, climate, operating hours, and the number of workers) site EUI from the building's energy consumption data and then benchmark the building against the normalized median (50th percentile) of peer buildings from data collected in CBECS and other sources (see Appendix X).

If it is desired to benchmark the building against EPA's building performance database to get an ENERGY STAR score, the EPA Portfolio Manager may be accessed at: https://openei.org/wiki/Environmental_Protection_Agency_(EPA)_Portfolio_Manager

The score is a benchmark that indicates how efficiently buildings use energy on a 1–100 scale. A score of 50 indicates that half the building's peers have an energy performance better than the building and half worse than the building. In other words, the building's performance is in the middle compared to its peers. A score of 75 or better indicates top performance (the building's performance is in the top quartile) and means the building may be eligible to earn ENERGY STAR certification.

ENERGY STAR scoring is available for the following property types:

- Bank/financial institution, including a bank branch
- College, university
- Convenience store with or without a gas station
- Courthouse
- Data center
- Distribution center
- Food sales
- Hospital (general medical and surgical)
- Hotel (with minimum 55% average annual occupancy)

- House of worship
- K–12 school
- Medical office
- Multifamily housing
- Office (with minimum 50% average annual occupancy)
- Parking area
- Residence hall/dormitory
- Retail store
- Senior care facility
- Supermarket/grocery store
- Warehouse (refrigerated and unrefrigerated)
- Wholesale club/supercenter.

The national median site EUI (refer to Appendix X) is a recommended benchmark metric for all buildings. The median value is the middle of the national population—where half of buildings use more energy and half use less. The median works better than the mean (arithmetic average) for comparing relative energy performance because it more accurately reflects the midpoint of energy use for most property types.

2.2.2.6 CEUS Data

The California Commercial End-Use Survey (CEUS) is a comprehensive study of commercial sector energy use. The California Energy Commission used the survey to collect commercial sector energy consumption and end-use profiles. The survey included a random sample of 2,790 commercial facilities collected from the service areas of Pacific Gas & Electric (PG&E), San Diego Gas & Electric (SDG&E), Southern California Edison (SCE), and the Sacramento Municipal Utility District (SMUD). For the electric-only utilities (SCE and SMUD), gas consumption was obtained separately from Southern California Gas (SCG) and PG&E. The sample was stratified by utility service area, climate region, building type, and energy consumption level. For each utility service area, electric and natural gas consumption, energy use intensities (EUI), and 16-day hourly end-use load profiles were estimated for 11 common commercial building type categories and 13 end-uses.[(7)]

The analysis to obtain weather-normalized energy consumption data consisted of three key steps:

1. Building survey data was collected in 2002, and initial building simulations were performed in a customized eQuest building energy simulation model using actual 2002 weather.
2. Simulation results were calibrated against actual building energy consumption data to obtain a calibrated simulation model.
3. Simulation results were weather-normalized by replacing the 2002 weather with normalized typical meteorological year (TMY) weather and re-running the simulations to obtain weather-normalized building EUI and end-use energy consumption.

The building type categories included in the CEUS study were:

- Small office (<30,000 SF)
- Large office (≥30,000 SF)
- Restaurant
- Retail
- Food store
- Refrigerated warehouse
- Unrefrigerated warehouse
- School
- College
- Healthcare
- Lodging.

The end-uses included:

- Space heating
- Space cooling
- Ventilation
- Water heating
- Interior lighting
- Exterior lighting
- Refrigeration
- Cooking
- Office equipment
- Air compressors
- Motors (non-HVAC)
- Process equipment
- Miscellaneous.

A map of the California climate zones is presented in Figure 2.4. PG&E coverage is principally in climate zones 1–5, SDG&E coverage in climate zone 13, SCE coverage in climate zones 7–10, and SMUD in climate zone 6. Average HDD and CDD for each zone are presented in Table 2.4.

For the SDG&E, SCE, PG&E, and SCG energy consumption data, 2002 weather data was used. For the SMUD energy consumption data, 2003 weather data was used. CEUS data for all utilities was then weather-normalized and includes:

- Whole building EUI by building type in each utility service area, kBtu/ft^2-yr.
- Whole building EUI by energy source (electricity/natural gas) for each building type in each utility service area, kWh/ft^2-yr and therms/ft^2-yr.
- End-use EUI by building type in each utility service area, kBtu/ft^2-yr.
- End-use EUI by energy source (electricity/natural gas) for each building type in each utility service area, kWh/ft^2-yr and therms/ft^2-yr.

Appendix VIII contains whole building and end-use energy consumption data (weather-normalized) for the four utility service areas surveyed.

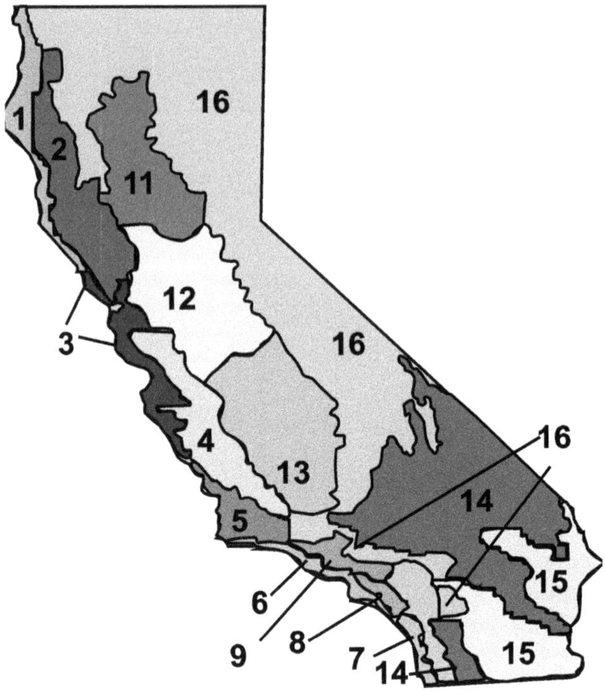

FIGURE 2.4 California climate zones.

TABLE 2.4
Average HDD and CDD in California Climate Zones

California Climate Zone	Average HDD	Average CDD
1	4,295	15
2	3,144	500
3	3,071	183
4	2,550	666
5	2,654	464
6	1,383	742
7	1,497	865
8	1,481	1,072
9	1,460	1,456
10	1,685	1,620
11	3,149	1,354
12	2,621	1,226
13	2,443	1,599
14	2,422	3,056
15	1,177	4,760
16	5,057	596

2.2.3 Weather-Normalization

The key to weather-normalization are the HDD and CDD. Building EUI data collected in a specific year must be normalized to remove the influence of unusual weather conditions (e.g., a warmer than normal winter or a colder than normal summer). This is best accomplished after evaluating the building's space heating end-use energy consumption and space cooling end-use energy consumption. The weather-normalized building EUI is most often used in benchmarking analysis, as well as in energy savings analysis.

2.2.3.1 Heating and Cooling Degree Days

The most common use of degree days is for tracking energy consumption in a building. An HDD is a measurement designed to quantify the demand for energy needed to heat a building. It is the number of degrees that a day's average temperature is below 65°F, which is the temperature below which buildings commonly need to be heated. Degree days may also be calculated based upon a standard temperature other than 65°F; however, throughout this book the 65°F standard temperature will be used. The daily average temperature is the mean of the maximum and minimum temperatures in a 24-hour period. To calculate the HDD, subtract the average of a day's high and low temperatures from 65. For example, if the day's average temperature is 50°F, its HDD is 15.

CDDs, on the other hand, are the number of degrees that the daily average temperature rises above 65°F. For example, if a weather station recorded an average daily temperature of 78°F, the CDD for that station would be 13.

The units of HDD and CDD are degree-days or °F-days.

Average HDD and CDD data for select cities in the U.S. are included in Appendix III. Appendix III also includes HDD and CDD data by census region in 2018 (latest CBECS) and 2020 (latest RECS), and for the U.S. nationally for the years 2018–2022.

The best source for HDD and CDD information is the National Weather Service (NWS). Thirty-year average HDD and CDD data (1991–2020) can be found by city at the NWS site (www.weather.gov/wrh/climate?wfo=box). Once at this site, first input the city of interest. When the page comes up, under Additional Forecasts and Information click on Past Weather Information (or Climate/Past Weather). On the next page that comes up, select Daily/Monthly Normals, and then under Variables select either HDD or CDD, depending upon what data you want to collect. Then click on GO, and the site will return average HDD or CDD data by month and summed for a year.

In reviewing utility data, you may need HDD and CDD monthly data for a specific period of time (e.g., March 2021 to February 2022). When there is a need for such data, this information may be obtained at www.degreedays.net. The historical time period covered is 36 months.

2.2.3.2 Methodology for Weather-Normalization

When weather-normalizing, it is generally assumed that only the space heating and cooling end-use energy consumptions will be impacted. As such, only these two components will be adjusted.

The annual energy consumption associated with the space cooling end-use may be weather-normalized using CDD as shown in Equation 2–2.

$$EUI_{N\text{-}C} = EUI_C * \left[CDD_{Avg}/CDD\right] \quad (2\text{–}2)$$

where EUI_{N-C} = normalized energy use intensity for space cooling, kBtu/ft²-yr
EUI_C = actual energy use intensity for space cooling during the period selected (e.g., recent 12 months), kBtu/ft²-yr
CDD = actual cooling degree days during the period selected (e.g., 12 months)
CDD_{Avg} = average long term CDD during the period selected (e.g., 12 months)

The annual energy consumption associated with the space heating end-use may be weather-normalized using HDD as shown in Equation 2–3.

$$EUI_{N\text{-}H} = EUI_H * \left[HDD_{Avg}/HDD\right] \quad (2\text{–}3)$$

where EUI_{N-H} = normalized energy use intensity for space heating, kBtu/ft²-yr
EUI_H = actual energy use intensity for space heating during the period selected (e.g., 12 months), kBtu/ft²-yr
HDD = actual heating degree days during the period selected (e.g., 12 months)
HDD_{Avg} = average long-term HDD during the period selected (e.g., 12 months)

EUI_{N-C} and EUI_{N-H} can then be added to the EUIs of the remaining end-uses not impacted by weather to get the building weather-normalized EUI that would be used for benchmarking.

Example Problem 2–4

Cooler Summer Than Usual (Fewer CDDs Than Usual) Case

Space cooling end-use energy consumption in a specific year has the following data:
Cooling EUI = 8.5 kBtu/SF-yr (actual)
CDD in the specific year = 400
CDD_{Avg} at the building location = 475

What is the weather-normalized cooling end-use energy consumption?

There are fewer CDDs than normal because it was a cooler summer. Therefore, the actual cooling EUI was lower than it normally would be.

To get the weather-normalized cooling EUI: EUI_{N-C} = (8.5)(475/400) = 10.1 kBtu/SF-yr

Example Problem 2–5

Warmer Summer Than Usual (More CDDs Than Usual) Case

Space cooling end-use energy consumption in a specific year has the following data: Cooling EUI = 8.5 kBtu/SF-yr (actual)

CDD in the specific year = 550

CDD_{Avg} at the location = 475

What is the weather-normalized cooling end-use energy consumption?

There are more CDDs than normal because it was a warmer summer. Therefore, the actual cooling EUI was higher than it normally would be.

To get a weather-normalized cooling EUI: $EUI_{N-C} = (8.5)(475/550) = 7.3$ kBtu/SF-yr

Example Problem 2–6

Warmer Winter Than Usual (Fewer HDDs Than Usual) Case

Space heating end-use energy consumption in a specific year with the following data: Heating EUI = 16.2 kBtu/SF-yr (actual)

HDD in the specific year = 840

HDD_{Avg} at the location = 975

What is the weather-normalized space heating end-use energy consumption?

There are fewer HDDs than normal because it was a warmer winter. Therefore, the actual space heating EUI was lower than it normally would be.

To get a weather-normalized space heating EUI: $EUI_{N-C} = (16.2)(975/840) = 18.8$ kBtu/SF-yr

Example Problem 2–7

Colder Winter Than Usual (More HDDs Than Usual) Case

Space heating end-use energy consumption in a specific year with the following data: Heating EUI = 16.2 kBtu/SF-yr (actual)

HDD in the specific year = 1,090

HDD_{Avg} at the location = 975

What is the weather-normalized space heating end-use energy consumption?

There are more HDDs than normal because it was a colder winter. Therefore, the actual space heating EUI was higher than it normally would be.

To get a weather-normalized space heating EUI: $EUI_{N-C} = (16.2)(975/1{,}090) = 14.5$ kBtu/SF-yr

Table 2.5 contains average HDD and CDD data for U.S. census divisions in select years. The years selected were designed to cover the CBECS and RECS survey years where whole building and/or end-use energy consumption data is provided but which may have to be adjusted (see Section 2.3.2).

TABLE 2.5
HDD and CDD for Census Divisions in Select Years*

Census Region	Census Division	HDD 2003	HDD 2009	HDD 2015	HDD 2018	HDD 2020	CDD 2003	CDD 2009	CDD 2015	CDD 2018	CDD 2020	States
Northeast	New England	6,908	6,646	6,525	6,326	5,822	502	362	557	667	644	CT, MA, VT, ME, NH, RI
	Mid-Atlantic	6,057	5,799	5,781	5,784	5,214	662	587	799	885	844	NJ, NY, PA
Mid-West	East North Central	6,423	6,540	6,168	6,433	5,854	633	547	728	972	831	IL, IN, MI, OH, WI
	West North Central	6,486	6,837	6,090	6,968	6,322	950	720	942	1,134	964	IA, KS, MN, MO, NE, ND, SD
South	South Atlantic	2,882	2,879	2,500	2,627	2,260	1,996	2,025	2,385	2,418	2,338	DE, FL, GA, MD (with DC), NC, SC, VA, WV
	East South Central	3,589	3,588	3,227	3,478	3,053	1,510	1,497	1,717	1,928	1,636	AL, KY, MS, TN
	West South Central	2,227	2,212	2,088	2,252	1,815	2,587	2,570	2,741	2,856	2,735	AK, LA, OK, TX
West	Mountain	4,633	5,016	4,595	4,803	4,807	1,603	1,504	1,483	1,573	1,665	AZ, CO, ID, MT, NV, NM, UT, WY
	Pacific	2,900**	3,150**	2,869	3,169	3,215	899**	884**	1,094	1,002	1,071	CA, OR, WA, AK, HI
U.S.	48 States**	4,495	4,482	4,128	4,291	3,967	1,280	1,230	1,457	1,579	1,476	National

*U.S. Energy Information Administration, Monthly Energy Reviews, Tables 1.9 and 1.10, Population-weighted Degree-day Averages (www.eia.gov/mer).
**Excludes AK and HI.

2.2.4 Building Simulation Modeling

If the resources (time and money) are available, the preferred way to get end-use energy consumption is building simulation modeling—although for equipment replacement in small to medium-sized buildings, rarely are the necessary resources available. Notwithstanding, if simulation modeling is to be done on an existing building, there is a considerable amount of data that must be collected (see Appendix XI), including the actual weather data for the period of time being modeling (which should coincide with the time period the existing building's energy consumption data was collected). Once the model is run, the energy consumption calculated by the model then needs to be calibrated against the building's actual energy consumption collected from utility bills. Unfortunately, choosing what building input data gets fine-tuned when the model is being calibrated can introduce error. Building energy model calibration is a technique that takes various types of measured performance data (e.g., energy use) and tunes key model parameters to match the simulated results with the actual measurements. Calibration methods can be labor-intensive and require considerable expertise.

Assuming the building simulation model has been calibrated sufficiently to be able to predict the existing building's energy consumption close to the building's actual energy consumption (in the baseline year), the calibrated model can then be run with typical meteorological year (TMY) weather to predict future building performance under various scenarios. (Note: TMY data contain one year of hourly data that best represents median weather conditions over a multiyear period. To calculate a TMY, a multiyear data set is analyzed, and 12 months of data are chosen from the time frames that best represent the median conditions. For example, a TMY developed from a set of data for the years 1998–2005 might use data from 2000 for January 2003 data, February 1999 data for March, and so on. The data are considered "typical" because the entirety of the meteorological data is condensed into one year's worth of the most usual conditions.)

Predicting future energy savings, however, presents new challenges. How the building will operate in the future can only be guessed. Will the occupancy change? Will mechanical equipment degrade over time? Will occupant behavior change? What kind of weather will be encountered in the future? How will utility rates behave in the future? It is obvious that answering these questions requires a considerable number of assumptions. In the final analysis, using sophisticated building simulation modeling can only provide a best guess estimate of what may happen in the future.

Example Problem 2–8

A building owner has authorized use of a building simulation model to assess future energy savings associated with an HVAC upgrade to an existing building. In view of the size of the project, the building owner believes this will enable a more informed decision on whether to proceed with this project.

What general strategy would you recommend for the building simulation modeling effort, and what would you say to the building owner to manage expectations?

Assuming all the required building characteristics and operational settings have been collected and input into the simulation model (understanding that this could be a time-consuming effort), the model should first be run with existing HVAC equipment using actual full year weather data associated with the utility bills collected and reviewed (this is the baseline year).

It is unlikely the energy consumption results from the model will equal the actual (baseline) energy consumption identified in the utility bills. As such, the model will need to be calibrated to better predict actual building energy consumption.

The calibrated model would then be re-run in the same year (baseline year) the utility bills had been reviewed, only this time with the high efficiency replacement equipment. At this point, the energy consumption difference between the two models would provide the building owner with a reasonable estimate of the energy cost savings associated with the energy efficiency upgrade had it been installed in the year the utility bills were reviewed (baseline year). However, this does not address the future energy savings. As such, the model should be re-run using typical meteorological year (TMY) data, and include an assumed annual energy escalation and annual equipment performance degradation. The building owner should then be advised that: "Assuming typical weather data exists over the lifetime of the high efficiency replacement equipment, the estimated present worth lifetime energy cost savings would likely be (insert)." To manage building owner expectations, it should also be pointed out that future energy cost savings in any single year can be influenced by the actual weather encountered which can be quite different from typical meteorological year weather.

2.2.5 Limitations

When utility bills are not available to review to establish whole building EUI, and benchmark-type data such as in CBECS or RECS or other such data are relied upon, limitations need to be recognized. CBECS data is based upon the principal building activity taking place in the building. Unfortunately, CBECS only supports certain building types. Building types not supported include industrial buildings, manufacturing plants, abandoned/vacant buildings, buildings undergoing renovation, and newly constructed buildings. In addition, the energy consumption of buildings undergoing repositioning (change of use) cannot be estimated since they lack the necessary historical energy consumption data. Moreover, buildings with mixed use add further complication (refer to Chapter 15, Section 15.2). There is also significant variability in energy consumption estimates provided in CBECS and RECS. This is evident when looking at the 25th and 75th percentile EUI estimates for each building type. Variability is also heavily influenced by the local climate where the building is located.

Finally, the impact of energy consumption associated with external features that may be associated with a building, such as parking areas or the presence of an outdoor pool, can further add complications. In ENERGY STAR Portfolio Manager, the energy consumption of these external features must be excluded to benchmark the building's energy consumption. ENERGY STAR Portfolio Manager provides an assessment of the building, not its parking area. As such, the parking area (ft^2) should not be included in the building ft^2. Fortunately, ENERGY STAR provides guidance on estimating energy consumption of parking areas when these are not separately metered.[8] Table 2.6 provides EPA methodology to estimate energy consumption of different types of parking areas in units of kBtu per ft^2 of parking area per year.

TABLE 2.6
Estimating Energy Consumption for Different Types of Parking Areas

Parking Type	Energy Need	Parking Area Site Energy, kBtu/ft^2-yr
Open	Lighting	2.989
Partially Enclosed	Lighting	8.967
Completely Enclosed	Lighting	8.967
	Ventilation	2.39

2.3 BUILDING END-USE ENERGY CONSUMPTION

The major commercial building end-uses include:

- Space heating
- Space cooling
- Domestic hot water (DHW) heating
- Lighting
- Ventilation.

Other end-uses may include:

- Cooking
- Refrigeration
- Office equipment
- Computers
- Miscellaneous.

Estimates of building end-use energy consumption may be obtained (assuming building simulation modeling is not being performed) as follows:

1. If utility data exists and the building uses natural gas or fuel oil for heating in the winter and air conditioning for cooling only in the summer, a graph of monthly electricity and monthly natural gas (or fuel oil) use can provide insight into space heating, DHW heating and space cooling energy consumption (see Section 2.3.1).
2. If utility energy consumption data does not exist, then end-use energy consumptions may be approximated from CBECS (see Section 2.3.2), RECS (see Section 2.3.3), CEUS (see Section 2.3.4), or other building end-use energy consumption data that might be available (see Section 2.3.6).
3. The local utility's technical resource manual (TRM) may provide insight into building end-use energy consumptions. For example, heating and cooling equivalent full load hours (EFLHs) (refer to Chapters 4 and 6), fan operating hours (refer to Chapter 7), and lighting operating hours (refer to Chapter 9) are often available in the TRM. When combined with knowledge of the existing equipment's capacity and energy efficiency, it may be possible to estimate space heating and cooling (see Section 2.3.5), ventilation (see Chapter 7) or lighting (see Chapter 9) end-use energy consumptions.

2.3.1 Analysis of Monthly Utility Data

Monthly building energy use data (e.g., electricity and natural gas) can be used to estimate select end-use energy consumptions. However, this will only work in those climates where buildings use natural gas for space heating in the winter months and where air conditioning is used only in the summer months. It is not applicable if the building is all electric (electric heating and cooling); or if the building gets all or some space heating via electricity (e.g., heat pump); or if the climate does not dictate any real need for space heating (e.g., South Florida), or where air conditioning is needed year-round.

The electricity (kWh) and fuel (CCF natural gas) consumed should first be calendarized and bar graphed. The average kWh of electricity used in the winter months when air conditioning is not called for would then be subtracted from the average kWh of electricity used in the summer months when air conditioning is used. The difference should approximate the kWh associated with space cooling.

In a similar manner, the average CCF used in the summer months would represent the amount of natural gas used for non-space heating purposes, such as for domestic hot water (DHW) heating. This would be subtracted from the average CCF of natural gas used in the winter months. The difference would approximate the space heating end-use energy consumption. If the only other natural gas use is principally for DHW heating, the DHW end-use energy consumption can also be estimated.

This approach can be done visually or more accurately by regression analysis (see Example Problem 2–9). The result would be building end-use energy consumption estimates for cooling, space heating, and DHW heating. The energy consumption of these three end-uses would be subtracted from the whole building energy consumption. The difference would be the energy consumption of the remaining building end-uses (e.g., lighting, ventilation; see Example Problem 2–10).

Example Problem 2–9

Given the following electricity and natural gas consumption data for a senior living home located in upstate New York, estimate the cooling and heating end-use energy consumption. Natural gas is principally used for space heating, DHW heating and cooking.

Nominal Utility Billing Period	No. Days in Billing Period	kWh in Billing Period	CDD in Billing Period	Therms in Billing Period	HDD in Billing Period
January	33	57,272	0	14,363	1,250.55
February	29	50,276	0	11,370	1,065.39
March	29	52,048	0	8,443	1,055.13
April	29	51,620	0.03	4,651	481.71
May	32	61,814	26.49	2,176	177.21
June	30	84,674	138.24	1,776	20.97
July	29	74,576	236.04	1,824	0
April	32	76,936	183.93	1,894	6.51
September	30	53,885	103.23	3,366	68.04
October	29	57,810	11.52	7,346	331.35
November	33	60,527	0	12,824	772.14
December	30	56,002	0	12,166	926.07
	365 days/yr	737,440 total kWh/yr	699.48 total CDD/yr	82,199 total therms/yr	6,155.07 total HDD/yr

The regression analysis function in Excel (under Data Analysis) was used to generate the following output.

Multiple Regression Output Summary

Cooling Regression Statistics

R Squared	0.697552712
Observations	12
CDD	107.66455 kWh/CDD
Space Cooling End-use:	75,365.19 kwh/yr or 10.22% of total electric usage

Heating Regression Statistics

R Squared	0.871135879
Observations	12
HDD	9.20738 therms/HDD
Space Heating End-use:	56,672 therms/yr or 68.94% of total gas usage

The regression equations were:

Whole Building kWh = 1,730.01555 (No. Days) + 107.6645486 (CDD) + 2,556.26045

Whole Building Therms = 456.0407608 (No. Days) + 9.20737957 (HDD) – 11,743.99529

Using the regression equations to see how well they predicted kWh and therm usage in each month:

	Actual		From Regression Equations	
Nominal Utility Billing Period	**kWh in Billing Period**	**Therms in Billing Period**	**kWh Calc with CDD**	**Therms Calc with HDD**
January	57,272	14,363	59,647	14,820
February	50,276	11,370	52,727	11,291
March	52,048	8,443	52,727	11,196
April	51,620	4,651	52,730	5,916
May	61,814	2,176	60,769	4,481
June	84,674	1,776	69,340	2,130
July	74,576	1,824	78,140	1,481
August	76,936	1,894	77,719	2,909
September	53,885	3,366	65,571	2,564
October	57,810	7,346	53,967	4,532
November	60,527	12,824	59,647	10,415
December	56,002	12,166	54,457	10,464

Let's now see how this compares with a quick analysis of the monthly energy use data, but *not* using regression analysis.

For electricity, the baseload kWh consumption when air conditioning was less likely to be used occurred in the 7 months of January to April and October to December. During these months, the kWh load was 385,555 kWh or an average of 55,079 kWh/month.

When air conditioning likely was used (in the 5 months from May to September), the kWh load was 351,885 kWh or an average of 70,377 kWh/month. The load difference with air conditioning in the warmer summer months was 15,298 kWh/month or 5 months × 15,298 kWh/month equal to 76,490 kWh/yr. This can be compared with the regression result of 75,365 kWh/yr.

For natural gas, the baseload therm consumption when space heating was less likely to be used occurred in the 6 months of April to September. During these months, the therm load was 15,687 therms or an average of 2,615 therms/month. When space heating likely was being used (in the 6 months from January to March and October to December), the therm load was 66,512 therms or an average of 11,085 therms/month. The load difference with space heating in the colder winter months was 8,470 therms/month or 6 months × 8,470 therms/month equal to 50,820 therms/yr. This can be compared with the regression result of 56,672 therms/yr.

Visually observing the data on a bar graph will provide similar results.

Example Problem 2–10

A 35,000 ft^2 office building has an EUI of 63.5 kBtu/ft^2-yr from utility bill review. The percentage end-use energy consumptions were estimated from local data on a similar building and were as follows:

End-Use	% of Building EUI (Similar Building)
Cooling	13
Space Heating	36
DHW Heating	1.5
Lighting	16
Ventilation	12
Other	21.5
Total	100

Data loggers on select equipment provided certain end-use energy consumptions at the building:

Space Heating	958 MM Btu/yr
Cooling	102,550 kWh/yr (or 350 MM Btu/yr)
DHW Heating	27 MM Btu/yr

Using the actual end-use energy consumptions above, what are the adjusted end-use energy consumptions attributed to all end-uses?

The whole building energy consumption = 63,500 Btu/yr × 35,000 ft^2 = 2,222.5 MM Btu/yr.

The adjusted space heating, cooling and DHW heating end-use percentages are:

End-use	MM Btu/yr	% of Total
Space Heating	958	43.1
Cooling	350	15.7
DHW Heating	27	1.2
Remaining Building Energy Consumption	887.5	40

The remaining end-uses will now have to be adjusted:

	Original % Bldg EUI
Lighting	16
Ventilation	12
Other	21.5
Total	49.5

Adjusted end-use calculations:

	MM Btu/yr	Adjusted % of Total Bldg EUI
Lighting = 16/49.5 × 100 = 32.3%	0.323 × 887.5 = 287	12.9
Ventilation = 12/49.5 × 100 = 24.2%	0.242 × 887.5 = 215	9.7
Other = 21.5/49.5 × 100 = 43.4%	0.434 × 887.5 = 385	17.3
Total	887	

The adjusted end-use energy consumptions become:

End-Use	MM Btu/yr	% of Total
Space Heating	958	43.1
Cooling	350	15.7
DHW Heating	27	1.2
Lighting	287	12.9
Ventilation	215	9.7
Other	385	17.3
Total	2,222	99.9

2.3.2 CBECS End-Use Data

The CBECS database (see Appendix V) provides average whole building energy and end-use energy consumption data for commercial real estate. As indicated previously, these data should be adjusted for the building location's climate zone rather than the very general CBECS climate zone. They also should be adjusted for the baseline year (assuming it is different than 2018 when CBECS was last updated). Finally, the data should be weather-normalized since predicting future end-use energy savings relies on weather-normalized end-use energy consumption.

2.3.2.1 Adjusting CBECS Data

Adjusting the 2018 CBECS whole building and end-use energy consumption data may be accomplished using heating and cooling degree days. An office building in Miami, Florida, will be used as an example.

1. Identify for an office building in the 2018 CBECS (CBECS Table E2) the whole building EUI and end-use energy consumptions (EUI in kBtu/ft^2-yr) with no reference to specific building location or climate region (in this table, CBECS looks at all office buildings in the U.S.). These are shown below:

	EUI	%
Space Heating	20.1	30.6
Space Cooling	5.1	7.8
Ventilation	12.9	19.7
DHW Heating	1.0	1.5
Lighting	7.8	11.9
Other	18.7	28.5
Total	65.6	100%

   These data are for all office buildings nationally. Unfortunately, end-use energy consumption for buildings in Miami is not provided. As such, it will be necessary to adjust these energy consumptions. (Note: What CBECS does provide is the whole building EUI for an office building in the South Atlantic census division, i.e., 66.2 kBtu/ft^2-yr. However, again, this is aggregated data from nine states, only one of which is Florida.)
2. Since the data in (1) is for all office buildings nationally, it will be necessary to obtain the 2018 HDD and CDD data for the U.S. (see Appendix III). The Annual CDD and HDD for the U.S. is 1,579 and 4,291, respectively.
3. Next, the HDD and CDD in Miami in 2018 will be needed. In 2018 Miami had an annual CDD and HDD of 4,814 and 78, respectively. Historic HDD and CDD data for cities across the country may be accessed at the National Weather Service site (www.weather.gov/wrh/climate). (*When the map comes up, click on the area of the country you are interested in. Then on the next view, select the specific location. Under "Product", check*

"Monthly Summarized Data". Under "Options", select the year range you want (you may want to select multiple years around the year you want). Under "Variables" select CDD or HDD, depending on what you want. Under "Summary" select "Sum". Click on "Go" to get your results.)

4. Assuming that only the space heating and space cooling end-use energy consumptions are impacted by weather (not an unreasonable approximation), they can now be adjusted to reflect the fact that the building is in Miami, Florida.
   CDD adjustment (more cooling energy): (4,814/1,579) × 5.1 = 15.5
   HDD adjustment (less heating energy): (78/4,291) × 20.1 = 0.4
   The adjusted end-use energy consumptions for 2018 would then be:

	Adjusted EUI	Adjusted%
Space Cooling	15.5	27.5
Space Heating	0.4	0.7
Ventilation	12.9	22.9
DHW Heating	1.0	1.8
Lighting	7.8	13.9
Other	18.7	33.2
Total	56.3	100%

5. These results will now have to be adjusted again to reflect whatever the baseline year is (e.g., the year utility bill data was collected). For example, let's assume that the baseline year is 2022. The end-use data in 2018 will need to be adjusted to 2022 using 2022 HDD and CDD data. This can be accomplished as shown below. The CDD for Miami in 2022 was 5,024 and HDD was 132.
   CDD adjustment (more cooling energy): (5,024/4,814) × 15.5 = 16.9
   HDD adjustment (more heating energy): (132/78) × 0.4 = 0.7

End-use	2018 EUI	2018%	2022 EUI	2022%
Space Cooling	15.5	27.5	16.9	29.1
Space Heating	0.4	0.7	0.7	1.2
Ventilation	12.9	22.9	12.9	22.2
DHW Heating	1.0	1.8	1.0	1.7
Lighting	7.8	13.9	7.8	13.5
Other	18.7	33.2	18.7	32.2
Total	56.3	100	58.0	99.9%

The 2022 end-use percentages would be multiplied by the whole building EUI (preferably obtained from utility data) to obtain the individual end-use energy consumptions that would be associated with the existing (baseline) equipment that is to be replaced. These would be the end-use energy consumptions associated with the whole building EUI data collected in 2022.

6. Finally, to predict future energy consumption saving, it is necessary to use weather-normalized end-use energy consumptions since it is not possible to predict future weather (or assume that it will be the same as encountered in 2022). In our case, the typical HDD and CDD for Miami for weather-normalization would be 130 and 4,458, respectively (see Appendix III). The methodology in step 5 would be repeated to obtain the weather-normalized end-use energy consumptions for use in energy savings calculations.

In summary then, the following adjustments were made in the above analysis:

- 2018 CBECS national office data was adjusted to Miami office data in 2018.
- Miami office data in 2018 was adjusted to Miami office data in baseline year 2022.
- Miami office data in baseline year 2022 was adjusted to weather-normalized data to make energy savings predictions.

Select modeled end-use percentage data for various building types is presented in Appendix VII. The models were developed in 2007–2012 and may be used to further assist in providing insight into end-use energy consumption estimations.

2.3.3 RECS End-Use Data

The RECS database (see Appendix VI) provides average whole building energy and end-use energy consumption data for multifamily property (five or more units). Appendix VI contains estimates for multifamily (five or more units) building types by census region and also state level data for all residential homes (not just multifamily).

2.3.4 CEUS End-Use Data

CEUS data (see Appendix VIII) includes end-use energy consumption (weather-normalized) for various commercial property types, including:

- Small office (<30,000 SF)
- Large office (≥30,000 SF)
- Restaurant
- Lodging
- Retail
- Food store
- Refrigerated warehouse
- Non-refrigerated warehouse
- School
- College
- Healthcare.

The data is provided statewide covering California's 16 climate zones (see Figure 2.4 and Table 2.4) by the service regions of the four major utilities: PG&E, SDG&E,

SCE, and SMUD. These utilities include the following major cities and California climate zones:

- PG&E major cities: San Francisco and San Jose—CA climate zones 1–5; principal CBECS climate zone: marine.
- SDG&E major cities: Fresno—CA climate zone 13; principal CBECS climate zone: hot-dry, mixed-dry.
- SCE major cities: San Diego, interior Los Angeles, Orange County—CA climate zones 7–10; principal CBECS climate zone: hot-dry, mixed-dry.
- SMUD major cities: Coastal Los Angeles, Santa Barbara—CA climate zone 6; principal CBECS climate zone: marine.

Example Problem 2–11

The following information on a 46,310 SF office building in Stockton, CA was collected in an energy audit for the period August 2018 through July 2019. The utility servicing the building is PG&E.

Electricity consumption	978,832	kWh
Natural gas consumption	3,313	CCF
Actual HDD	3,035	"Colder winter than normal"
Actual CDD	1,525	"Warmer summer than normal"
Long-term Avg. HDD	2,621	
Long-term Avg. CDD	1,226	

The energy auditor wants to estimate the energy savings for an upgrade to the building but needs the weather normalized baseline building end-use energy consumptions.
The actual energy consumption of the building in the specified 12 month period is:

Electricity:	(978,832 kWh/yr)*(3.412 kBtu/kWh) / 46,310 SF =	72.1	kBtu/SF-yr
Gas:	(3,313 CCF/yr)*(102.9 kBtu/CCF) / 46,310 SF =	7.4	kBtu/SF-yr
	Total	79.5	kBtu/SF-yr

The weather normalized end-use percentages can be obtained from Appendix VIII for a large office building. The electricity and natural gas end-uses are combined to get end-use as a percent of whole building energy consumption.

End-use	Normalized % of Building EUI
Space heating	27.44
Cooling	13.03

DWH heating		3.67
Ventilation		12.53
Lighting		19.17
Other		24.16
	Total	100%

While we have the weather normalized end-use percentages, we do not have the weather normalized whole building EUI.

The weather normalized whole building EUI can be obtained as follows (assuming that weather principally impacts space heating and cooling):

End-use	Actual Total EUI	Normalized Total EUI	Normalized CEUS Total Percentages
Cooling	x	(1,226/1,525)*x	13.03
Space heating	y	(2,621/3,035)*y	27.44
Other	z	z	59.53
Total	79.5	T	

1,226/1,525 = 0.8

2,621/3,035= 0.86

Normalized Total EUI = T = (0.8*x) + (0.86 * y) + z

$0.1303*T = 0.8 * x$ so $T = 6.14 * x$

$0.2744 * T = 0.86 * y$ so $T = 3.13 * y$

$0.5953 * T = z$ so $T = 1.68 * z$

$6.14 * x = 3.13 * y$

$x = 0.51 * y$

since T = T,

$1.68 * z = 3.13 * y$

$z = 1.863 * y$

Also: $79.5 = x + y + z$

$79.5 = (0.51 * y) + y + (1.863 * y)$

$79.5 = 3.373 * y$

$y = 23.57$

and $x = 12.02$

$z = 43.91$

$T = 73.8$

In summary:

End-use	Actual Total EUI	Normalized Total EUI	Normalized CEUS Total Percentages
Cooling	12.02	9.62	13.03
Space heating	23.57	20.27	27.44
Other	43,91	43.91	59.53
Total	79.5	73.8	100

The "Other" end-uses would then be as follows:

DHW Heating		2.71	3.67
Lighting		14.15	19.17
Ventilation		9.25	12.53
Other		17.83	24.16
	Sub-total	43.9	59.53

2.3.5 EFLH Data for End-Use Estimation

Estimating cooling and heating energy requirements is an integral part of designing and managing buildings. Where complex hourly simulation models are not favored, estimates may be derived by simplified methods that are less computationally intensive but still provide reasonable estimates compared to those obtained from the more complicated approaches. The equivalent full load hours (EFLH) method is one such simplified energy consumption estimation method. The EFLH is the ratio of the annual load delivered to the space to be conditioned to the installed hourly capacity. There are EFLHs for cooling and space heating loads in different types of buildings in different geographies (refer to Tables 4.4, 6.1, and 6.2). Various TRMs also include estimations for cooling and space heating EFLHs.

The EFLH equations to estimate the input cooling and space heating end-use energy consumptions are given by Equations 2–4 through 2–7.

$$E_{Out\text{-}C} = CAP_C * EFLH_C \tag{2–4}$$

$$E_{In\text{-}C} = \left(E_{Out\text{-}C} / EER_C\right) * 3.412 \tag{2–5}$$

$$E_{Out\text{-}H} = CAP_H * EFLH_H \tag{2–6}$$

$$E_{In\text{-}H} = E_{Out\text{-}H} / \eta_{Baseline} \tag{2–7}$$

where

$E_{Out\text{-}C}$ = output (delivered) cooling end-use energy consumption, Btu/yr
CAP_C = output (delivered) air conditioning cooling capacity, Btu/hr
$EFLH_C$ = equivalent full load hours for space cooling, hr/yr
$E_{In\text{-}C}$ = input cooling end-use energy consumption, Btu/yr
EER_C = energy efficiency ratio of cooling equipment, Btu output/W-hr input (see Chapter 4)
3.412 = btu/W-hr
$E_{Out\text{-}H}$ = output (delivered) space heating end-use energy consumption, Btu/yr
CAP_H = output (delivered) space heating capacity, Btu/hr
$EFLH_H$ = equivalent full load hours for space heating, hr/yr
$E_{In\text{-}H}$ = input heating end-use energy consumption, Btu/yr
$\eta_{Baseline}$ = heating system thermal efficiency, fraction (see Chapter 6).

The $E_{In\text{-}C}$ and $E_{In\text{-}H}$ would provide an estimate of the cooling and space heating end-use energy consumptions. However, this requires knowledge of the energy efficiency metrics of the existing heating (CAP_H and $\eta_{Baseline}$) and cooling (CAP_C and EER) equipment that will be replaced, as well as knowledge of the $EFLH_C$ and $EFLH_H$.

2.3.6 Other Sources for End-Use Energy Consumption Data

There are a number of other reference sources that might be used to provide estimates of building end-use energy consumption.[9–11] The goal would be to get the end-use data most appropriate for your building type and location (climate zone).

2.3.7 Limitations

Just as whole building energy use estimation has limitations, so do the energy consumptions associated with the various building end-uses. There may be considerable variability in end-use energy consumption for different property types in different climate zones. This may be due among other things, to:

- The variability in building characteristics, including size, orientation, occupancy, etc.
- The variability in building operation, including set-points.
- The variability in HVAC systems, their configuration and operation for different property types in different climate zones.
- The variability in weather.

Short of monitoring energy-consuming equipment with data loggers, end-use energy consumptions can only be estimated. Data loggers can measure and record specific energy consumption data over a certain period of time and can be used when the most accurate energy consumption data is required. Unfortunately, these are rarely used except in investment grade energy audits performed when energy savings guarantees are required.

REFERENCES

1. U.S. Energy Information Agency (EIA), Commercial Building Energy Consumption Survey (CBECS), Table B6, building size, number of buildings, 2018, https://www.eia.gov/consumption/commercial/data/2018/bc/pdf/b6.pdf
2. U.S. Energy Information Agency (EIA), Commercial Building Energy Consumption Survey (CBECS), Building type definitions, 2018, https://www.eia.gov/consumption/commercial/building-type-definitions.php
3. U.S. Energy Information Agency (EIA), Residential Energy Consumption Survey (RECS), 2009, https://www.eia.gov/consumption/residential/data/2009/
4. U.S. Energy Information Agency (EIA), Residential Energy Consumption Survey (RECS), 2015, https://www.eia.gov/consumption/residential/data/2015/
5. ASTM Standard E3224, Standard guide for building energy performance and improvement evaluation in the assessment of property condition, 2019, https://www.astm.org/e3224-19.html

6. Pacific Northwest National Laboratory, Guide to determining climate regions by county, PNNL-17211 Rev. 3, August 2015, https://www.energy.gov/eere/buildings/articles/building-america-best-practices-series-volume-73-guide-determining-climate
7. California Energy Commission, California Commercial End-Use Survey (CEUS), Report CEC-400-2006-005, March 2006, https://www.energy.ca.gov/data-reports/surveys/california-commercial-end-use-survey/2006-california-commercial-end-use-survey
8. ENERGY STAR Portfolio Manager, Technical reference, parking and the ENERGY STAR Score in the U.S. and Canada, February 2018, www.energystar.gov/sites/default/files/tools/Parking_August_2018_EN_508.pdf
9. The National Trust for Historic Preservation, Quantifying the value of building reuse, Quantis International Report, January 2012, https://cdn.savingplaces.org/2023/05/24/11/14/36/697/The_Greenest_Building_Full.pdf
10. Heller, J., Heater, M. and Frankel, M., Sensitivity analysis: Comparing the impact of design, operation, and tenant behavior on building energy performance, Report, New Buildings Institute White Paper, July 2011.
11. Konopacki, S.J. and Akbari, H., Estimates of energy consumption by building type and end use at U.S. army installations, LBL-38577, August 1996, https://eta-publications.lbl.gov/sites/default/files/lbl-38577.pdf

3 Establishing Baseline Energy Efficiency Metrics

3.1 EQUIPMENT ENERGY PERFORMANCE

To estimate the energy savings associated with replacing an existing energy-consuming piece of equipment with new, high efficiency equipment, it is essential to have some knowledge related to the current energy performance of the existing equipment. Assuming a data logger will not be installed to monitor the equipment's energy consumption, estimating its current energy performance can represent a challenge. Such equipment may, for example, no longer have an observable and readable identification tag (or nameplate tag), or the building owner/manager may not have access to the operating manual, original purchase order or proposal, or any other supporting documentation that might provide insight into the existing equipment's energy efficiency rating when installed. An identification tag on the equipment would likely identify the manufacturer and model number which can then be further researched. However, finding specifications for equipment that may be more than a decade old can be challenging. Moreover, even if the research can successfully identify the equipment's original energy efficiency metrics, the equipment's current efficiency would unlikely be at the same performance level as when it was newly installed.

3.1.1 Building Energy Codes

If research does not yield the needed information, based upon the approximate age of the equipment, the building energy code in effect when the equipment was installed may provide insight into its energy efficiency metrics. Unfortunately, the U.S. does not have a national energy code or standard, so energy codes are adopted at the state and local levels of government. Because of this, the local codes and editions in place vary widely, but they generally follow the International Energy Conservation Code (IECC) or ASHRAE 90.1 Energy Standard for Buildings. Table 3.1 provides a summary of state energy codes in effect as of March 2023.(1) As such, it can be assumed that the equipment's energy efficiency metrics likely complied with the IECC or ASHRAE 90.1 standards in effect at that time as most manufacturers design their equipment to comply with the latest ASHRAE 90.1 standard.

3.1.2 ASHRAE 90.1 Building Energy Standard

The first ASHRAE 90 Building Energy Standard was published in 1975. There have been multiple editions to it since. In 1999 the ASHRAE board of directors voted to place the standard on continuous maintenance, based on rapid changes in energy

DOI: 10.1201/9781032692777-3

TABLE 3.1
State Commercial Energy Codes

State	State Current Code (as of March 2023)
Alabama	90.1–2013
Alaska	None statewide
Arizona	Home rule*
Arkansas	2009 IECC and 90.1–2007
California	2022 Building Energy Efficiency Standards
Colorado	Home rule
Connecticut	2021 IECC and 90.1–2019
Delaware	2018 IECC and 90.1–2016
District of Columbia	90.1–2013
Florida	2018 IECC and 90.1–2016
Georgia	2015 IECC and 90.1–2013
Hawaii	Home rule
Idaho	2018 IECC and 90.1–2016
Illinois	2018 IECC and 90.1–2016
Indiana	90.1–2007
Iowa	2012 IECC and 90.1–2010
Kansas	Home rule
Kentucky	2012 IECC and 90.1–2010
Louisiana	90.1–2007
Maine	2015 IECC and 90.1–2013
Maryland	2018 IECC and 90.1–2016
Massachusetts	2018 IECC and 90.1–2016
Michigan	2015 IECC and 90.1.2013
Minnesota	2018 IECC and 90.1–2016
Mississippi	None statewide
Missouri	Home rule
Montana	2021 IECC and 90.1–2019
Nebraska	2018 IECC and 90.1–2016
Nevada	2018 IECC and 90.1–2016
New Hampshire	2018 IECC and 90.1–2016
New Jersey	90.1–2019
New Mexico	2018 IECC and 90.1–2016
New York	2018 IECC and 90.1–2016
North Carolina	2015 IECC and 90.1–2013
North Dakota	Home rule
Ohio	2012 IECC and 90.1–2010
Oklahoma	2006 IECC and 90.1–2004
Oregon	90.1–2019
Pennsylvania	2018 IECC and 90.1–2016
Rhode Island	2018 IECC and 90.1–2016
South Carolina	2009 IECC and 90.1–2007
South Dakota	Home rule

(*Continued*)

TABLE 3.1 (*Continued*)
State Commercial Energy Codes

State	State Current Code (as of March 2023)
Tennessee	2012 IECC and 90.1–2010
Texas	2015 IECC and 90.1–2013
Utah	2018 IECC and 90.1–2016
Vermont	2018 IECC and 90.1–2016
Virginia	2018 IECC and 90.1–2016
Washington	2018 Washington State Energy Code
West Virginia	90.1–2013
Wisconsin	2015 IECC and 90.1–2013
Wyoming	Home rule

*A home rule state is one where codes are adopted and enforced at the local level. Some home rule states will have a mandate that jurisdictions can go above code but also have to meet a certain minimum code. In general terms, the ideal of home rule is defined as the ability of a local government to act and make policy in all areas that have not been designated to be of statewide interest through general law, state constitutional provisions, or initiatives and referenda.

technology and energy prices. The standard was renamed ASHRAE 90.1 in 2001. It has since been updated in 2004, 2007, 2010, 2013, 2016, 2019, and 2022 to reflect newer and more efficient technologies.

In general, there are two paths for building designers to comply with ASHRAE 90.1:

1. A *prescriptive* path designed to be a fast, definitive, and conservative approach to code compliance. Materials and equipment must meet certain minimum levels of stringency, which are quantified in tables.
2. A *performance* path designed to achieve specific results, rather than meeting prescribed requirements for individual building components. Performance paths typically are based on the anticipated results from application of the prescriptive path. Performance-based approaches use an established baseline measurement from which certain systems must perform. This path requires more detail regarding building design, materials, and systems; however, it is a more flexible approach than the prescriptive path. The approach is particularly desirable for larger buildings, as it provides opportunities for trade-offs across energy-influencing systems to come up with the most cost-effective means for achieving compliance. Performance-based codes are technology neutral, thus enabling quicker incorporation of energy saving technologies and practices into the marketplace.

To estimate the energy efficiency metrics of existing equipment when such information is not available from other sources, the 90.1 prescriptive levels can be used. If you know the age of the equipment but have no information on its energy efficiency metrics, it would be reasonable to assume its energy performance when initially installed would likely have met the ASHRAE 90.1 standard in effect closest to the year the equipment was installed. For example, if no data is available on an existing 10-ton roof-top air conditioner installed in 2000, the ASHRAE 90.1 standard in effect at that time (i.e., 90.1–1999) would have required an energy efficiency (EER) of 8.9. As we shall see later, this energy efficiency metric will then have to be adjusted to reflect any performance degradation from the time it was installed to the present (see Section 3.3).

An excellent resource for previous ASHRAE 90.1 standards may be found at www.ashrae.org/technical-resources/standards-and-guidelines/read-only-versions-of-ashrae-standards. At this website the following 90.1 standards can be reviewed (read only): 1989, 1999, 2001, 2004, 2007, 2010, 2013, 2016, and 2019.

3.1.3 IECC Standard

The International Energy Conservation Code (IECC) is a building code created by the International Code Council in 2000. It is a model code adopted by many states and municipal governments in the U.S. for the establishment of minimum design and construction requirements for energy efficiency. The code is updated every 3 years to provide an ongoing standard of best practices for energy efficiency. IECCs have been published in 2000, 2003, 2006, 2009, 2012, 2015, 2018, and 2021.

There is a difference between the ASHRAE 90.1 standard and IECC. First, ASHRAE 90.1 is a minimum standard of energy efficiency, not a code. The IECC is a model energy code that references the ASHRAE 90.1 standard. Another difference is that ASHRAE 90.1 and IECC have different 3-year cycles. IECC follows behind ASHRAE 90.1 by 2 years. Hence, for example, IECC 2021 follows ASHRAE 90.1–2019.

3.2 ESTABLISHING BASELINE ENERGY EFFICIENCY METRICS FOR EXISTING EQUIPMENT

A general approach to identify the energy efficiency metrics of an existing piece of equipment is as follows.

1. If the existing equipment was initially installed by the contractor asked to replace it, information on the existing equipment may be available in the contractor's files. Whether or not equipment energy efficiency metrics can be found in these files is another matter. Useful information may also be found on the equipment's identification tag (or nameplate), assuming it still exists and is readable. At a minimum, the manufacturer, make and model number (see Figure 3.1) hopefully will be available because internet research might then provide the needed information. If it does not, the manufacturer, assuming the manufacturer is known, can be contacted to see if the needed information might be available.
2. If (1) does not yield the needed information, based upon the approximate age of the equipment, the building energy code in effect when the equipment was installed may provide insight into its energy efficiency metrics. If the local building energy code in effect at that time is not known, it can reasonably be

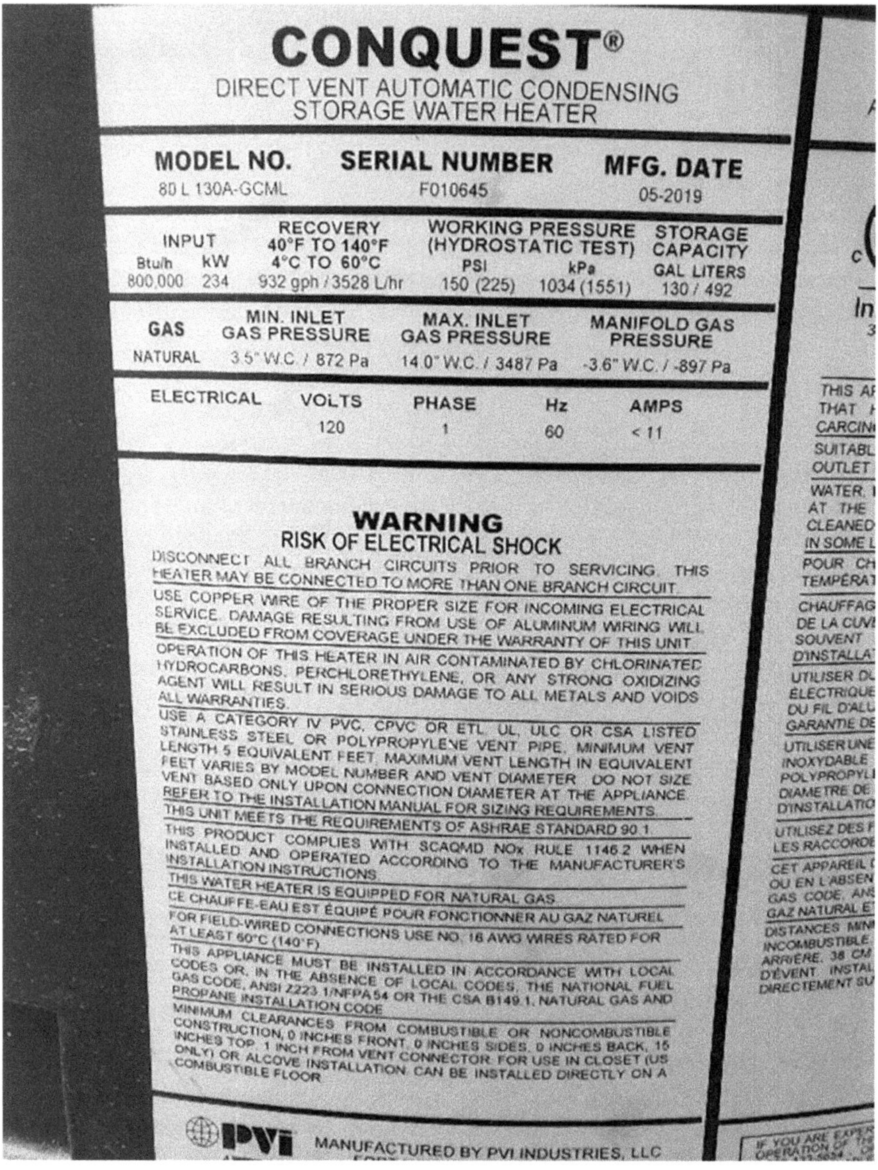

FIGURE 3.1 DHW water heater nameplate.

assumed that the equipment's energy efficiency metrics likely complied with the IECC or ASHRAE 90.1 building energy performance standards in effect at that time. (It should be noted that local building codes may not follow the latest IECC or ASHRAE 90.1 standards. However, most manufacturers design their equipment to comply with the latest ASHRAE 90.1 standard.)

3. Assuming the energy efficiency metrics of the existing equipment when initially installed can be established, a factor to represent performance degradation from the time the equipment was installed to the present should

next be considered. Any performance degradation factor will most likely be dependent upon how the equipment was maintained and usually relies on the contractor's experience. A default annual energy efficiency metric degradation rate for most HVAC equipment between 0.25% and 1.0% is a reasonable estimate in lieu of manufacturer data (see Section 3.3).

4. Once the baseline energy efficiency metrics can be established for the existing equipment and the energy efficiency metrics are obtained for the replacement equipment, an estimated percent energy consumption savings can be determined (refer to later chapters in this book on specific pieces of equipment). The percent energy consumption savings can then be multiplied by the baseline weather-normalized impacted end-use energy consumption (e.g., cooling, space heating) to predict annual energy consumption savings. The estimated energy cost savings can then be calculated by applying the local utility rate schedule(s).

3.3 PERFORMANCE DEGRADATION

All energy-consuming systems in a building have an estimated useful life (refer to Appendix IX). However, these values should be used only as a guide since actual lifetimes will vary depending on many things such as the type of equipment, its use, and how it has been maintained. For example, different practices of HVAC system maintenance can result in substantial differences in energy consumption. If a piece of equipment is not well maintained, its performance will degrade. For example, with packaged roof-top air conditioning units (RTUs), performance degradation might experience an annual performance degradation as suggested in Table 3.2. Well-maintained RTUs typically have an EUL of 15–20 years. However, equipment that is left to run until a failure is encountered (reactive maintenance) might last only 10–12 years or less.

More specific performance degradation default values used by the author are summarized in Tables 3.3 and 3.4. To estimate the performance degradation of HVAC

TABLE 3.2
General Degradation of Cooling Systems under Difference Maintenance Scenarios

Condition of Equipment	Example	Performance Degradation
Well-maintained maintenance contract	Equipment is under a maintenance contract with a focus on proactive, preventive maintenance; no deterioration evident	Minimal, less than 0.25% per year
Minimal level of maintenance	Equipment inspected and maintained to keep it running (not necessarily running well); scheduled maintenance; some evidence of deterioration	Some, perhaps 1% per year
Poorly maintained	Deferred maintenance is the modus operandi; significant deterioration	More, greater than 2% per year

equipment from the time the existing equipment was installed to the time it is being replaced, Equation 3–1 can be used.

$$EEPM_D = EEPM_{NoD} * (1 - M)^A \tag{3–1}$$

where $EEPM_D$ = energy efficiency performance metric with performance degradation (e.g., EER/IEER for RTUs, split system A/Cs, COP for heat pumps; AFUE for boilers/furnaces, full load kW/ton for chillers, etc.)

$EEPM_{NoD}$ = energy efficiency performance metric without consideration of performance degradation

A = age of equipment, years

M = maintenance factor, where for HVAC equipment (air conditioning systems—packaged RTUs and split systems, heat pumps, chillers, boilers, furnaces):

- M = 0.01 for systems with deferred maintenance
- M = 0.03 for systems that are not routinely maintained
- M = 0.0025–0.005 for systems maintained under a maintenance contract
- M = 0.005 default

At the default M = 0.005, the ratio of the degraded energy efficiency metric to the non-degraded is shown in Table 3.3.

TABLE 3.3
Ratio of Degraded to Non-Degraded Energy Efficiency Metric as a Function of Equipment Age for M = 0.005

Age of Equipment	Ratio Degraded to Non-Degraded
10	0.9511
15	0.9276
20	0.9046
25	0.8822
30	0.8604

TABLE 3.4
Annual Performance Degradation as a Function of Maintenance Practices

	Performance Degradation, Percent (%) per Year	
Equipment	Scheduled Maintenance	Deferred Maintenance
DX Coil A/C	0.25	1.0
Chiller	0.25	1.0
Boiler	0.20	0.5
Heat Pump	0.25	1.0

Example Problem 3–1

An existing roof-top packaged air conditioning unit was installed 18 years ago and is being replaced. When originally installed, the unit's energy efficiency metric, EER, was 8.9. The unit experienced only a minimum level of maintenance annually (i.e., deferred maintenance). Estimate the existing unit's energy efficiency today for use as a baseline in calculating the energy savings.
From Equation 3–1:

$$EEPM_D = EEPM_{NoD} * (1-M)^A$$

where $EEPM_{NoD} = 8.9$
$A = 18$

The only unknown in this equation is M. Based on only a minimum level of maintenance annually (deferred maintenance), it is not unreasonable to assume that the performance degradation might be 1% per year (M = 0.01). Therefore:

$$EEPM_D = 8.9 * (1-0.01)\wedge 18 = 7.4$$

Hence, when doing the energy savings calculations, the baseline energy efficiency metric today would be 7.4 rather than 8.9 when the unit was first installed 18 years ago.

REFERENCE

1. U.S. DOE, Office of Energy Efficiency and Renewable Energy, Building energy codes program, www.energycodes.gov/state-portal.

4 Air Conditioning Systems

4.1 INTRODUCTION

The heating, ventilating, and air conditioning (HVAC) system for a building is the equipment that delivers heated or cooled air to various parts of the building. Its purpose is to add or remove heat and moisture where needed (the conditioned space) and remove undesirable air components to maintain the desired environmental conditions for building occupants. As such, the HVAC system must provide an acceptable level of occupancy comfort, maintain good indoor air quality (IAQ), and keep system costs and energy requirements to a minimum.

The HVAC system is responsible for a significant portion of the energy use and energy cost in most commercial buildings. Thus, looking for ways to save on the energy costs of operating a building's HVAC system is an important part of any energy management program.

One of the ways to reduce costs in HVAC systems is to reduce the amount of energy that must be added to or extracted from an area to bring the area to the desired temperature range. Two major strategies for accomplishing this are available:

1. *Reduce the heating or cooling load*: the amount of cooling needed in an area can be reduced, for example, by reducing the amount of heat brought into the area through the building envelope, by reducing the number of heat sources such as lighting, or by modifying the energy consumption characteristics of the equipment. The amount of heating needed can be reduced, for example, by reducing infiltration of cold outside air, or by insulating roofs or ceilings so that less heat escapes.
2. *Change the targeted (set-point) temperature range*: this means changing the temperature limit above which air cooling occurs—the upper set-point, and the temperature limit below which heating occurs—the lower set-point. If heating does not start until the temperature is 55°F or lower, less heat will be used than if the set-point temperature for heating is 65°F. Similarly, an upper control limit of 75°F for cooling to be actuated is more economical than a set-point of 70°F. Changing these temperatures has the effect of changing the heating or cooling load imposed upon the HVAC system.

4.2 EQUIPMENT DESCRIPTION

Air conditioning systems can generally be classified as follows:

1. Based on the cooling/heating fluid that is used:
   a. All-air systems that use only air for cooling or heating.
   b. All-water (hydronic) systems that use only water for cooling or heating.

DOI: 10.1201/9781032692777-4

   c. Air-water combination systems that use both water and air for cooling or heating.
2. Based on the system used (i.e., packaged or split systems). A packaged system is an all-in-one system where air conditioning components are factory selected and assembled in a package, such as packaged roof-top air conditioning units (RTUs) or packaged terminal air conditioning units (PTACs). A split system separates the condenser/compressor, typically outside the building, and evaporator, typically inside the building.
3. Based on interior space zoning (i.e., single zone or multiple zone systems). A single zone system can satisfactorily air condition only one zone in a building. A multiple zone system can satisfactorily air condition many different zones.

The typical components found in HVAC systems include dampers, grilles and filters, condenser and evaporator coils, fans, ductwork, terminal units, and a control system. Each of these contributes to the operation of the HVAC system.

Dampers

A damper controls the flow of air. Dampers are used to regulate the flow of outside air into a system or to control the flow from one part of the building to another part.

Grills and Filters

A screen—or grille—is typically placed upstream of a damper to catch larger contaminants such as bugs, lint, and debris before they enter the air distribution system. Filters are typically used to remove smaller contaminants such as dust and dirt.

Condenser and Evaporator Coils (Heat Exchange Surfaces)

The air that circulates through the HVAC system usually needs to be heated or cooled to be useful. This heating or cooling takes place when air is forced around a coil or finned surface containing hot or cold fluid. If these heat exchange surfaces are fouled with dirt, grease, or other materials with poor heat conduction properties, heat exchange will be inefficient, and more heat or cooling energy will have to be used to heat or cool air to the desired temperature.

Fans

Fans provide the motive power to move air through the air distribution system.

Ductwork

Ductwork directs and conducts air from the heat exchange surfaces to the rooms where the hot or cold air is desired, and it conducts the exhaust air from these rooms back to the mixing plenum and to the outside.

Terminal Units

A terminal unit, or terminal, is a device or equipment installed directly in or above the conditioned space to cool, heat, filter, and mix outdoor air with recirculating air. Fan coil units and variable air volume boxes are common terminal units.

Fan Coil Units

A fan coil unit (FCU) is installed directly inside the conditioned space or in the ceiling plenum just above the conditioned space. The water systems used for the fan coils are either two-pipe or four-pipe systems. A four-pipe fan coil system is equipped with two supply mains: a chilled water supply and hot water supply. There are also two return mains: a chilled water return and a hot water return. A two-pipe system is equipped with a supply pipe and a return pipe. Chilled water is supplied to the coil to cool and dehumidify the air during cooling mode operation. In heating mode operation, chilled water is changed over to hot water and then supplied to the coil to heat the air. In locations with a moderate winter, an electric heating coil in the FCU is sometimes used for winter heating, simplifying the operation of the fan-coil system.

Outdoor air is often conditioned (cooled, heated, dehumidified) in a separate outdoor air unit or makeup air handling unit (MAU). The outdoor air is then transported to the fan coils where outdoor air is mixed with recirculated air and conditioned as needed. If the conditioned outdoor air is supplied directly to the conditioned space via supply ducts in a dedicated outside air ventilation system (DOAS), then it is mixed with the conditioned recirculating air from the fan coil unit serving the conditioned space.

Variable Air Volume Boxes

In variable air volume (VAV) systems (see Section 4.2.1.1), the supply air temperature is constant, and the supply airflow rate to the zones is varied. Each VAV box serves a certain zone and is connected via cable to a thermostat located within the zone. The thermostat indicates to the VAV box if it needs to increase or decrease the flow rate to meet the demand.

HVAC Control System

An HVAC control system transforms the operating instructions for desired environmental conditions into the air temperatures and ventilation volumes desired in the working environment. The control system has the task of regulating the HVAC system so that these instructions are met as close as possible. The control system accomplishes its function through a system of sensors, actuators, and communication links. The sensors send appropriate electrical or pneumatic signals when some temperature, pressure, or humidity threshold has been crossed. Upon reaching their destinations, the control signals are then translated into additional pneumatic or electrical signals that are used to open or close dampers, to regulate fans, and to initiate or stop the source of heating or cooling. Dampers can be opened or closed to regulate the amount of incoming air, the amount of exhaust air that is mixed with fresh air, and the amount of air that is introduced into an area. Fans can be turned on or off to increase air coming into a room or being exhausted from it, or their speed can be regulated so that the amount of air coming into a room is no more than is needed for proper ventilation and temperature control. Boilers or chillers can be turned on to provide heat or cooling for the heat exchangers.

Air Conditioning System Sizing Rules of Thumb

If the output cooling capacity of an air conditioning system is not available, a rule of thumb on estimating capacity is to divide the building's total square feet (ft^2) that is conditioned by somewhere between 150 and 550 ft^2/ton, depending on the type building

TABLE 4.1
Cooling Capacity General Rules of Thumb for Different Property Types*

Building Type	Building ft^2/ton
Apartment	450
Office	400
Public Assembly	
Auditorium, Church, Theater	400
Library, Museums	300
Schools	250
Hospital	350
Lodging	350
Retail	300
Restaurants	150
Shopping Centers	400

*A building's climate zone may have a noticeable impact on these values.

and climate zone (location), to get an approximate cooling capacity in tons. For different building types, common values used are presented in Table 4.1. These are only rough estimates as they do not reflect the building's climate zone. As such, every effort should be made to obtain the existing equipment's output capacity (refer to Chapter 3).

It should be noted that the terms "output cooling capacity" and "refrigeration capacity" are often used interchangeably. Also, you may see "RT" or "refrigeration ton" used, which is equal to 12,000 Btu/hour.

4.2.1 Packaged RTUs/Split Systems/PTACs

Unitary heating, ventilation, and air conditioning systems, which combine heating and cooling in one unit (packaged) or a few sections (split), are one of the most common technologies used for space conditioning in commercial buildings. They are designed to be flexible so that the equipment can be installed indoors (e.g., in a mechanical room, attics, ceilings) or outdoors (on roof-tops or on the ground).

The most common HVAC systems designed to satisfy building cooling requirements are direct expansion (DX) systems, in which there is direct heat exchange between the building air and the refrigerant. The basic vapor compression cycle, when applied directly to the job of building cooling, is referred to as a DX refrigeration system (see Figure 4.1). In Figure 4.1, the compressor constricts the refrigerant vapor and moves it towards the condenser. The heat of compression raises the temperature of the refrigerant vapor causing it to be a high pressure superheated vapor. As the refrigerant moves into the condenser, the condenser rejects the heat in the refrigerant, causing it to change state and condense into a high pressure, high temperature liquid. As the refrigerant passes through the metering device,

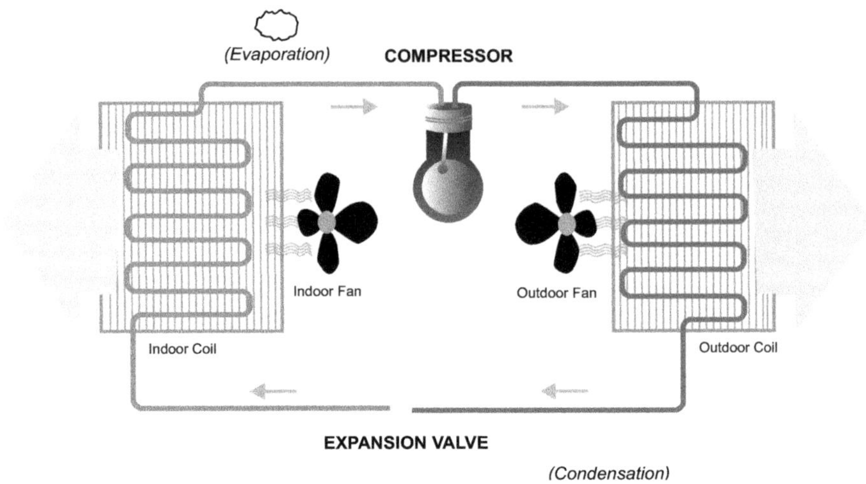

FIGURE 4.1 DX air conditioning system schematic. (*Source: Courtesy of Carrier.*)

its temperature, pressure and state change once again. Some of the low-pressure liquid refrigerant instantly boils off, forming "flash gas." As this mixture of liquid and gas passes through the evaporator, heat is absorbed, and the remaining liquid refrigerant changes its state back into a vapor. At the outlet of the evaporator, 100% of the low-pressure vapor flows back through the suction line to the compressor.

In the DX system, a direct-expansion coil is used as the evaporator where the heat of vaporization is removed from the air to be cooled and conducted to spaces requiring conditioning. The condenser releases the heat of condensation into the incoming air. The condenser is often combined with the reciprocating compressor(s) to form the condensing unit that is generally located outdoors. Either the DX coil is installed in a roof-top packaged unit consisting of fan(s), filters, condensing unit, and controls, mounted on the roof of the building; or it is a component in a split (separate condenser and evaporator units) system. Air is forced through the DX coil by a fan for better heat transfer.

Packaged RTUs

Most packaged roof-top units (see Figures 4.2 and 4.3) are small to medium-sized systems in which the bulk of the refrigeration machinery is located on the roof. These usually have reciprocating, scroll, or rotary compressors with air-cooled condensers. Packaged RTUs are usually applied to low-rise buildings, such as small office buildings or shopping malls. In a unitary system all components of the system, including the air handlers, are located on the roof. In a split system the compressor and condenser are located on the roof (or on the ground adjacent to the building), while the evaporator(s) and air handler(s) are located inside the building. Sophisticated roof-top systems may use single zone, multizone, or VAV arrangements. They also may be equipped with a natural gas or electric heating section.

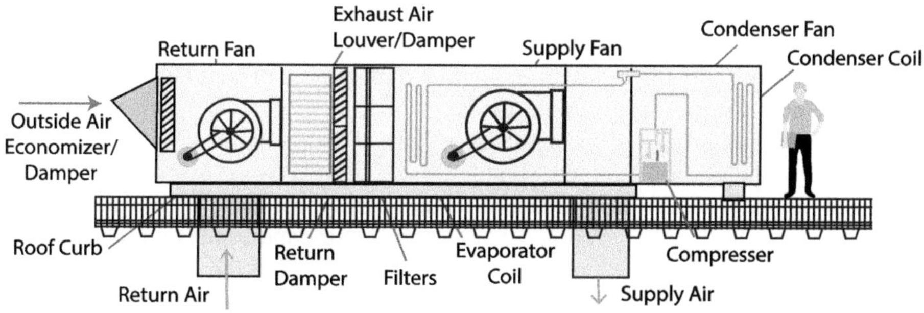

FIGURE 4.2 Typical packaged RTU schematic. (*Source: Courtesy MEPAcademy.*)

FIGURE 4.3 Packaged RTU. (*Source: Courtesy of Carrier.*)

There are many reasons for the popularity of packaged roof-top units including:

- Relatively low first cost.
- Ease of installation.
- Elimination of the need for a machinery room in the building.
- Maintenance can be conducted without disrupting activities within the building and can be carried out without access to the building when the building is closed.
- Placing equipment on a roof does not reduce the rentable floor area, as may be the case with equipment located inside the building enclosure (e.g., in a machinery room).
- Each tenant can own and operate a separate unit, relieving the landlord of the need to service and pay for equipment operation.

There are some disadvantages, including the following:

- Critical roof-top units must be maintained regardless of weather conditions (e.g., heavy rain, snow or high winds).
- Units are not protected from outside weather conditions.

Packaged air conditioners are often divided into two types: those with water-cooled condensers and those with air-cooled condensers. In air conditioners with air-cooled condensers, the condenser is cooled by atmospheric air. The fan located inside the unit sucks the outside air and blows it over the condenser coil, cooling it in the process. In air conditioners with water-cooled condensers, the condenser is cooled by water. The condenser is of the shell and tube type, with refrigerant flowing along the tube side and cooling water flowing along the shell side. Water is supplied continuously in these systems to maintain operation of the air conditioning system. Packaged systems with air-cooled condensers are used more commonly than those with water-cooled condensers since air is freely available.

Split Systems

Split systems are DX systems in which the evaporator and condenser are separated and placed at different locations (see Figures 4.4 and 4.5), with the compressor located near the condenser. They are connected by refrigerant lines to the air handling system, which includes the evaporator. The refrigerant piping, even with its insulation, is only inches in diameter, compared to ducts that typically are feet in diameter. Ducts are notorious for losing energy due to their construction, with up to

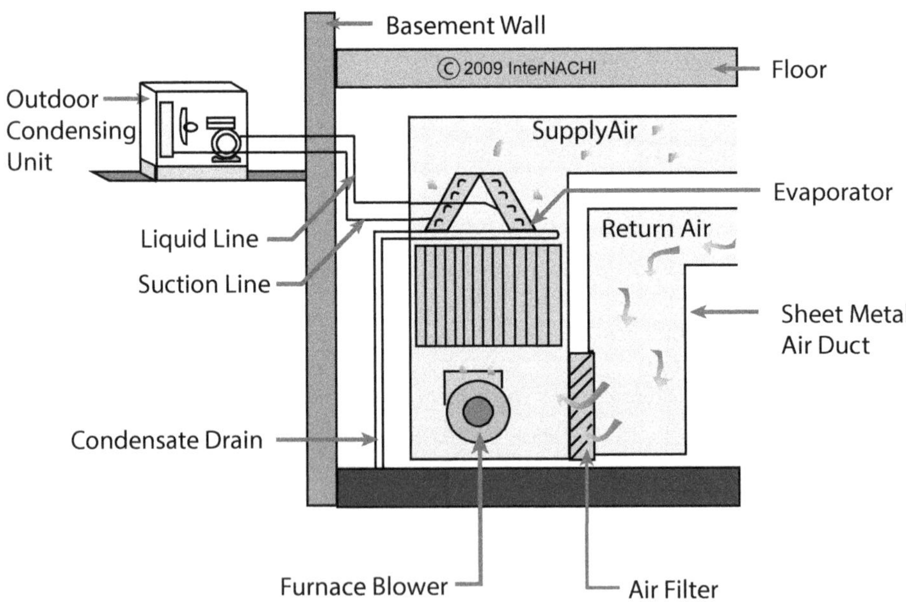

FIGURE 4.4 Split system schematic. (*Source: Courtesy InterNACHI.*)

FIGURE 4.5 Split system air-cooled condenser. (*Source: Courtesy of Carrier.*)

30% of generated energy dissipating while making its way through ductwork. The main advantage of the split system is that it allows the air handling part of the unit to be indoors, where it is easier to maintain and does not need to be weatherproofed. The noise of the compressor is outside and can be located at some distance from the air-handling unit.

PTACs

A through-the-wall air-cooled room air conditioner (see Figure 4.6), known as a packaged terminal air conditioner (PTAC), is designed to cool or heat individual room spaces, such as in motels, hotels, and apartments. They have the following advantages:

- There are no supply, return, or exhaust ducts.
- They are compact, flexible, and have low initial cost.
- Building space is saved.
- It is easier to match the requirements of an individual control zone.
- They are quick to install.

They have the following disadvantages:

- Temperature control is usually on/off, resulting in space temperature swings.
- Air filters are limited to coarse or low-efficiency filters.
- Local outdoor ventilation air intake is often affected by wind speed and wind direction.

FIGURE 4.6 PTAC unit.

- Noise level may not be suitable for certain applications.
- More regular maintenance of coils and filters is required than for packaged and central systems.

4.2.1.1 Variable Air Volume Systems

A common upgrade to the single zone packaged RTU is implementation of a VAV system (see Figures 4.7 and 4.8). Unlike constant air volume (CAV) systems, which supply a constant airflow at a variable temperature, VAV systems vary the airflow at a constant or varying temperature. Compared to CAV systems, VAV systems provide more precise temperature control, reduced compressor wear, lower energy consumption by system fans, and less fan noise. Instead of a motor starter, a variable speed drive (VSD) has been added to the supply air fan. Instead of one space temperature sensor, there are multiple space temperature sensors serving multiple zones. For each zone there is an associated control damper (known as a VAV box or VAV terminal unit). The zone damper, serving its individual space(s), modulates open or closed based on input from the space temperature sensor. As the zone dampers open and close, the static pressure in the supply ductwork changes. A static pressure sensor detects these changes and sends a signal to the VSD on the supply fan to change its speed/airflow higher or lower. With VAV systems, each individual space almost never experiences situations where its temperature cannot be satisfied. VAV system upgrades to packaged RTUs are common today in commercial buildings.

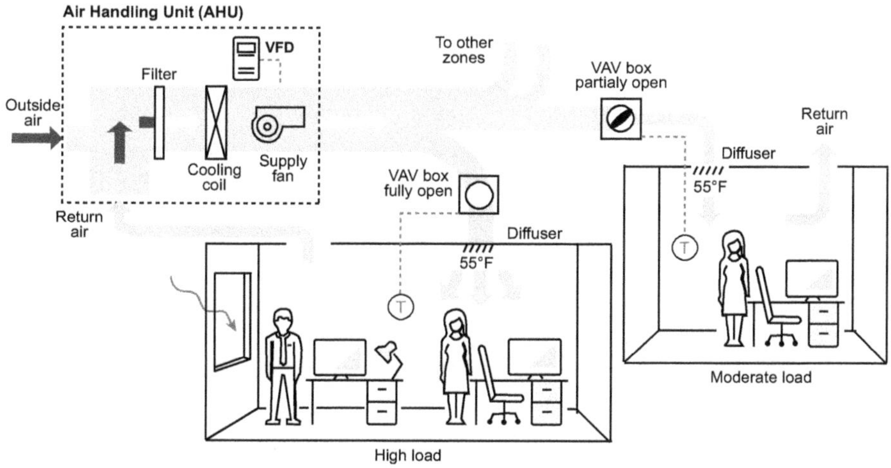

FIGURE 4.7 VAV schematic. (*Source: Credit: U.S. DOE, PNNL, VAV Systems O&M.*)

FIGURE 4.8 VAV terminal box with reheat coils.

4.2.2 Chiller Plants

Chilled water systems are the heart of central HVAC cooling, providing cooling throughout a building or group of buildings from a single source. Centralized cooling offers numerous operating, reliability, and efficiency advantages over individual DX systems and, on a life-cycle basis, can have significantly lower total cost. Chilled water from a chiller is sent to cooling coils where it subsequently cools air. Typical applications for chilled water systems include large buildings (offices, laboratories, etc.) or multi-building campuses where it is desirable to provide cooling from a central facility.

Air-Cooled Chillers

Air-cooled chillers (see Figures 4.9 and 4.10) are very common, especially in smaller commercial and office-type buildings. They are usually located externally, either up on the roof or at ground level. This is because air-cooled chillers do not use cooling towers; instead they dump their heat into the ambient air and therefore need access to a lot of fresh air to reject the unwanted heat from the building. The chillers produce "chilled water," which is pumped out around the building to the air handling units (AHUs) and fan coil units (FCUs).

The cycle starts with the evaporator, which has a liquid refrigerant that flows over the evaporator tube bundle where it evaporates. In the process, heat is absorbed from the chilled water circulating through the bundle. The compressor then pulls the refrigerant vapor from the evaporator and pumps it to the condenser. The refrigerant condenses when it is in the condenser tubes, releasing its internal heat to the air. The high-pressure liquid then moves through the expansion device and into the evaporator. To complete the continuous cycle, the refrigerant flows back over the chilled water coils and absorbs more heat.

Centrifugal and screw compressors are widely used in large chillers, whereas screw, scroll, and reciprocating compressors are widely used in small and medium-sized chillers. Screw and scroll types are also gradually replacing the reciprocating compressors in small and medium-sized chillers.

Water-Cooled Chillers

Water-cooled chillers (see Figures 4.11 and 4.12) are almost always located inside the building. They work almost the same way as air-cooled chillers. The difference is that they remove heat from chilled water by exhausting the heat to a second isolated water line called the condenser water line.

The condenser water flows through the chiller and picks up heat. The condenser water then returns to the cooling tower located outside of the building where heat is removed from the condenser water by evaporating some of the condenser water into the atmosphere. As some of the condenser water evaporates, heat is removed from the condenser water and the cool condenser water flows back to the chiller.

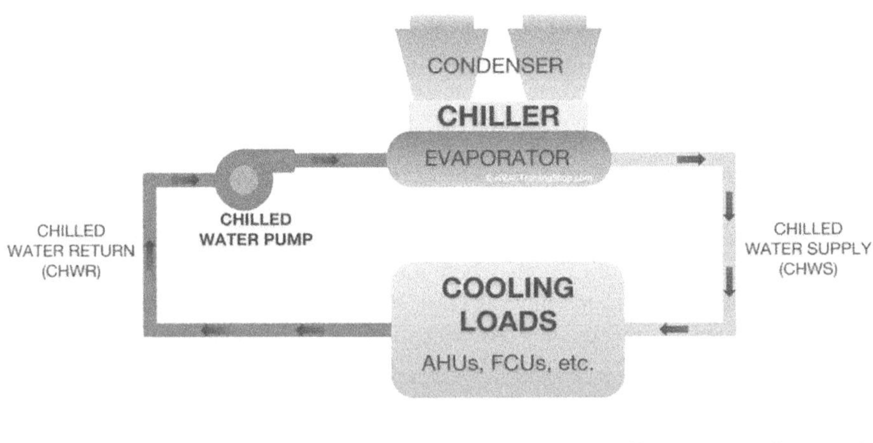

FIGURE 4.9 Air-cooled chiller schematic. (*Source: Courtesy of hvactrainingshop.com.*)

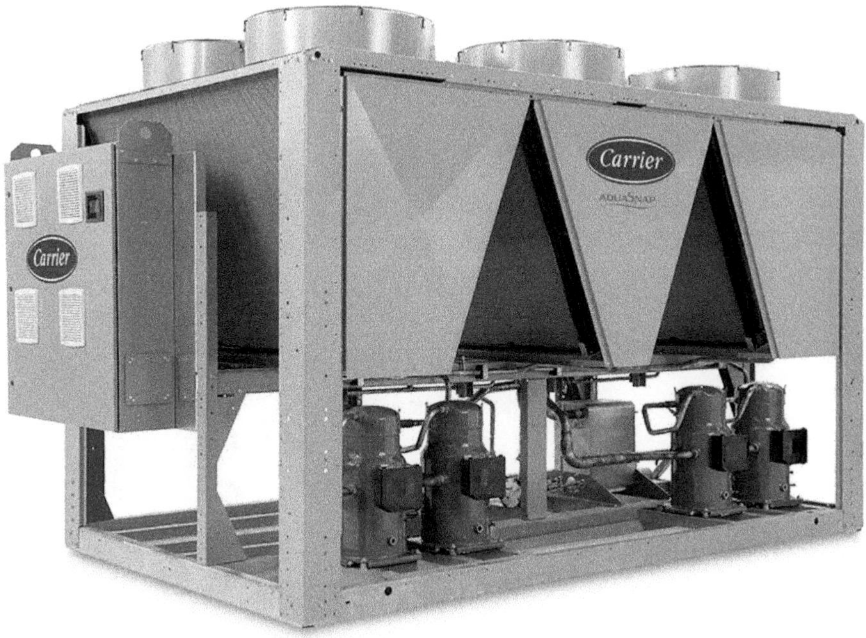

FIGURE 4.10 Air-cooled scroll chiller. (*Source: Courtesy of Carrier.*)

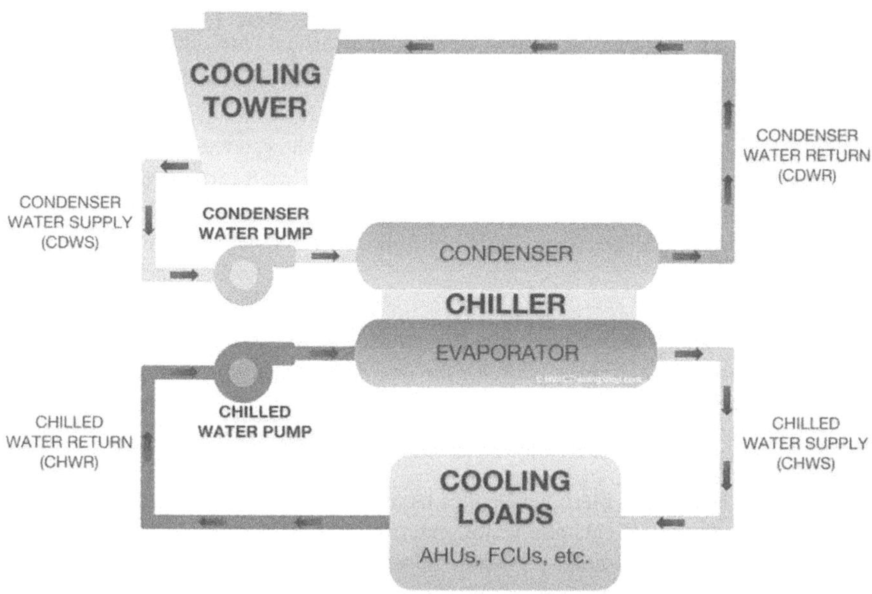

FIGURE 4.11 Water-cooled chiller schematic. (*Source: Courtesy of hvactrainingshop.com.*)

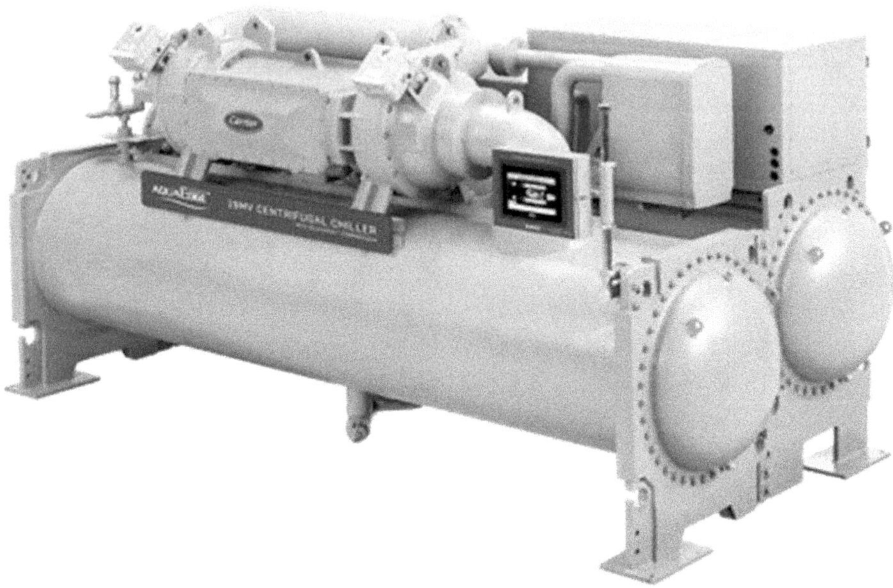

FIGURE 4.12 Water-cooled centrifugal chiller. (*Source: Courtesy of Carrier.*)

Cooling Towers

Cooling towers (see Figures 4.13 and 4.14) are used to reject heat from a condenser to the outdoor air via the use of condenser water. More than 1,000 Btu of heat can be released to the atmosphere for each pound of water evaporated. A condenser water pump circulates condenser water from the cooling tower basin, through the piping system and the condenser, finally returning the water to the cooling tower wet deck. For large building air conditioning systems, vertical counterflow cooling towers are the most prevalent. They are equipped with either induced draft or forced draft fans.

Every cooling tower, no matter how configured or constructed, includes the following components:

1. *Fill*—the heat transfer media in the cooling tower designed to maximize contact between the return condenser water and the ambient air.
2. *Hot water distribution ("wet deck")*—pans or basins with spray nozzles designed to provide an even distribution of the return condenser water entering the fill.
3. *Cold water basin*—the basin, either as an integral part of the tower or as a separate sump, collects the water passing through the tower for supply to the system by the condenser water pump. The basin must also be sized to contain enough water to supply the condenser water system until the pump returns water to the tower.
4. *Fan(s)*—all cooling towers used for HVAC applications are mechanical draft towers that use one or more fans to provide airflow through the tower.

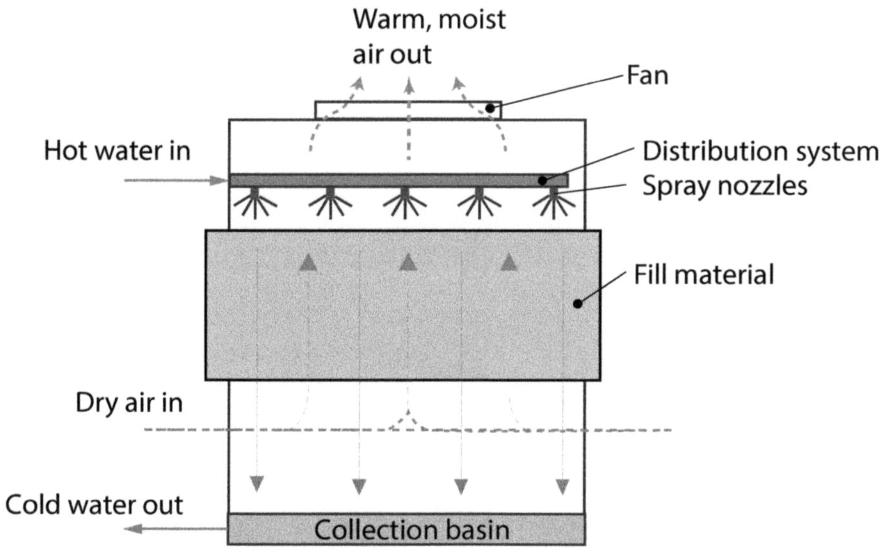

FIGURE 4.13 Cooling tower schematic.

5. *Inlet louvers and drift eliminators*—inlet louvers act to force the air entering the tower into as straight and even flow pattern as possible, while the drift eliminators are designed to trap and remove any entrained water droplets that may exit the tower.

In counterflow towers (as shown in Figure 4.13), the water and air move in opposite directions; that is, the condenser water flows vertically downward and the air flows vertically upward. In crossflow towers, the two flow streams are arranged at 90 degrees to each other: the water flows vertically downward through the fill, while the air flows horizontally through it. One or more fans, connected to one or more motors via a drive assembly, provide the motive power for airflow through the cooling tower. Two types of fans are commonly used: centrifugal fans and axial propeller fans. Both types can be used with forced draft cooling towers, but propeller fans typically are only used with induced draft towers.

Variable frequency drives (VFDs) have been successfully applied to cooling tower fan motors so that condenser water supply temperature can be controlled by changing the fan speed in direct proportion to the change in condenser water supply temperature.

4.2.3 Evaporative Pre-Cooling

Depending upon a building's operating hours and location, evaporative cooling/pre-cooling (see Figures 4.15 and 4.16) can be applied effectively from both a cost and performance perspective. Evaporative cooling works best in hot, dry climates, such as in the Southwestern U.S. In low-humidity areas, evaporating water into the air provides a natural and energy-efficient means of cooling. Evaporative coolers,

FIGURE 4.14 Cooling tower installation.

also called swamp coolers, rely on this principle. By passing outdoor air over water-saturated pads, the water in the pads evaporates, typically reducing the air temperature by 15° to 30°F before it is directed into the main air conditioning unit or directly into the space to be cooled.

Two general types of evaporative cooling equipment are available: spray-type air washers and wetted-media–type coolers. Spray-type air washers are versatile in that

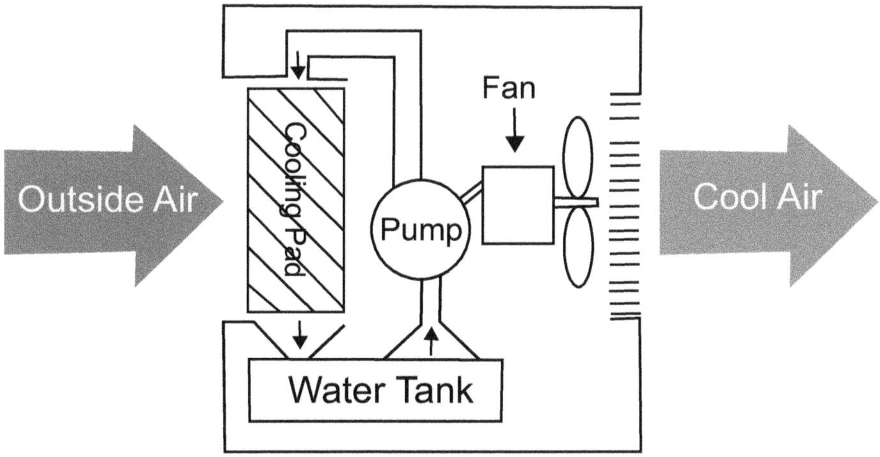

FIGURE 4.15 Evaporative cooling schematic.

FIGURE 4.16 Evaporative (or swamp) cooler installation.

they can use chilled water from a refrigeration system when the outdoor air wet bulb temperature is high, but they can also provide cooling without chilled water when the outdoor air is suitable for conventional evaporative cooling. Water reservoirs require constant maintenance to prevent corrosion and the buildup of bacteria and fungi that would otherwise be dispersed into the airstream during the evaporative cooling process. When this is not done, air washers can become a potential source of biological contaminants and a cause of poor indoor air quality and/or sick building syndrome. Because wetted-media coolers use nonrefrigerated water, they are adaptable to air-conditioning systems only in those climate zones where very low ambient wet bulb temperatures prevail year-round. In a climate such as that of Phoenix, Arizona, a dual air conditioning system may make sense—wetted media for 10 months of the year when dry ambient conditions are prevalent and compression refrigeration during the 2-month humid season. The extremely low operating cost of wetted media evaporative coolers often justifies the higher initial investment of a dual air conditioning system.

The systems described thus far are direct evaporative coolers because water is injected into the primary (supply) airstream, which is cooled adiabatically. Indirect evaporative cooling can take advantage of the evaporative cooling process but without increasing humidity. The main difference with an indirect system is that a heat exchanger is used to cool the air supplied to the building space. The evaporative cooling cycle occurs in the heat exchanger. Hot outside air is blown through a heat exchanger that is supplied with water. One design for this type of heat exchanger features a series of metal tubes that are kept wet on their outside surfaces. As hot air passes over these tubes, the water evaporates, and the tubes are cooled. After passing over the tubes, the cool moist air is exhausted to the outside. As cooling happens on the heat exchanger's exterior surfaces, hot exterior air is drawn through the tube interiors. This air is cooled, but without gaining any extra humidity, before it is blown through ductwork to the building interior. This cooling method is more suitable for areas where additional humidity isn't desirable for interior air. Because indirect evaporative cooling requires two fans rather than one—the supply fan and a smaller secondary fan for the heat exchanger's airflow motor, it consumes more electricity than direct evaporative cooling.

By their nature, evaporative coolers also continually use water. In areas with limited water supply, building owners may be concerned about the water-use impact of adding an evaporative cooler.

4.3 ENERGY RECOVERY

When evaluating how much energy, Q, is available for recovery, there are two equations that can be used, depending on the medium that is carrying the energy.

If the medium is air:

$$Q_A = \rho_A * c_{pA} * 60 * CFM * \Delta T$$

$$= (0.075) * (0.24) * 60 * CFM * \Delta T$$

$$Q_A = 1.08 * CFM * \Delta T \tag{4–1}$$

where Q_A = energy in air for a specific air ΔT, Btu/hr
ρ_A = density of air, lb/ft^3
c_{pA} = specific heat capacity of air, Btu/lb °F
CFM = airflow rate, ft^3/min
ΔT = temperature difference in air, °F
60 = 60 minutes/hr

If the medium is water:

$$Q_W = \rho_W * c_{pW} * 60 * GPM * \Delta T$$

$$= (8.33) * (1) * 60 * GPM * \Delta T$$

$$Q_W = 500 * GPM * \Delta T \quad (4\text{–}2)$$

where Q_W = energy in water for a specific water ΔT, Btu/hr
ρ_W = density of water, lb/gal
c_{pW} = specific heat capacity of water, Btu/lb °F
GPM = water flow rate, gal/min
ΔT = temperature difference in water, °F
60 = 60 minutes/hr

4.3.1 Economizers

An economizer (see Figures 4.17 and 4.18) uses cool outdoor air to cool the building instead of operating the air conditioner's compressor. An economizer is an add-on feature to an HVAC air handler that draws in outdoor air and mixes it with return air from indoors. A controller in the economizer measures the temperature and humidity of the outdoor air. When the outdoor temperature is favorable, the control switches the air conditioner's compressor off and cools the building directly with outdoor air. The economizer cools the building using far less energy compared to compressor air conditioning.

Economizers often operate at night when the outdoor air is cooler than the indoor air in a process referred to as "free cooling." Most commercial buildings have some spaces that need cooling all year long. If it is colder outdoors than indoors, it often makes energy sense to "economize" by bringing in outdoor air to reduce the hours of mechanical cooling system operation. Economizers mix enough outdoor air with the indoor air to meet the thermostat set-point without using the compressor.

Economizer equipment costs extra and, depending on the climate, the hours when the economizer is saving cooling energy may be limited. In climates that are warm and humid, the number of hours when the outside air has a lower enthalpy than the return air enthalpy may be very few. In southern states such as Florida, for example, economizers are seldom used. This is because the number of hours during which the outside enthalpy falls below the controlled space temperature is insufficient to justify the investment.

Economizers may be categorized as dry bulb economizers, single enthalpy economizers, differential enthalpy economizers, or integrated differential enthalpy

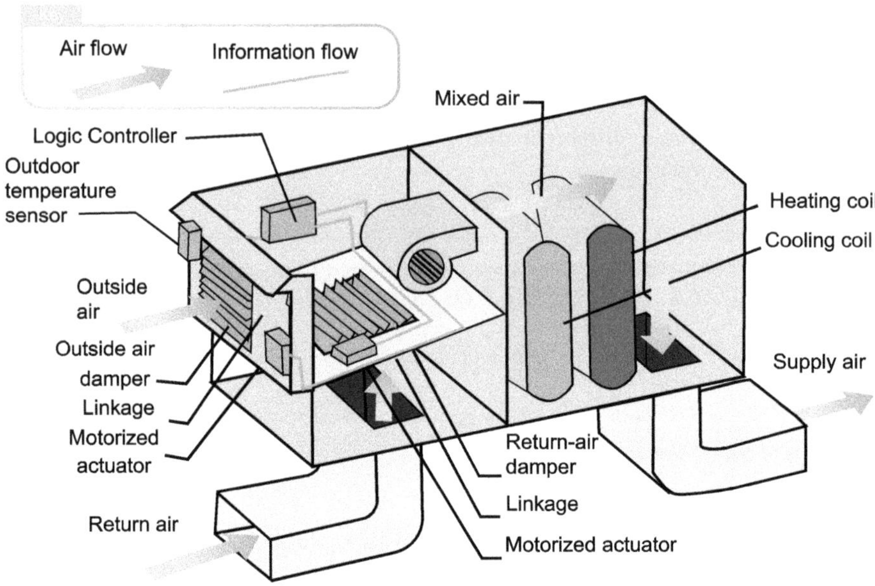

FIGURE 4.17 Air-side economizer schematic. (*Source: Credit: U.S. EPA, ENERGY STAR.*)

FIGURE 4.18 Economizer installation.

economizers. Dry bulb economizers have sensors that detect if outdoor air is a certain temperature, which will then open a damper to allow outdoor air in. There is no detection for humidity, which could make for unfavorable conditions inside the conditioned space. Single enthalpy economizers have sensors to determine if the humidity and heat levels are below presets before using outdoor air to cool a building. Differential enthalpy economizers have two sensors, one to measure indoor air enthalpy and one to measure outdoor air enthalpy. Integrated differential enthalpy economizers have sensors that track return and outdoor air enthalpy, plus there are controls that communicate with an indoor thermostat. If the outdoor air temperature and enthalpy are low enough, cooling will start with the economizer, and if not, the compressor will cool the building.

Allowing some of the exhaust air to be recirculated back into the fresh air intake to offset the heating or cooling demand will also necessitate that you evaluate what you are doing to be sure it is safe and healthy. As such, it may be necessary to use CO_2 sensors to ensure the exhaust air has a low CO_2 level. If the CO_2 level is too high, then the air cannot be re-used, the economizer mixing damper will close, and then all the return air will be rejected from the building.

4.3.2 Thermal and Desiccant Wheels

A thermal wheel (or rotary heat exchanger) is a slowly rotating matrix of finely corrugated metal, operating in both opposing airstreams (see Figures 4.19 and 4.20).

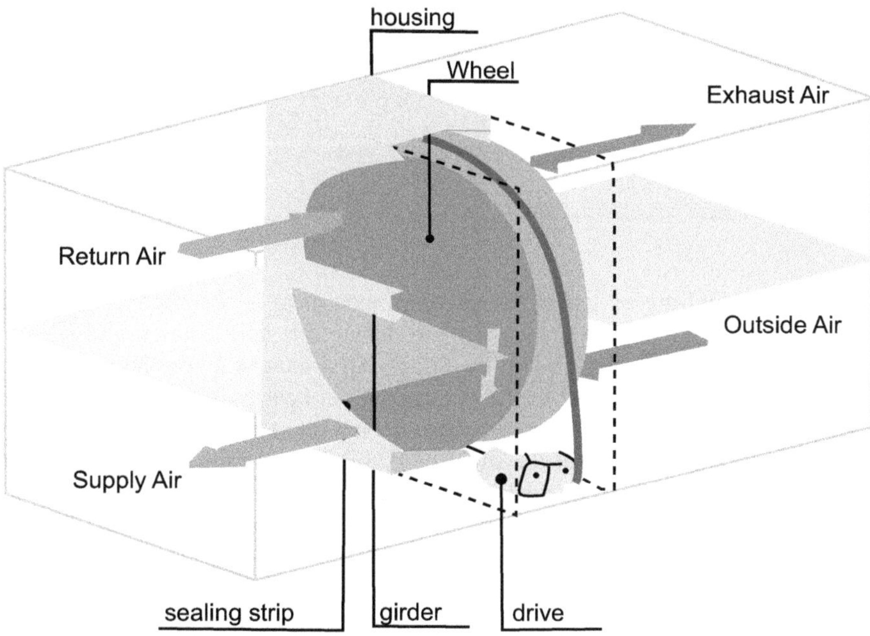

FIGURE 4.19 Rotary heat exchanger schematic.

FIGURE 4.20 Heat wheel installation.

When the air handling unit is in the heating mode, heat is absorbed as air passes through the matrix in the exhaust airstream during one-half rotation and released during the second half rotation into the supply airstream in a continuous process. When the air handling unit is in the cooling mode, heat is released as air passes through the matrix in the exhaust airstream during one half rotation and absorbed in the second half rotation into the supply airstream. Heat recovery efficiencies up to 85% are possible. Wheels are also available with a hydroscopic coating to provide latent heat transfer and the drying or humidification of airstreams.

Desiccant wheels contain chemicals that are quick to pick up heat and moisture, and quick to give them up again if exposed to a cooler, drier atmosphere. In use, the supply air is ducted through one-half of the wheel and the exhaust air through the other half. Suppose it is a hot summer day, so the exhaust is cooler and drier than the supply of outside air. The chemical coating in the section of the coil in the exhaust stream becomes relatively cool and dry. Now the wheel is slowly rotated and the cool, dry section moves into the incoming hot, humid air, drying and cooling the air. Similarly, a section is moving from hot and humid into cool and dry, where it gives up moisture and becomes cooler. The wheel speed, a few revolutions per minute, is adjusted to maximize the transfer of heat and moisture. Control of wheel speed to maximize savings is a complex issue since the transfer of sensible and of latent heat do not vary in direct relation to each other. Some cross-contamination is possible, but there are ways of minimizing (though not eliminating) it.

4.3.3 Heat Exchangers and Heat Pipes

The plate-type heat exchanger uses thin sheets of metal to separate the two streams of air, so they do not come into contact or mix at all. The temperature difference between the two airstreams will cause heat to transfer over from the hot exhaust stream through the metal walls of the heat exchanger and into the cold intake stream. Heat recovery efficiencies up to 70% are possible.

Another energy savings solution is the heat pipe (see Figure 4.21). A heat pipe is a metal tube that is filled with an evaporative fluid (refrigerant) and then sealed at both ends. When heat is applied to one end of the tube, the refrigerant inside evaporates and the vapor moves to the other, cooler end. The vapor then condenses in the cooler end and returns to the warmer end by gravity to repeat the process. The heat pipe is activated by the temperature difference between the two ends and does not use energy to operate.

When used in an air conditioning system, one end of the heat pipe is placed in the return air system and is heated by the warm return air. That heat is "free" because no additional energy is expended to generate the heat. Because the heat pipe absorbs heat from the return air, that air is cooler when it goes to the cooling coil so the cooling coil can work at lower temperature. This means that the cooling coil can remove more moisture from the air; it also means that the cooling load on the compressor

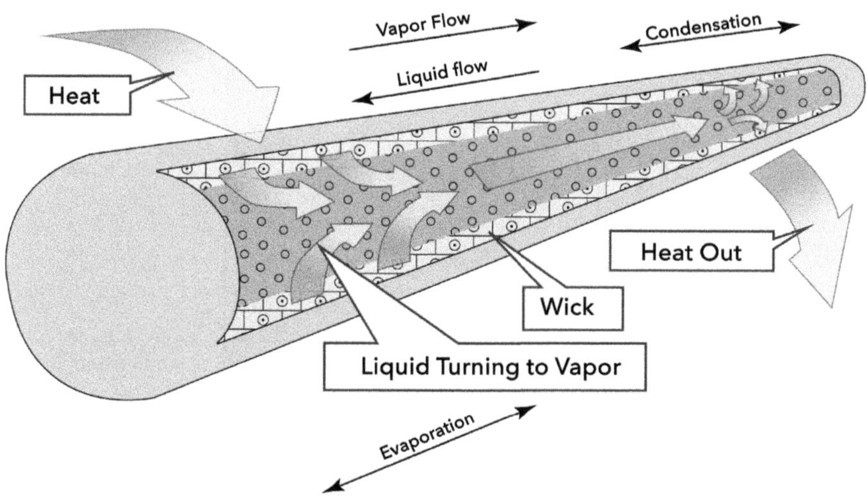

FIGURE 4.21 Heat pipe schematic.

is lower. The other end of the heat pipe is placed in the supply, or conditioned air system. The heat from the first section in the return airstream is transferred to this section which is used to reheat the chilled air and lower the humidity before the air is distributed throughout the system. The process of reheating the air causes that end of the heat pipe to become chilled.

4.3.4 Energy Recovery Ventilators

An energy recovery ventilator (ERV) is an air-to-air heat exchanger that acts as an additional heat recovery agent when coupled with a dedicated outdoor air system (DOAS). The energy recovery ventilator works by pre-tempering the incoming air with heat that is being exhausted from the building, resulting in additional and often significant operational savings. An energy recovery ventilator is ideal for more severe climates with extreme design conditions.

Energy can be recovered either in its sensible (temperature only) or latent (moisture) form, or a combination of both from multiple sources. Sensible energy can be extracted, for example, from outgoing airstreams to heat supply air. Units used for this purpose are called sensible heat exchange devices or heat recovery ventilators (HRVs). Devices that transfer both sensible heat and moisture (latent heat) are known as energy or enthalpy devices or ERVs (see Figures 4.22 and 4.23). Air conditioners use considerable energy to dehumidify moist airstreams. ERVs are designed to capture cooling and heating energy and can enhance dehumidification. They typically recover sensible and latent heat energy from building exhaust air by indirectly crossing airstreams through a rotating wheel (see Section 4.3.2) or fixed plate heat exchanger (see Section 4.3.3). ERV systems are a more popular choice in hot-humid climate zones such as in the Mid-West and Southern regions of the U.S., where reducing humidity in the incoming air is important for occupant comfort. HRV systems typically are more popular in colder, northern climates with longer heating seasons.

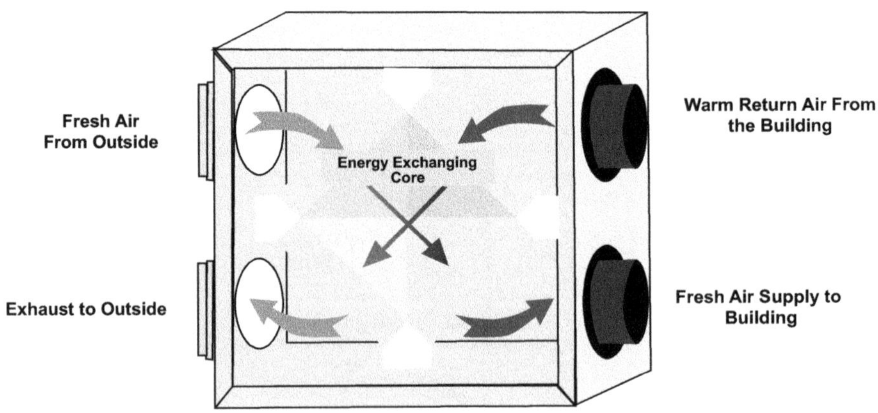

FIGURE 4.22 Energy recovery ventilator schematic. (*Source: Courtesy of Carrier.*)

FIGURE 4.23 ERV-DOAS installation.

Introducing outdoor or ventilation air is the primary means of diluting air contaminants to achieve acceptable indoor air quality. ERVs can cost-effectively provide large amounts of outdoor air to meet a commercial building's minimum ventilation requirements as prescribed in ASHRAE Standard 62.1.[1]

4.4 ENERGY EFFICIENCY METRICS

4.4.1 Air Conditioning Systems

The efficiencies of air conditioners are usually measured in terms of their energy efficiency ratio (EER) and integrated energy efficiency ratio (IEER) in units of Btu of cooling output (or cooling delivered to the space to meet cooling load requirements) per watt-hour of electricity input (energy consumed as reflected in utility bills). EER/IEER units can also be expressed as the Btu/hr of cooling output to the watts of electric power input. EER represents the unit efficiency at maximum (full) load capacity (at a single outside air temperature). However, operation at full load typically represents operation only a small percentage of the time, typically about 2% of a given unit's operating hours. Accordingly, while it is useful to know the unit's EER, it can be misleading since it only accounts for a small fraction of the unit's overall hours. With this limitation in mind, the IEER was developed. IEER reflects the efficiency of HVAC equipment operating at different load levels. This ratio accounts for both full load and part load efficiencies and does so by weighting EER values at different load capacities and adding them together. The IEER weightings in the formula are based on the approximate number of hours a unit spends at each load point throughout its life. The IEER encompasses EER ratings at four different load percentages: 2% at 100% load, 61.7% at 75% load, 23.8% at 50% load, and 12.5% at 25% load. This is shown in Equation 4–3.

$$\text{IEER} = 0.02*\text{A} + 0.617*\text{B} + 0.238*\text{C} + 0.125*\text{D} \tag{4–3}$$

where IEER = integrated energy efficiency ratio, Btu output/W-hr input
A = EER for the A/C running at 100% capacity (2% of the time)
B = EER for the A/C running at 75% capacity (61.7% of the time)
C = EER for the A/C running at 50% capacity (23.8% of the time)
D = EER for the A/C running at 25% capacity (12.5% of the time).

For small units, the efficiency might be represented by the seasonal energy efficiency ratio (SEER). The SEER is calculated using the output Btu of the air conditioner over the cooling season (e.g., summer) divided by the total W-hr of electric energy input over the same cooling season. (*N.B.* Starting in 2023, the SEER2 standard replaced the SEER rating with the main difference being the testing conditions.)

New high efficiency unit EERs are typically in the 12.6–14.3 range (or IEERs in the 14.8–18.5 range) depending on size (tons capacity); older units might have EERs in the 6.8–10.3 range. EERs/IEERs will be somewhat less for large capacity units (e.g., greater than 10 tons) compared to smaller units.

4.4.2 Chillers

Chiller efficiency is usually measured in terms of a coefficient of performance (COP). COP is the cooling output produced by the chiller divided by the amount of electrical energy consumed or input by the chiller to produce this cooling.

Chiller efficiencies depend on the type compressor used in the chiller, and whether air or water cooling is used. COPs for air-cooled chillers installed between the years 2000 and 2010, for example, can range between 2.5–2.8. Water-cooled, centrifugal, or screw compressor chillers installed in this same time period can have COPs in the 3.8–5.5 range.

Chiller efficiencies may also be expressed as EERs, where:

$$\text{EER} = \text{COP} \times 3.412 \tag{4–4}$$

where EER = energy efficiency ratio, Btu output/W-hr input
COP = coefficient of performance, W-hr output/W-hr input
3.412 = conversion factor, (3,412 Btu/kWh) × (kW/1,000 W)

Chiller efficiencies are greatly affected by the load on the system. Conventional single-speed chillers will typically reach their highest COP at, or close to, maximum load but most chillers will run at part load most of the year. Part load COPs can be considerably less than full load COPs (i.e., they are less efficient). Typically, a chiller must operate at 70% or greater load to be close to its full load COP. (*N.B. If a variable frequency drive [refer to Chapter 14] can be retrofitted to the motor controls, this will reduce the rotation speed of the compressor, which in turn decreases the electrical demand and increases the COP at part load.*)

Single chillers typically do not operate for long at maximum capacity. To provide data to evaluate a single chiller part load performance, the integrated part load value (IPLV) is used. The IPLV rating method requires manufacturers to determine the input energy requirement at four operating conditions with a "weighing factor" representing the percentage of the hours of the cooling season at which the chiller is expected to operate at the defined rating point. The IPLV encompasses COP (or EER) ratings at four different load percentages: 1% at 100% load, 42% at 75% load, 45% at 50% load, and 12% at 25% load. Efficiencies for air-cooled electric chillers are typically measured using the units of EER, whereas water-cooled electric chiller efficiencies are typically measured in units of kW/ton. Efficiencies expressed in EER can be converted to kW/ton using the relationship: kW/ton = 12 / EER. When COP is used in the IPLV equation, for example, the IPLV units are kW output/kW input or (cooling capacity output of the chiller in kW)/(kW input to the chiller).[2]

$$\text{IPLV} = 0.01\text{A} + 0.42\text{B} + 0.45\text{C} + 0.12\text{D} \tag{4–5}$$

where IPLV = integrated part load value, kW output/kW input (for COP) or Btu output/W-hr input (for EER)
A = COP or EER for the chiller running at 100% capacity (1% of the time)
B = COP or EER for the chiller running at 75% capacity (42% of the time)
C = COP or EER for the chiller running at 50% capacity (45% of the time)
D = COP or EER for the chiller running at 25% capacity (12% of the time)

A more common way of calculating IPLV is using the kW/ton rating in Equation 4–6.[2]

$$\text{IPLV} = 1/\left[(0.01/\text{A}) + (0.42/\text{B}) + (0.45/\text{C}) + (0.12/\text{D})\right] \tag{4–6}$$

where IPLV = integrated part load value, kW input/ton output
A = kW/ton for the chiller running at 100% capacity (1% of the time)
B = kW/ton for the chiller running at 75% capacity (42% of the time)
C = kW/ton for the chiller running at 50% capacity (45% of the time)
D = kW/ton for the chiller running at 25% capacity (12% of the time)

For chillers running at maximum (100%) capacity for long periods of time, the full load kW/ton best represents the operational efficiency.

Example Problem 4–1

The spec sheet on a chiller has the following efficiencies at different loads:

Load	kW/ton
100% Load:	0.57
75% Load:	0.59
50% Load:	0.66
25% Load:	0.77

What is the IPLV?

Since the efficiency units are kW/ton, Equation 4–6 is used for the IPLV calculation.

$$IPLV = 1 / ((0.01/A) + (0.42/B) + (0.45/C) + (0.12/D))$$

$$IPLV = 1/((0.01/0.57) + (0.42/0.59) + (0.45/0.66) + (0.12/0.77))$$

$$IPLV = 0.64 \text{ kW/ton}$$

Let's calculate the IPLV again only using units of COP. We first must convert from kW/ton units to COP units (kW output/kW input) using the following conversion factor:

$$COP = 3.517 / (\text{kw/ton})$$

The calculations for COP are shown below:

Load	kW/ton	COP
100% Load:	0.57	6.17
75% Load:	0.59	5.96
50% Load:	0.66	5.33
25% Load:	0.77	4.57

$$IPLV = 0.01*6.17 + 0.42 *5.96 + 0.45 * 5.33 + 0.12 * 4.57$$

$$IPLV = 5.5 \text{ kW out/kW input}$$

Converting to kW input/ton output:

$$(5.5 \text{ kW out / kW in})*(3{,}412 \text{ Btu/kW})/(12{,}000 \text{ Btu/ton})$$
$$= 1.56 \text{ tons output / kW input}$$

$$\text{or } 1/1.56 = 0.64 \text{ kW input/ton output}$$

Note: kW/ton can be used in either equation and the results will be the same.

$$IPLV = 0.01*0.57 + 0.42*0.59+0.45 * 0.66 + 0.12 * 0.77$$

$$IPLV = 0.64 \text{ kW/ton}$$

The IPLV rating is only directly applicable to the single chiller installation where the condenser water supply temperature is allowed to fall as the outdoor wet bulb temperature falls. Anytime multiple chillers are used, particularly if careful control allows the chillers to be used at or near their optimum efficiency range of 40%–80% of rated capacity, use of the IPLV may not be valid. Therefore, to compare the energy performance in multiple chiller installations, more comprehensive energy analysis is required.

TABLE 4.2
Baseline Chiller Efficiency Values (kW/ton) over Time for Different Types of Chillers[3]

	Screw/Scroll/Reciprocating Water-Cooled Chillers					
	IPLV kW/ton			Full Load kW/ton		
Year	≤150 tons	150–300 tons	≥300 tons	≤150 tons	150–300 tons	≥300 tons
1990–2001	0.815	0.737	0.658	0.926	0.837	0.748
2002–2012	0.695	0.632	0.562	0.79	0.718	0.639
	Centrifugal Water-Cooled Chillers					
	IPLV kW/ton			Full Load kW/ton		
Year	≤150 tons	150–300 tons	≥300 tons	≤150 tons	150–300 tons	≥300 tons
1990–2001	0.815	0.737	0.658	0.926	0.837	0.748
2002–2012	0.619	0.558	0.508	0.703	0.634	0.577
	Air-Cooled Chillers					
	Part Load EER Btu/hr/Watt			Full Load EER Btu/hr/Watt		
Year	≤150 tons	150–300 tons	≥300 tons	≤150 tons	150–300 tons	≥300 tons
1990–2001	12.157	11.258	11.258	9.21	8.529	8.529
2002–2011	12.611	12.611	12.611	9.554	9.554	9.554
2012	12.622	12.622	12.622	9.562	9.562	9.562

Note: Chillers in this table do not have variable speed drives.

New high efficiency air-cooled chiller systems can have full load kW/ton in the 0.8–1.1 range and IPLVs in the 0.65–0.85 range, while new water-cooled systems may have full load kW/ton in the 0.48–0.66 range and IPLVs as low as 0.3, depending on size (tons) and type; older air-cooled chiller systems can have full load kW/ton in the 1.26–1.4 range and IPLVs in the 0.94–1.1 range, while older water-cooled systems may have full load kW/ton in the 0.72–0.84 range and IPLVs in the 0.51–0.74 range, depending on size. Table 4.2 provides chiller efficiency values for older chillers that might be due for replacement today.

Note: The IPLV rating is targeted to a very specific situation; that is, when the project's design conditions are equal to AHRI standard conditions. The non-standard part load value (NPLV) is a newer rating for chillers that provides a weighted average year-round efficiency considering two main parameters: the percent of operating hours at different loads and the different ambient conditions that can affect the entering water temperature. The NPLV rating therefore allows for efficiency measurements at a wider range of conditions. NPLV is calculated using the same equation as the IPLV.

4.4.3 Cooling Towers

Cooling towers are normally rated in tons of cooling, with a nominal cooling tower ton typically equal to the amount of power to cool 3 gpm of condenser water from

95°F to 85°F (a ΔT of 10°F). Alternate ratings are in gallons per minute with a specific temperature difference between water inlet and outlet temperatures (ΔT). For example, if the ΔT is 7°F, the gpm/ton would be 4.28. If it is 15°F, the gpm/ton would be 2.0. ΔT is typically in the 10–15°F range.

A rule of thumb is that cooling tower tons are approximately equal to 1.25 times the chiller tons since heat from associated equipment such as the compressor, pumps, and fans contribute additional heat load. Therefore, if chiller tons are known, that tonnage can be multiplied by 1.25 to obtain the approximate size of the cooling tower in tons. Rules of thumb used for cooling towers include:

- 3 gpm condenser water (10°F ΔT) per ton of cooling.
- 300 cfm airflow per ton of cooling.
- 0.067 hp/ton fan energy for induced draft (draw-through) towers.
- 0.143 hp/ton fan energy for forced draft (blow-through) towers.

4.4.4 Evaporative Coolers

Evaporative coolers are rated by the cubic feet per minute (cfm) of air that they deliver. Most models range from 3,000 to 25,000 cfm. Manufacturers recommend providing enough air-moving capacity for 20 to 40 air changes per hour, depending on the climate.[(4)] It takes almost 8,700 Btu of heat to evaporate 1 gallon of water. For evaporative cooling, this heat is taken from the air, cooling the air as the water evaporates. Electricity is used by a supply fan motor and a small sump pump.

4.5 ENERGY SAVINGS

4.5.1 Packaged RTUs/Split Systems/PTAC Units

Air conditioning equipment replacements can often result in a 10%–30% savings in space cooling end-use energy consumption. When replacing an existing inefficient cooling system with a new higher efficiency system, the energy savings can be estimated using Equation 4–7.

$$ES_{C\text{-}kWh} = CAP_C * EFLH_C * \left(\left(1/EER_{Baseline}\right) - \left(1/EER_{new}\right)\right) * \left(kW/1000\ W\right) \quad (4\text{–}7)$$

where $ES_{C\text{-}kWh}$ = space cooling energy consumption savings, kWh/yr
CAP_C = nominal rated output capacity of the unit, Btu/hr
$EFLH_C$ = cooling equivalent full load hours (see Table 4.3)
$EER_{Baseline}$ = EER (or IEER) of the existing unit (baseline), Btu output/W-hr input
EER_{new} = EER (or IEER) of the proposed new unit, Btu output/W-hr input

CAP_C is typically the rated output capacity (in Btu/hr or tons) for the specific piece of equipment. This is usually known or available in the manufacturer's spec sheets. The $EFLH_C$ can be estimated using Table 4.3[(5)] or the state technical resource manuals

(TRMs).[3,6–9] For example, the EFLH for cooling by property type in the Connecticut TRM is presented in Table 4.4.[10]

When using Equation 4–7 (or for that matter any of the equations in this book), be sure to check the units of each term in the equation to insure dimensional consistency; that

TABLE 4.3
Equivalent Full Load Hours for Cooling by Location and Building Type*[5]

Location	School	Office	Retail	Hospital
Atlanta, GA	690–830	1,080–1,360	1,380–1,860	2,010–2,850
Baltimore, MD	500–610	690–1,080	880–1,480	1,340–2,340
Bismarck, ND	150–250	250–540	340–780	540–1,290
Boston, MA	300–510	450–970	610–1,380	1,020–2,330
Charleston, WV	430–570	620–1,140	820–1,600	1,260–2,560
Charlotte, NC	650–730	1,080–1,340	1,350–1,830	1,990–2,820
Chicago, IL	280–410	420–780	550–1,090	870–1,780
Dallas, TX	830–890	1,350–1,580	1,680–2,090	2,320–3,100
Detroit, MI	230–360	390–820	530–1,170	870–1,950
Fairbanks, AK	26–54	64–200	110–320	210–600
Great Falls, MT	130–220	210–490	290–710	500–1,210
Houston, TX	940–1,000	1,550–1,770	1,870–2,290	2,540–3,320
Indianapolis, IN	380–560	560–1,000	730–1,410	1,120–2,250
Los Angeles, CA	780–910	1,280–1,670	1,740–2,350	2,740–3,770
Louisville, KY	550–670	770–1,250	1,000–1,720	1,480–2,690
Madison, WI	210–310	320–640	420–900	680–1,490
Memphis, TN	700–830	1,090–1,350	1,350–1,780	1,910–2,680
Miami, FL	1,260–1,300	1,980–2,150	2,350–2,740	3,110–3,890
Minneapolis, MN	200–300	320–610	430–870	680–1,420
Montgomery, AL	840–910	1,260–1,510	1,550–1,990	2,170–2,950
Nashville, TN	570–740	830–1,280	1,030–1,710	1,490–2,620
New Orleans, LA	920–990	1,500–1,720	1,820–2,240	2,500–3,280
New York City, NY	360–550	540–1,040	720–1,480	1,160–2,440
Omaha, NE	310–440	480–820	610–1,130	920–1,780
Phoenix, AZ	950–1,000	1,340–1,610	1,630–2,090	2,220–3,040
Pittsburgh, PA	300–530	440–920	600–1,310	960–2,160
Portland, ME	190–300	310–630	410–900	700–1,520
Richmond, VA	630–730	880–1,310	1,110–1,770	1,650–2,760
Sacramento, CA	680–850	1,080–1,430	1,460–2,020	2,250–3,180
Salt Lake City, UT	410–710	510–1,090	660–1,520	1,060–2,470
Seattle, WA	260–460	440–1,200	710–1,860	1,340–3,270
St. Louis, MO	460–550	680–1,100	850–1,500	1,260–2,330
Tampa, FL	1,050–1,110	1,800–2,000	2,170–2,580	2,910–3,710
Tulsa, OK	580–770	830–1,300	1,030–1,730	1,470–2,630

*TMY2 weather data utilized in simulation modeling for normalization.

TABLE 4.4
Cooling EFLH in Connecticut by Property Type

Property Type	Cooling EFLHs
Banks, Financial Centers	797
Religious Worship/Church	564
College/University	594
College—Cafeteria	1,139
College—Classes/Administrative	646
College—Dormitory	709
Commercial Condos	837
Convenience Stores	1,139
Convention Center	564
Dining—Bar Lounge/Leisure	854
Dining—Cafeteria/Fast Food	1,149
Dining—Family	854
Entertainment	564
Exercise Center	1,069
Fast Food Restaurant	1,139
Fire Station (unmanned)	564
Food Stores	837
Gym	646
Hospital	1,308
Hospital/Healthcare	1,307
Laundromat	837
Library	797
Lodging	708
Mall Concourse	938
Medical Offices	797
Motion Picture Theater	564
Multifamily	1,306
Museum	797
Nursing Homes	1,069
Office	797
Office/Retail	797
Performing Arts Theaters	646
Police/Fire Station	1,306
Post Office	797
Refrigerated Warehouse	648
Restaurants	854
Retail	837
Schools	594
Sports Arena	564
Warehouse—Non-Refrigerated	648

is, that the units on the right side of the equation equal the units of the left side of the equation. For Equation 4–7, our dimensional consistency check would result in:

$$\text{kWh/yr input} = (\text{Btu/hr output}) * (\text{hr/yr}) * (\text{W-hr input/Btu output}) * (\text{kW/W})$$

It is important to note that there is a difference between energy output units (e.g., Btu/yr output or kWh/yr output) versus energy input units, e.g., Btu/yr input or kWh/yr input. Input units refer to the energy consumed; that is, the energy that is purchased as reflected in utility bills. Output units refer to the energy delivered to the space to satisfy the load to meet space conditioning requirements. When used in equations, input energy does not equal output energy (despite the fact that the "nominal" units are the same), but rather must incorporate an energy efficiency metric. For example, Btu/yr output (delivered energy to the space) must be divided by the energy efficiency metric, EER or IEER, with appropriate conversion factors to get the Btu/yr input energy.

The percent cooling energy savings is given by taking Equation 4–7 and dividing by:

$$CAP_C * EFLH_C * (1/EER_{Baseline}) * (kW/1000\ W)$$

The result will give the cooling end-use energy consumption savings:

$$\%ES_{C\text{-}kWh} = \left[1 - (EER_{Baseline}/EER_{new})\right] * 100 \qquad (4\text{–}8)$$

where $\%ES_{C\text{-}kWh}$ = percent cooling end-use energy consumption savings.

Multiplying $\%ES_{C\text{-}kWh}$ by the baseline cooling end-use input energy consumption in kWh/yr will give the kWh/yr saved (which is the $ES_{C\text{-}kWh}$ in Equation 4–7). Hence, in using Equation 4–8 all that is needed are the energy efficiency metrics; that is, for the existing baseline equipment and new high efficiency equipment, and the baseline cooling end-use input energy consumption. The latter can be obtained from the whole building energy consumption (which can be calculated from the building's utility bills; refer to Chapter 2, Section 2.2.1) and the percent of the whole building energy consumption associated with the cooling end-use (refer to Chapter 2, Section 2.3).

Peak electric demand for a commercial building is the maximum kW used typically in a 15-minute or 30-minute interval in the billing period. On-peak electric demand savings (kW) represents the reduction in the demand for electricity that coincides with periods of peak demand for electricity. Due to high temperatures requiring air conditioning, summertime usually creates peak demand for electricity.

Peak demand savings can be estimated using Equation 4–9.

$$DS_{kW} = CAP_C * ((1/EER_{Baseline}) - (1/EER_{new})) * (kW/1000\ W) * CF_{summer} \qquad (4\text{–}9)$$

where DS_{kW} = peak demand savings, kW

CF_{summer} = utility coincidence factor in summer (0.84 default)

Equation 4–9 provides the peak demand savings. All things being equal, assuming no ratchet provisions and that billed demand equals read demand, the peak demand savings would occur in the summer months when cooling occurs (i.e., the summer season). If this is the case, then the annual peak demand savings would be the summer peak demand savings multiplied by the number of billing periods constituting the summer season. Hence, if the peak demand savings is 5 kW and the summer season when cooling occurs comprises 4 months, the annual demand savings would be 20 kW. It is important to understand how the utility charges for demand and whether ratchet provisions exist before estimating demand cost savings.

The coincidence factor (CF) represents the fraction of the connected load reduction expected to occur at the same time as a particular system's peak period. This fraction is multiplied by the gross demand energy savings (e.g., kW) determined by the energy efficiency measure (EEM) to calculate the portion of this gross demand coincident with the on-peak period. If the entire gross demand energy savings is coincident with the on-peak period, the CF would be equal to one. Coincidence factor(s) are often available in state utility technical resource manuals (refer to Chapter 1, Section 1.3). For space cooling, a summer coincidence factor of 0.84 for all property types would not be unreasonable, but it is always best to check the state TRM.

The IEER can also be used in Equations 4–7, 4–8 and 4–9 to replace EER. Energy professionals often prefer to use the IEER as it is more realistic and will often produce more attractive energy savings. If you only have an IEER for a piece of equipment and need the EER, it can be estimated using Equation 4–10.

$$\text{EER} \sim (1.12 * \text{IEER}) - \left(0.02 * \text{IEER}^2\right) \tag{4–10}$$

Typical energy savings realized by upgrading packaged air conditioning, split system and PTAC units often results in 10%–30% savings in space cooling end-use energy consumption.

Example Problem 4–2

An office building plans to replace an existing 10-ton roof-top air conditioner with an EER of 8.5. The new unit will have an EER of 12.5. Estimate the kWh/yr in energy savings and peak demand (kW) savings, and the percent of the cooling end-use energy consumption that would be saved. Assume the cooling EFLH is 797 hr (TMY) and the utility summer coincidence factor is 0.84.

Unit Output Capacity = 10 tons * 12,000 Btu/hr/ton=12,000 Btu/hr

From Equation 4–7:

$$ES_{C\text{-}kWh} = CAP_C * EFLH_C * \left((1/EER_{Baseline}) - (1/EER_{new})\right) * (kW/1000\ W)$$

$$= 120{,}000 * 797 * \left((1/8.5) - (1/12.5)\right) * (1/1000) = 3{,}596\ \text{kWh/yr}$$

From Equation 4–9:

$$DS_{kW} = CAP_C * \left(\left(1/EER_{Baseline}\right) - \left(1/EER_{new}\right)\right) * \left(kW/1000\ W\right) * CF_{summer}$$

$$= 120{,}000 * \left(\left(1/8.5\right) - \left(1/12.5\right)\right) * \left(1/1000\right) * 0.84 = 3.8\ kW$$

The percent cooling end-use energy consumption that would be saved is:

Cooling energy saved = 3,596 kWh/yr saved

Baseline cooling end use energy consumption
= 120,000/1000 * 797 * (1/8.5) = 11,252 kWh/yr

The percent energy savings = 3,596/11,252 = 0.3196 or 32%.
Or by using Equation 4–8:

$$\%\ ES_{C\text{-}kWh} = \left(1 - \left(EER_{Baseline}/EER_{New}\right)\right) * 100$$

$$= \left(1-\left(8.5/12.5\right)\right) = 32\%$$

Example Problem 4–3

A 10-ton packaged air conditioning unit on an office building in Hartford, Connecticut is replacing an existing unit that was installed in 2010. The new high efficiency unit to be installed in 2023 has an EER of 14.2. What is the expected cooling end-use energy consumption savings?

Existing unit:

Assuming the EER of the existing 10-ton unit installed in 2010 complied with ASHRAE Standard 90.1–2010:

$$EER = 11.2$$

Estimating the performance degradation that might have taken place over 13 years using Equation 3–1 in Chapter 3:

$$EEPM_D = EEPM_{NoD} * \left(1-M\right)^{A}$$

The only unknown in this equation is M. Assume that the performance degradation is 0.5% per year:

$$EEPM_D = 11.2 * (1-0.005)\text{^}3 = 10.49$$

$$EFLH_{cooling}\left(TMY\right) = 797 \text{ hrs from Table 4.4}$$

Estimated space cooling output (delivered to the space) = (10 tons) × (12,000 Btu/hr per ton) × (1,306 hr/yr) = 95,640,000 Btu/yr

The input energy of the existing unit in kWh/yr would then be:

$$\left[\left(95{,}640{,}000 \text{ Btu/yr output}\right) / \left(10.49 \text{ Btu output/ watt-hr input}\right)\right] * \left(\text{kw/1000 W}\right)$$
$$= 9{,}117 \text{ kWh/yr}$$

New unit:

$$\left[\left(95{,}640{,}000 \text{ Btu/yr output}\right) / \left(14.2 \text{ Btu output/watt-hr input}\right)\right] * \left(\text{kw/1000 W}\right)$$
$$= 6{,}735 \text{ kWh input}$$

$$\text{Savings} = 9{,}117 - 6{,}735 = 2{,}382 \text{ kWh/yr}$$

$$\%\text{savings} = \left(2{,}382/9{,}117\right)*100 = 26.1\%$$

or alternately, using Equation 4–8:

$$\%\ ES_{C\text{-}kWh} = \left(1 - \left(EER_{Baseline}/EER_{New}\right)\right) * 100$$

$$= \left(1 - \left(10.49/14.2\right)\right) * 100 = 26.1\%$$

and the kWh/yr savings would be:

$$= 0.261 * 9{,}117 \text{ kWh input} = 2{,}382 \text{ kWh/yr}$$

Example Problem 4–4

The cooling end-use energy consumption for an office building is 75,000 kWh/yr. What is the cooling end-use energy delivered to the building space annually? The air conditioning equipment has an EER of 8.2.

End-use energy consumption input, $E_{IN\text{-}C}$ =	75,000 kWh/yr input
Air conditioner efficiency (EER) =	8.2 Btu output (delivered to the space)/W-hr energy input
	or 0.0024033 kWh output/W-hr energy input
	or 2.4032825 kWh output/kWh energy input
	or 2.4 COP

Using Equation 4–12, solving for $E_{Out\text{-}C}$

$E_{Out\text{-}C} = E_{In\text{-}C} * COP$ = 75,000 kWh input/yr × 2.4 kWh output/kWh input / (293 kWh/MM Btu)
= 615 MM Btu/yr output (to meet the building space conditioning requirement)

Example Problem 4–5

An office building owner wants to replace an old 10-ton roof-top air conditioning unit. To be competitive, Contractor A proposes the least costly unit meeting the current energy code. Contractor B believes proposing a higher efficiency unit exceeding code, while initially more expensive, will provide the building owner with the most attractive solution and win the project. Which proposal is more attractive? Assume a unity coincidence factor.

The following information is available:

	Existing Unit	New Unit at Code	New Unit above Code
Size, tons	10	10	10
IEER	10.1	12.9	14.8
Installed Cost		$15,800	$17,300
Utility Incentive, $/ton			$50
EFLH Cooling (TMY)	800	800	800
Cooling Months	6	6	6
MM Btu/yr Delivered to space	96	96	96
$/kWh	$0.21	$0.21	$0.21
$/kW	$17	$17	$17

The energy savings calculations for each scenario are provided below:

% Energy Savings		21.7%	31.8%
MM Btu/yr Delivered to space	96	96	96
kWh/yr input	9,505	7,442	6,486
Electricity Cost/yr	$1,996	$1,563	$1,362
Peak Demand kW/month	11.9	9.3	8.1

kW peak demand in Cooling Season*	71.3	55.8	48.6
Peak Demand Cost/yr	$1,212	$949	$827
Savings in Yr. 1 vs. Existing		$696	$1,019
Savings in Yr. 1 Above Code vs. Code			$322
Above code incentive			$500
Installed Cost difference, above code v. at code			$1,000
Simple Payback, years			3.1

* Assumes billed demand equals read demand, and no ratchet provisions.

The simple payback time is based solely on the first-year savings. However, the savings will accrue each year over the 15-year estimated useful life of the equipment, and assuming electricity costs will escalate in future years, the savings will more than justify spending the extra money for the above-code unit. As such, the building owner will likely be willing to spend a few extra dollars on purchasing the above-code equipment (see Chapter 16 to develop life-cycle savings).

Example Problem 4–6

You are asked to perform a preliminary desktop estimate of the energy savings associated installing new, high efficiency RTUs on a 39,500 ft^2, all-electric office building constructed in 2011 in Orange County, California, assuming an electricity cost of $0.16/kWh. No other information has been provided.

Since no other information is available on this office building, we will have to estimate all the data that is needed:

EUI data of the building:

Whole building EUI:	74.1	kBtu/ft^2-yr	(Appendix VIII, Orange County)
Cooling end-use EUI:	13.52	kBtu/ft^2-yr	(Appendix VIII, Orange County)

To get the EER/IEER of the existing air conditioners, the capacity (tons) will need to be estimated:

ft^2/ton:	400 (Table 4.1)
RTU Capacity:	39,500 SF / 400 ft^2/ton = 99 tons, assume 100 tons

Assume five 20-ton units are being installed.
Based on the age of the building:

Building age:	12 years
Applicable 90.1 code:	ASHRAE 90.1–2011 (Chapter 3, Section 3.1.2)
IEER for 20-ton units:	10.1

Factoring in performance degradation using Equation 3–1 in Chapter 3, using the IEER metric, and assuming M = 0.005:

$$EEPM_D = EEPM_{NoD} * (1-M)^A$$

$$= 10.1 * (0.995) \wedge 12 = 9.5$$

IEER revised: 9.5

High efficiency RTU performance metrics:

IEER:	12.4 (Internet search of various manufacturers' high efficiency units)

Using Equation 4–8 with the IEERs:

$$\% ES_C = \left[1 - \left(IEER_{Baseline} / IEER_{new}\right)\right] * 100$$

$$= \left[1 - (9.5/12.4)\right] * 100 = 23.4\%$$

The energy consumption savings would be:

$$ES_C = 0.234 * 13.52 * 39{,}500 = 124{,}965 \text{ kBtu/yr} = 36{,}615 \text{ kWh/yr}$$

Peak demand savings can be estimated from Equation 4–9:

$$DS_{kW} = CAP_C * \left(\left(1/IEER_{Baseline}\right) - \left(1/IEER_{new}\right)\right) * \left(kW/1000\ W\right) * CF_{summer}$$

$$DS_{kW} = (100 \text{ tons} * 12{,}000 \text{ Btu/hr/ton}) * \left((1/9.5) - (1/12.4)\right) * (1/1000) * 0.84$$
$$= 24.9 \text{ kW}$$

The first-year energy cost savings due to more efficient cooling is then estimated as:

$$36{,}615 \text{ kWh/yr} * \$0.16/\text{kWh} = \$5{,}858 /\text{yr}$$

plus peak demand savings.

The life-cycle energy cost savings over the equipment's estimated useful life will be more meaningful. Chapter 16 provides this calculation methodology (see Example Problem 16–4) and does a cash flow assessment assuming the estimated total installed cost of the new equipment is available.

A Word About $EFLH_C$ . . .

$EFLH_C$ is calculated using Equation 4–11:

$$EFLH_C = E_{Out\text{-}C} / CAP_C \tag{4–11}$$

where $EFLH_C$ = equivalent full load hours for space cooling, hr/yr
$E_{Out\text{-}C}$ = output (delivered) cooling end-use energy consumption, Btu/yr
CAP_C = output (delivered) air conditioning cooling capacity, Btu/hr

The total energy Btu/yr delivered to the space, that is, output from the air conditioning unit over the cooling season, is not the total energy input consumed by the equipment ($E_{In\text{-}C}$), the latter being the cooling end-use energy consumption as discussed in Chapter 2. $E_{In\text{-}C}$ is estimated from the COP_C energy efficiency metric and the output cooling end-use energy consumption, $E_{Out\text{-}C}$.

$$E_{In\text{-}C} = E_{Out\text{-}C} / COP_C \tag{4–12}$$

where $E_{In\text{-}C}$ = total energy input used by the equipment, Btu/yr
COP_C = coefficient of performance for cooling, Btu output/Btu input

The theoretical $EFLH_C$ can be calculated from outdoor air conditions and the building HVAC design set-point using Equation 4–13.

$$EFLH_C = \left[CDD * 24\right] / \left[T_{Design} - 65\right] \tag{4–13}$$

where CDD = cooling degree days, °F-day
T_{Design} = outside air design temperature, °F
24 = 24 hr/day
65 = reference temperature, °F

The design outdoor air temperatures (OATs) for select cities in the U.S. are included in Table 4.5, and CDD can be obtained at www.degreedays.net for your building's location for a 12-month period.

Unfortunately, the theoretical $EFLH_C$ ignores the property type and specific HVAC system characteristics, both of which can have a significant impact on the building's cooling load. TRM manuals, on the other hand, typically have $EFLH_C$ by building type, location, and select HVAC characteristics such as whether or not an economizer is used in the building. Interestingly, ASHRAE found that variations in

TABLE 4.5

Design Outdoor Air Temperature by City for Cooling and Heating[11]

Location	1% Cooling Temperature	99% Heating Temperature
Atlanta, GA	92	20
Baltimore, MD	94	12
Bismarck, ND	90	−14
Boston, MA	90	1
Charleston, WV	89	15
Charlotte, NC	94	22
Chicago, IL	91	−3
Dallas, TX	101	22
Detroit, MI	90	4
Fairbanks, AK	80	−40
Great Falls, MT	91	−11
Hartford, CT	88	7
Houston, TX	99	27
Indianapolis, IN	92	3
Los Angeles, CA	102	14
Louisville, KY	93	10
Madison, WI	88	−6
Memphis, TN	97	20
Miami, FL	92	48
Minneapolis, MN	89	−11
Montgomery, AL	96	27
Nashville, TN	97	14
New Orleans, LA	92	34
New York City, NY	92	9
Omaha, NE	92	−3
Phoenix, AZ	111	24
Pittsburgh, PA	87	5
Portland, ME	87	−4
Richmond, VA	93	20
Sacramento, CA	103	29
Salt Lake City, UT	96	11
Seattle, WA	83	11
St. Louis, MO	94	2
Tampa, FL	95	33
Tulsa, OK	100	17

operating hours produced large variations in EFLH. As such, operating hours may be a key determinant in defining EFLH for the different building types.[5]

It also should be noted that the $EFLH_C$ equation is based on the air conditioner running at full capacity when the outside air temperature reaches the design temperature. However, most air conditioners do not operate at full capacity when the outside temperature reaches the design temperature. An HVAC contractor may purposely oversize a system simply because it is better to have excessive cooling capacity than insufficient cooling capacity. Or an HVAC contractor may not correctly estimate the equipment size because of the complexity of estimating cooling load for a particular building (e.g., the thermal characteristic of the building may be difficult to quantify). The general consensus about air conditioner oversizing is that it probably is prevalent.[12] In general, the greater the oversizing, the fewer the operating hours, and the less efficiently a unit operates.

Example Problem 4–7

What is the theoretical cooling $EFLH_C$ for Bridgeport, Connecticut? Assume the following:

CDD for Bridgeport =	814
Design temperature for Bridgeport =	85°F

Using Equation 4–13:

$$EFLH_C = [CDD * 24]/[T_{Design} - 65]$$

$$EFLH_C = (814)*24/(85-65) = 977 \text{ hr/yr}$$

Example Problem 4–8

Estimate the $EFLH_C$ knowing the space cooling end-use energy consumption (from utility bill data and end-use analysis in Chapter 2) for an office building in Chicago, Illinois, with a 110-ton packaged RTU having an IEER of 12. Assume the space cooling end-use energy consumption (weather-normalized) for the unit is 75,000 kWh/yr.

From Equation 4–7:

$$ES_{C\text{-}kWh} = CAP_C * EFLH_C * ((1/IEER_{Baseline}) - (1/IEER_{new})) * (kW/1000\ W)$$

Or using Equation 4–7 to separate out the Btu/yr of the existing equipment:

$$E_{In\text{-}C\text{-}Existing} = (CAP_C * EFLH_C/IEER_{Baseline}) * (1/1{,}000)$$

Everything is known in this equation except for $EFLH_C$. Hence,

$$E_{In\text{-}C\text{-}Existing} = 75{,}000 \text{ kWh/yr}$$

$$= [(110 \text{ ton} * 12{,}000 \text{ Btu/hr/ton}) * EFLH_C * (kW/1000W)]/$$
$$(12 \text{ Btu output/W-hr input})$$

Solving for $EFLH_C$:

$$75{,}000 \text{ kWh/yr input} = (110 \text{ tons} * 12{,}000 \text{ Btu/hr/ton})_{output} * (EFLH_C \text{in hr/yr}) * (\text{kW}/1000 \text{ W})/(12 \text{ Btu output/W-hr input})$$

$$EFLH_C = \left[(75{,}000 \text{ kWh/yr}) * (12 \text{ Btu out/W-hr in}) * (1000 \text{ W/kW})\right] / \left[1{,}320{,}000 \text{ btu/hr}\right] = 682 \text{ hr/yr}$$

Table 4.2 has the EFLH for cooling in an office building in Chicago as 420–780 hr/yr.

Example Problem 4–9

If the air conditioner in the office building in Example Problem 4–4 was located where the $EFLH_C$ = 683 hr/yr, estimate the size of the air conditioner. What would the cooling end-use energy consumption savings be if the new air conditioner has an EER = 11.2?

End-use energy consumption input, $E_{In\text{-}C}$ =	75,000 kWh/yr input
From Equation 4–12:	$E_{Out\text{-}C} = E_{In\text{-}C} * COP$ COP = 2.4 $E_{Out\text{-}C}$ = 180,246 kWh/yr output = 615 MM Btu/yr $E_{Out\text{-}C}$ = (tons capacity) × (12,000 Btu/hr/ton) × ($EFLH_C$ hr/yr) tons capacity = (615 MM Btu/yr)/(12,000 × 683) = 75 tons
If the new air conditioner has an EER = 11.2 and the $E_{Out\text{-}Baseline} = E_{Out\text{-}New}$ (building demand is unchanged):	$E_{Out\text{-}New}$ = 615 MM Btu/yr output $E_{In\text{-}New}$ = (615 MM Btu output/yr)/[(11.2 Btu output/W-hr Input) × (1,000 W/kW)] = 54,911 kWh/yr Energy savings = 75,000 – 54,911 = 20,089 kWh/yr (or 26.8% energy savings)

The energy savings could also be obtained directly from Equation 4–18 as $(1 - (EER_{Existing}/EER_{New}))$= (1 – (8.2/11.2)) = 0.268 = 26.8% or the savings in kWh/yr = 0.268 × 75,000 = 20,089 kWh/yr.

Another way of estimating energy savings when the cooling end-use energy consumption of the baseline existing equipment, $E_{In\text{-}C\text{-}Baseline}$, is known or can be estimated, is by using Equation 4–14:

$$ES_{C\text{-}Btu} = E_{In\text{-}C\text{-}Baseline} - E_{In\text{-}C\text{-}New} \quad (4\text{–}14)$$

where $ES_{C\text{-}Btu}$ = space cooling end-use energy consumption savings, Btu/yr

However, we still need to get $E_{In\text{-}C\text{-}New}$. This can be done as follows:

$$E_{Out\text{-}C\text{-}Baseline} = E_{In\text{-}C\text{-}Baseline} * IEER_{Baseline} * \left(1000/3{,}412\right) \quad (4\text{–}15)$$

where $E_{Out\text{-}C\text{-}Baseline}$ = existing equipment cooling energy delivered (output), Btu/yr
$E_{In\text{-}C\text{-}Baseline}$ = existing equipment cooling energy consumed (input), Btu/yr
$IEER_{Baseline}$ = integrated energy efficiency ratio of existing equipment Btu output/W-hr input
1,000 = conversion factor, 1,000 W/kW
3,412 = conversion factor, 3,412 Btu/kWh

Assuming the cooling energy load (what is delivered to the space to satisfy its demand), does not change for the new system, then:

$$E_{Out\text{-}C\text{-}Baseline} = E_{Out\text{-}C\text{-}New} \quad (4\text{–}16)$$

and

$$E_{In\text{-}C\text{-}New} = (E_{Out\text{-}C\text{-}New} / IEER_{New}) * \left(3{,}412/1{,}000\right) \quad (4\text{–}17)$$

where $E_{In\text{-}C\text{-}New}$ = new equipment cooling energy consumed (input), Btu/yr
$E_{Out\text{-}C\text{-}New}$ = new equipment cooling energy delivered (output) to the space, Btu/yr (assumed equal to $E_{Out\text{-}C\text{-}Baseline}$)
$IEER_{New}$ = integrated energy efficiency ratio of new equipment, Btu output/W-hr input

To convert $ES_{C\text{-}Btu}$ (in Btu/yr) to $ES_{C\text{-}kWh}$ (in kWh/yr), divide by 3,412 Btu/kWh.

Alternately, to directly estimate the energy savings, Equation 4–18 can be used:

$$ES_{C\text{-}Btu} = E_{In\text{-}C\text{-}Baseline} * \left(1 - \left(IEER_{Baseline}/IEER_{New}\right)\right) \quad (4\text{–}18)$$

4.5.2 Chillers

For kWh energy savings, use IPLV in units of kW consumed (input)/ton of cooling delivered (output). For kW demand savings, use the full load efficiency in units of kW/ton. If the cooling efficiency is expressed as an EER, it can be converted to kW/ton using Equation 4–19.

$$\left(kW/ton\right) = 12 / EER \quad (4\text{–}19)$$

The energy savings associated with replacing an existing inefficient chiller with a new higher efficiency chiller can be estimated using Equation 4–20.

$$ES_{C\text{-}kWh} = CAP_C * EFLH_C * \left[\left(1/IPLV_{Baseline}\right) - \left(1/IPLV_{New}\right)\right] / 1{,}000 \qquad (4\text{–}20)$$

where $ES_{C\text{-}kWh}$ = space cooling end-use energy consumption savings, kWh/yr
CAP_C = output (delivered) cooling capacity, Btu/hr
$IPLV_{Baseline}$ = baseline integrated part load value, Btu output/W-hr input
$IPLV_{New}$ = integrated part load value of new chiller, Btu output/W-hr input
1,000 = conversion factor, 1,000 W/kW

In Equation 4–20, the dimensional unit consistency check would result in:

$$\left(\text{kWh input/yr}\right) = \left(\text{Btu/hr output}\right)\left(\text{hr/yr cooling}\right) * \left(\text{W-hr input/Btu output}\right)\left(\text{kW/W}\right)$$

If the IPLV units in Equation 4–20 are in kW input/tons output:

$$ES_{C\text{-}kWh} = \left(CAP_C / 12{,}000 \text{ Btu/hr/ton}\right) * EFLH_C * \left(IPLV_{Baseline\text{-}Alt} - IPLV_{New\text{-}Alt}\right) \qquad (4\text{–}21)$$

where CAP_C = cooling capacity delivered, Btu/hr
$IPLV_{Baseline\text{-}Alt}$ = baseline integrated part load value in kW consumed/ton delivered
$IPLV_{New\text{-}Alt}$ = integrated part load value of new chiller in kW consumed/ton delivered
12,000 = conversion factor, 12,000 Btu/hr per ton

In Equation 4–21, the dimensional unit consistency check would result in:

$$\left(\text{kWh input/yr}\right) = \left(\text{Btu/hr output}\right) * \left(\text{ton/Btu/hr}\right) * \left(\text{hrs/yr cooling}\right) * \left(\text{kW input/ton output}\right)$$

The percent savings is given by taking Equation 4–20 and dividing by:

$$\left(CAP_C / 12{,}000\right) * EFLH_C * \left(1/IPLV_{Baseline\text{-}Alt}\right)$$

The result will give the percent energy savings as:

$$\%\, ES_{C\text{-}kWh} = \left[1 - \left(IPLV_{New\text{-}Alt} / IPLV_{Baseline\text{-}Alt}\right)\right] * 100 \qquad (4\text{–}22)$$

Multiplying the percent energy savings by the baseline input energy (i.e., energy consumed), in kWh/yr will give the kWh/yr saved.

The peak demand savings due to the energy consumption savings can then be estimated. Using the full load value (FL) energy efficiency metric in kW input/ton output, the peak demand savings would be:

$$DS_{kW} = \left(CAP_C / 12{,}000\right) * \left[\left(FL_{Baseline}\right) - \left(FL_{New}\right)\right] * CF_{summer} \qquad (4\text{–}23)$$

where DS_{kW} = peak demand energy savings, kW
$FL_{Baseline}$ = full load efficiency of existing equipment, kW input/ton output

FL_{New} = full load efficiency of new equipment, kW input/ton output
CF_{summer} = coincidence factor for cooling (e.g., summer)

In Equation 4–23, the dimensional unit consistency check results in:

$$(\text{kW}) = \left[(\text{Btu/hr})/(\text{Btu/hr/ton})\right] * (\text{kW/ton})$$

Equation 4–23 provides the peak demand savings. All things being equal, assuming no ratchet provisions and that the billed demand equals the read demand, the peak demand savings would occur in the summer months when cooling occurs (i.e., the summer season). If this is the case, then the annual peak demand savings would be the summer peak demand savings multiplied by the number of billing periods constituting the summer season. Hence, if the peak demand savings is 10 kW and the summer season when cooling occurs comprises 4 months, the annual peak demand savings would be 40 kW. It is important to understand how the utility charges for demand and whether ratchet provisions exist before estimating demand cost savings.

Typical space cooling energy consumption savings for chiller upgrades are in the 10%–25% range.

Example Problem 4–10

A new 950,000 Btu/hr chiller (output) is replacing an existing chiller in an office building located in Connecticut. The baseline IPLV is 0.61 kW/ton and the new chiller has an IPLV of 0.55 kW/ton. What is the annual energy consumption and peak demand savings? Assume $CF_{cooling} = 0.8$.

From Table 4.4, $EFLH_{Cooling} = 797$ hr/yr

$$ES_{cooling} = \left(\text{Output } CAP_{cooling}/(12{,}000 \text{ Btu/ton})\right) * EFLH_{cooling} * \left(IPLV_{Baseline} - IPLV_{New}\right) \text{kW input/ton output}$$

$$= (950{,}000/12{,}000) * 797 * (0.61 - 0.55)$$

$$= 3{,}786 \text{ kWh/yr}$$

$$\text{Input power, kWh} = IPLV_{Baseline} * \text{Tons} * EFLH$$

$$= 38{,}488 \text{kWh/yr}$$

$$\% \text{ savings/yr} = (3{,}786 \text{ kWh}/38{,}488 \text{ kWh}) * 100$$

$$= 9.8\%$$

or

$$\% \text{ savings} = \left(1 - \left(IPLV_{new}/IPLV_{baseline}\right)\right) = (1 - 0.9016) = 9.8\%$$

For the calculation of peak demand savings:

$$DS_{cooling} = \left(CAP_{cooling}/(12{,}000\ \text{Btu/ton})\right) * \left(IPLV_{Baseline} - IPLV_{New}\right) * CF_{summer}$$

$$= (950{,}000/12{,}000)\ \text{tons} * 0.06\ \text{kW/ton} * 0.8$$

$$= 3.8\ \text{kW peak demand savings}$$

Example Problem 4–11

A chiller produces 5,000 kW of cooling using 920 kW of energy. What is the COP?

COP = 5,000 / 920 = 5.4 kW output per kW input

Another way to look at the efficiency rating is in kW/ton:

5,000 kW *3,412 Btu/hr/kw = 17.06 MM Btu/hr = 1,422 tons output

920 kw used (input) / 1,422 tons = 0.65 kW input/ton output

And still another way to look at the efficiency rating is using EER:

EER = 3.412* COP = 18.4 Btu output per W-hr input

4.5.3 Cooling Towers

The information that must first be obtained in pumping to a tower is to determine the minimum and maximum flow rates required by the chiller condensers. In the past, there were simple rules such as 3 gal/min per ton of cooling for an electric chiller. This assumed that the heat of rejection included the 12,000 Btu/hr/ton plus 3,000 Btu/hr/ton for the chiller compressor and auxiliaries for a total of 15,000 Btu/hr/ton of cooling. Today, with improved chiller performance, the water rates are calculated more carefully. The flow rates should be checked at specific design water temperatures such as 95°F return and 85°F sump water temperatures. If the heat of rejection from the chiller is expressed in Btu/hr at a specific temperature difference, Equation 4–24 can be used to determine the water rate:

$$\text{GPM-T} = (H_R) * (\text{Tons})/(500 * \Delta T) \qquad (4\text{–}24)$$

where GPM-T = cooling tower water flow rate, gal/min per ton
H_R = total heat of rejection, Btu/hr per ton (see Table 4.6)

ΔT = temperature difference (inlet-outlet), °F
500 = (density of water, 8.33 lb/gal) × (heat capacity of water, 1 Btu/lb °F) × (60 min/hr)

Cooling tower efficiency can be expressed as:

$$\eta_{CT} = 100 * \left(T_{inlet} - T_{outlet}\right) / \left(T_{inlet} - T_{WB}\right) \quad (4\text{–}25)$$

where η_{CT} = cooling tower efficiency, % (typically 70%–75%)
T_{inlet} = inlet temp. of water to the tower, °F
T_{outlet} = outlet temp. of water from the tower, °F
T_{WB} = wet bulb temp. of air, °F
($T_{inlet} - T_{outlet}$) is typically in the range of 10–15°F

Energy savings (kWh) for cooling towers are included in certain state technical resource manuals such as New York State's, where calculations are provided when retrofitting with close approach (≤6°F) cooling towers.[(6)]

4.5.4 Evaporative Coolers

For an existing commercial building in an applicable climate, the energy efficiency upgrade might include a direct evaporative pre-cooler upstream of the air conditioning system. This process reduces the temperature of the outside air before it is used to cool the condenser coil for DX units or air-cooled chillers. The temperature reduction is achieved by having the incoming air pass through a saturated media or mist wall which would increase the humidity ratio under adiabatic conditions. This allows the dry bulb temperature to decrease while the wet bulb temperature remains constant, effectively increasing the heat rejection capacity from the condenser coils into the air.

Since the evaporation of 1 gallon of water will absorb 8,652 Btu of heat energy, and 1 ton of air conditioning is the equivalent of 12,000 Btu of energy, it takes the evaporation of less than 1.5 gallons of water to provide the equivalent cooling effect

TABLE 4.6
Chiller System Total Heat of Rejection

Compressor Input Power, kW/ton	Approximate Heat of Compression, % (including 2% for condenser water pump(s))	Total Heat Rejection, Btu/hr/ton
0.8	25	15,000
0.7	22	14,640
0.6	19	14,280
0.5	16	13,920
0.4	14	13,680

of 1 ton of air conditioning. Since the cost of 1.5 gallons of water is considerably less than the cost to run 1 ton of air conditioning, savings can be realized. As such, every ton of heat load removed by an evaporative cooling system means 1 less ton of cooling load placed on the building's HVAC system.

The energy savings associated with installing an evaporative pre-cooler upstream of the DX cooling system or air-cooled chiller in dry climates such as the Southwestern U.S. can be estimated by Equation 4–26.[13]

$$ES_{C\text{-kwh}} = CAP_C * EFLH_{Reduction} / (EER * 1{,}000) \qquad (4\text{–}26)$$

where $ES_{C\text{-kWh}}$ = cooling energy savings, kWh
$EFLH_{Reduction}$ = annual equivalent full load cooling hour reduction (see Table 4.7 for Texas)
CAP_C = air conditioning system rated cooling output capacity, Btu/hr
EER_C = cooling energy efficiency ratio, output Btu/hr/W input
1,000 = 1,000 W/kW

Peak demand reduction can be estimated using Equation 4–27.

$$DS_{kW} = CAP_C * DRF/(EER * 1{,}000) \qquad (4\text{–}27)$$

where DS_{kW} = peak demand savings, kW
DRF = demand reduction factor, dimensionless (see Table 4.7 for Texas)

Evaporative cooling can complement (and in some cases replace) traditional direct expansion air conditioning in some climates. In drier climates, evaporative cooling can save 30% to 70% of the energy required by an equivalent direct expansion (DX) system. For example, in one particular project, installing an evaporative precooler upstream of a 50-ton chiller in California (climate zone 12—hot dry, 1,332 CDD) saved 22% of the space cooling (chiller) end-use energy consumption and approximately 20% of the peak demand associated with the chiller energy consumption savings.

4.5.5 Dual Enthalpy Economizers

Dual enthalpy economizers regulate the amount of outside air introduced into the ventilation system based on the relative temperature and humidity of the outside and return air. If the enthalpy (latent and sensible heat) of the outside air is less than that of the return air when space cooling is required, then outside air is allowed in to reduce or eliminate the cooling requirement of the air conditioning equipment.

The energy savings associated with the installation of dual enthalpy control air-side economizers integrated into a central air handling system on packaged units can be estimated from the tons of air conditioning and a factor representing the economizer kWh/ton savings. The economizer kWh/ton savings is determined by modeling for a specific building type, location, operating hours, and cooling system

TABLE 4.7
EFLH Reduction and Demand Reduction Values for Evaporative Pre-Cooling in Texas

Building Type	Principal Activity	Ft. Worth				Houston				Corpus Christi				El Paso			
		DX		Air-Cooled Chiller		DX		Air-Cooled Chiller		DX		Air-Cooled Chiller		DX		Air-Cooled Chiller	
		DRF	EFLHRed.	DRF	EFLHRed.	DRF	EFLHRed.	DRF	EFLHRed.	DRF	EFLHRed.	DRF	EFLHRed.	DRF	EFLHRed.	DRF	EFLHRed.
Education	College	0.21	192	0.19	195	0.20	173	0.17	175	0.13	161	0.11	160	0.27	240	0.22	254
	Primary School	0.24	120	0.12	80	0.21	118	0.10	74	0.14	113	0.07	68	0.30	161	0.17	120
	Secondary school	0.21	131	0.19	132	0.20	118	0.17	119	0.13	110	0.11	109	0.27	163	0.22	172
Food Sales	Convenience Store	0.24	214			0.22	193			0.14	188			0.25	232		
	Supermarket	0.15	78			0.14	76			0.08	74			0.12	76		
Food Service	Full-Service Restaurant	0.23	194			0.21	171			0.13	157			0.25	223		
	Quick Service Restaurant	0.24	185			0.22	167			0.14	162			0.25	201		
Healthcare	Hospital	0.24	230	0.22	216	0.21	202	0.19	187	0.15	199	0.09	169	0.26	273	0.20	247
	Outpatient Care	0.19	174			0.18	157			0.12	150			0.23	259		
Multifamily	Mid-Rise	0.16	230	0.15	120	0.17	257	0.14	105	0.14	181	0.09	104	0.28	264	0.15	140
Lodging	Large Hotel	0.15	137	0.18	212	0.14	120	0.14	193	0.08	116	0.10	179	0.19	201	0.19	300
	Nursing Home	0.16	234	0.15	122	0.17	261	0.14	107	0.14	183	0.09	106	0.28	268	0.15	142
	Small Hotel/Motel	0.15	133			0.13	113			0.08	109			0.17	193		
Mercantile	Stand-Alone Retail	0.24	158	0.19	120	0.22	152	0.19	126	0.14	148	0.12	120	0.25	198	0.18	131
	Strip Mall	0.23	156			0.21	152			0.13	146			0.26	207		
Office	Large Office	0.26	220	0.23	231	0.24	203	0.23	150	0.16	192	0.13	137	0.32	314	0.22	199
	Medium Office	0.20	102			0.19	94			0.11	90			0.25	137		
	Small Office	0.22	156			0.20	138			0.13	131			0.26	215		
Public Assembly		0.24	161	0.12	108	0.21	159	0.10	99	0.14	152	0.07	92	0.30	217	0.17	162
Religious Worship		0.24	95	0.19	72	0.22	92	0.19	77	0.14	89	0.12	72	0.25	119	0.18	79
Service		0.23	150			0.21	132			0.13	122			0.25	173		
Warehouse		0.20	93			0.18	81			0.12	74			0.25	82		

Source: Texas Technical Resource Manual, volume 3, pp. 2–101 ff, October 30, 2017.

type and operation. Economizer kWh/ton savings factors are often included in select state technical resource manuals, including those of New York State,[14] Delaware,[15] Ohio,[7] Massachusetts,[16] and Vermont.[17]

Energy savings can also be estimated knowing the cooling system capacity (tons cooling output), the efficiency of the cooling system (IEER), and the annual hours when "free cooling" occurs.[18]

Economizers can reduce energy use by anywhere from 15% to as much as 70%, depending on site-specific conditions, including building location, building use, occupancy, HVAC system, and controls. In northern states, for example, air-side dual enthalpy economizers might save up to 45% of mechanical refrigeration (space cooling) end-use energy consumption compared to installations without economizers. Since economizers will not likely be saving energy during peak hours, peak demand savings are not expected.

Example Problem 4–12

A heating system uses 15,000 cfm of ventilation and exhaust air. The exhaust air temperature is 72°F and the average intake air temperature is 40°F. The system operates 1,950 hours a season. The building owner is considering saving energy by using a heat wheel to heat the intake air with exhaust air. The heat wheel manufacturer says the wheel will heat the intake air to 60°F. How much energy would be saved if the heat wheel is installed?

The energy saved is found from the sensible heat equation for preheating the intake air—Equation 4–1:

$$Q_A = 1.08 * CFM * \Delta T$$

To get the energy saved over the season, multiply Q_A by the hr/season):

$$= Q_A * \text{ (operating hours per season)}$$

$$= 1.08 * 15{,}000 * 20 * 1{,}950$$

$$= 631.8 \text{ MM Btu/season}$$

It should be noted that a small amount of energy will be required by the fan to move the air through the wheel (see Chapter 7). If the pressure drop across the wheel is not available, assume 0.5 in. W.C. across the wheel.

4.5.6 Thermal and Desiccant Wheels

Heat recovery efficiencies up to 85% are possible.

4.5.7 Heat Exchangers and Heat Pipes

Heat recovery efficiencies up to 70% are possible.

4.5.8 Energy Recovery Ventilators

To assess product efficiency, the sensible recovery efficiency (SRE) and the total recovery efficiency (TRE) metrics are commonly used to compare products on a level playing field. The higher the rating, the more efficient the product. SRE is the net sensible energy recovered by the supply airstream as a percentage of the potential sensible energy that could be recovered plus the exhaust fan energy. TRE is the net total energy recovery (sensible plus latent) as a percentage of the potential sensible energy that could be recovered plus the exhaust fan energy. SRE and TRE include the HRV and ERV fan energy in the energy performance data. SREs for ERV and HRV generally are in the 65%–81% range.

The adjusted SRE or ASRE and adjusted TRE or ATRE metrics are utilized where HRV and ERV fan energy is removed and considered separately. This may be done to provide a more accurate assessment of ventilator system energy performance, showing the benefits of balanced heat and energy recovery ventilation during the heating season.

ERVs can recover about 70% of sensible energy and 60% of latent energy. Performance is typically characterized by effectiveness; the pressure drop, pumping, or fan power of fluids; cross airflow (i.e., the amount of air leakage from one stream to the other); and frost control (used to prevent frosting on the heat exchanger).

An ERV will typically handle approximately half of the load required to bring outside ventilation air to the required temperature. In other words, buildings with a ventilation system that does not include ERV capabilities could require double the tonnage of mechanical cooling and heating.

The general rule of thumb for ERV pretreatment of ventilation air is that there can be an HVAC equipment reduction of approximately 2–2.5 tons for each 1,000 cfm of outdoor air handled by an ERV. For example, instead of installing a 75-ton cooling system for a project that required 40% outdoor air (12,000 cfm outdoor air and 18,000 cfm return air), it may be possible to install a 12,000 cfm ERV and a 45- to 50-ton cooling system unit. This reduction in the size of the cooling equipment can help offset the initial cost of the ERV. Incorporating the energy savings of the ERV can often result in simple payback times of 3 years or less for most systems depending on the building's geographic location. In addition, an ERV can reduce summer electrical peak load demand.

Energy savings are typically estimated using bin methodology where for each temperature interval the kWh/yr saved (when in the cooling mode) or Btu/yr saved (when in the heating mode) can be determined knowing the supply airflow rate (cfm), heat exchanger efficiency, the temperature and enthalpy of outside and inside air, cooling system efficiency (kW/ton), heating system efficiency (%), operating hours, and fan kW in each bin temperature range. The total savings would then be the sum of all the bin interval savings.

REFERENCES

1. ASHRAE 62.1–2022 Standard, *Ventilation for Acceptable Indoor Air Quality*, Atlanta, GA: American Society of Heating, Refrigerating and Air-Conditioning Engineers, Inc., 2022.

2. AHRI Standard 550/590–2020 (I/P) with Addendum 1, 2020 standard for performance rating of water-chilling and heat pump water-heating packages using the vapor compression cycle, Appendix D, February 2022, https://www.ahrinet.org/system/files/2023-06/AHRI%20Standard%20550-590%202020%20%28I-P%29%20with%20Addendum%201.pdf
3. Arkansas Technical Reference Manual, *Deemed Savings, Installation and Efficiency* (Version 9.1, Vol. 2, Tables 272–277, pp. 288–293), August 31, 2022, https://apsc.arkansas.gov/wp-content/uploads/AR_TRM_V9.1_Volume_1_2_and_3_on_8-31-22.pdf
4. U.S. DOE, see www.energy.gov/energysaver/evaporative-coolers.
5. Carlson, S., ASHRAE research project report RP-1120: Development of equivalent full load heating and cooling hours for GCHPs applied in various building types and locations, March 2001, https://www.techstreet.com/ashrae/searches/39741364
6. New York standard approach for estimating energy savings from energy efficiency programs, Version 7, pp. 366–367 and Appendix J, also Appendix G for EFLH, April 15, 2019, https://dps.ny.gov/system/files/documents/2022/11/technical-resource-manual-version-7-filed-april-15-2019-effective-january-1-2020.pdf
7. State of Ohio energy efficiency technical reference manual, pp. 201–203, also pp. 147, 192, 196 and 199 for EFLH, August 6, 2010, https://resourceinsight.com/wp-content/uploads/2022/01/VEIC-ohio-TRM.pdf
8. United Illuminating and Eversource, Connecticut Program Savings Document, 12th edition, Appendix 5, 2017, https://www.puc.nh.gov/EESE%20Board/EERS_WG/ct_trm.pdf
9. Illinois Statewide Technical Reference Manual, Version 5.0, pp. 158, 165, 171, 184 and 189, 2016.
10. CT Program Savings Document (TRM), 12th Edition, Appendix 5, 2017, https://www.puc.nh.gov/EESE%20Board/EERS_WG/ct_trm.pdf
11. EPA ENERGY STAR, Design temperature limit reference guide, 2019 edition, see www.energystar.gov/sites/default/files/asset/document/Design%20Temperature%20Limit%20Reference%20Guide%20%282019%20Ed%29%20-%20ENERGY%20STAR%20SFNH%20Rev11_MFNC%20Rev02.pdf
12. National Renewable Energy Laboratory (NREL), Right-size heating and cooling equipment, report issued in 2002, see www.nrel.gov/docs/fy02osti/31318.pdf
13. Texas Technical Resource Manual, Version 5, Vol. 3, pp. 2–101 ff, 2018, https://texasefficiency.com/wp-content/uploads/2023/11/trm5v3.pdf
14. New York standard approach for estimating energy savings from energy efficiency programs, Version 7, pp. 373–374 and Appendix J, April 15, 2019, https://dps.ny.gov/system/files/documents/2022/11/technical-resource-manual-version-7-filed-april-15-2019-effective-january-1-2020.pdf
15. Delaware Technical Resource Manual, pp. 431–433, July 1, 2016, https://regulations.delaware.gov/register/august2016/proposed/DE%20TRM%20complete%202016.pdf
16. Massachusetts Technical Reference Manual, pp. 203–204, Section 3.39, pp. 555–557, November 2021, https://masssavedata.com/TRL/Technical%20Reference%20Manual%202022-2024%20Final.pdf
17. Efficiency Vermont, Technical Reference Manual (TRM), pp. 45–47, March 16, 2015, https://www.puc.nh.gov/EESE%20Board/EERS_WG/vt_trm.pdf
18. State of Pennsylvania Technical Reference Manual, Section 3.2.9, June 2015, https://www.puc.pa.gov/docket/M-2012-2313373
19. SPEED Statewide Partnership for Energy Efficient Demonstrations, *Condenser Air Pre-Cooler Retrofits, HVAC Energy Efficiency Case Study* (Beale Air Force Base, Marysville, CA), Davis, CA: University of California, Davis, 2012, https://wcec.ucdavis.edu/wp-content/uploads/2013/11/EvaporCool_Beale_ProjectReport_Final.pdf

5 Heat Pumps

5.1 EQUIPMENT DESCRIPTION

A heat pump is a packaged air conditioner or a packaged unit with a reversing valve. Heat pumps transfer heat extracted from one medium (air, water) to another medium via the vapor-compression refrigeration cycle. Heat pump systems are generally classified as either air source heat pumps or water source heat pumps.

A heat pump has all the main components of an air conditioner or packaged unit: fan, filters, compressor, evaporator, condenser, and controls. The apparatus for changing from cooling to heating or vice versa is a reversing valve, in which the refrigerant flow to the condenser is changed to the evaporator. Alternatively, air passage through the evaporator may be changed over to passage through the condenser. A supplementary heater is often provided when the heat pump capacity does not meet the required output during low outdoor temperatures.

5.1.1 Air Source Heat Pumps

In an air source heat pump (ASHP) system, outdoor air acts as a heat source from which heat is extracted during heating, and as a heat sink to which heat is rejected during cooling (see Figure 5.1). Since air is readily available, air source heat pumps are the most widely used heat pumps in commercial buildings. The cooling capacity of most air source heat pumps is between 1 and 30 tons. Air source heat pumps can be classified as individual unit heat pumps and packaged heat pumps. Individual unit heat pumps serve only one unit. Packaged heat pumps can be subdivided into roof-top heat pumps and split heat pumps.

Most air source heat pumps consist of single or multiple compressors, indoor coils through which air is conditioned, outdoor coils where heat is extracted from or rejected to the outdoor air, capillary tubes, reversing valves that change the heating operation to a cooling operation and vice versa, an accumulator to store liquid refrigerant, and other accessories.

All air source heat pumps can accumulate frost on their outdoor coils under certain conditions. The outdoor coil is always colder than the outdoor air, and when the coil temperature is below the ambient dew point, moisture in the air condenses on the coil. When the coil temperature is below freezing, this condensate turns to ice on the coil. The accumulated ice reduces heat transfer from the air to the refrigerant in the coil. If not removed the ice would eventually block airflow through the coil. To maintain efficiency, the heat pump must periodically get rid of any accumulated ice by "defrosting." It can do so by temporarily turning into an air conditioner by removing heat from the indoors and pushing it into the outdoor coil. This melts the ice off the coil. The defrosting process unavoidably decreases the heat pump's efficiency, so the defrost cycle needs to be kept as brief as possible.

DOI: 10.1201/9781032692777-5

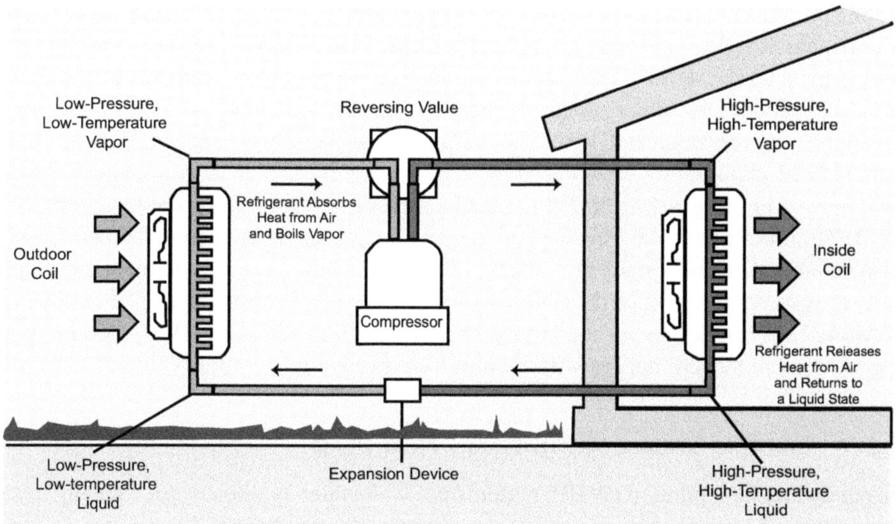

FIGURE 5.1 Air source heat pump heating cycle operation.

5.1.2 Water Source Heat Pumps

Water source heat pumps (WSHPs) can take the heat out of an area that is rejecting it and use it in another area that is demanding it. The Btus of cooling load and the heat of compression in the refrigeration circuit of a traditional ASHP is rejected to the atmosphere through the outdoor condensing unit. The quality of heat is low, and it is not economical in an ASHP to recover this energy. This energy is blown into the atmosphere and wasted. In a WSHP, these Btus are rejected into a common water loop that acts as a reserve of this energy, which can be easily transported to a space that is demanding heat. For example, some water source heat pumps located in the shady side of the perimeter zone of a building may extract heat from a water loop to heat the supply air, and other water source heat pumps in the core part of the building may reject heat to the water loop to cool the supply air. Excess heat is therefore transferred from the core of the building to the perimeter zone of the building. As water is physically moved through a water pump to different areas of the same building or a group of buildings, thermal energy is also being simultaneously transported. Even though the quality of heat is low, it is efficiently captured and transported to where it is needed due to the high heat-transfer coefficients achievable with water and due to its high specific heat. The energy that is normally wasted in an ASHP is now recovered and used somewhere else, thus offsetting new energy demand. The overall building energy consumption is decreased, and Btus are used more effectively.

A WSHP usually consists of an air coil that is a finned coil to condition the air; a double-tube water coil to reject or extract heat from the water loop; a forward-curved centrifugal fan, which is often located downstream of the air coil; single or twin hermetic compressors; a short capillary tube; a reversing valve; an outer casing; controls; and accessories. A two-speed fan motor is often used for better capacity control.

WSHP equipment is more efficient and smaller in size than ASHP equipment. Traditional ASHPs can require each air handling unit to have a separate condensing unit. For a large, multi-unit system, which is common in a commercial building, multiple condensing units would be needed that are not only noisy but also may present a challenge to install since they require considerable free space. With a multi-unit WSHP installation, heat exchange can be accomplished with a single, central evaporative cooling tower or dry cooler located on the ground or the roof-top. The WSHP units can be placed in dropped ceilings or hidden away from occupied spaces in mechanical rooms or utility closets. Placing the units in ceilings, near the point of use, also results in less ductwork and less fan-energy consumption. Fan-energy consumption can be among the largest energy components of an HVAC system, and a good overall system design will attempt to minimize it.

5.1.3 Ground Source (Geothermal) Heat Pumps

Groundwater heat pump (GWHP) systems use well water as a heat source during heating and as a heat sink during cooling. In summer, the geothermal heat pump extracts heat from a building and transfers it to the ground for cooling. In winter, the geothermal heat pump takes natural heat from the ground and transfers it to the building for heating. When the groundwater is more than 30 ft deep, its year-round temperature is relatively constant. Groundwater heat pump systems are usually open-loop systems. They are mainly used in northern climates, particularly in low-rise commercial buildings where groundwater is readily available and local codes permit such usage.

GWHP configurations include:

1. Water to air, where water from a ground source is circulated through the unit, and air ducted to and from the space to be conditioned is either heated or cooled.
2. Water to water, where water from a ground source is circulated through the unit, and chilled or hot water is circulated to fan coil units for cooling, and to heating elements or fan coils for heating.

On average, GWHPs will yield savings that are about 40% more than would be provided by ASHPs. This is because underground temperatures are higher in winter than air temperatures. As a result, GWHPs can provide more heat over the course of the winter than ASHPs.

GWHPs would not typically be included in an energy efficiency upgrade to an existing commercial building; however, they are being considered in new construction projects where conditions will allow it.

5.1.4 VRF Heat Pumps

A variable refrigerant flow (VRF) system (also referred to as a variable refrigerant volume [VRV] system which is trademarked by Daikin Industries that developed the technology) uses enhanced heat pumps to provide space heating and cooling to

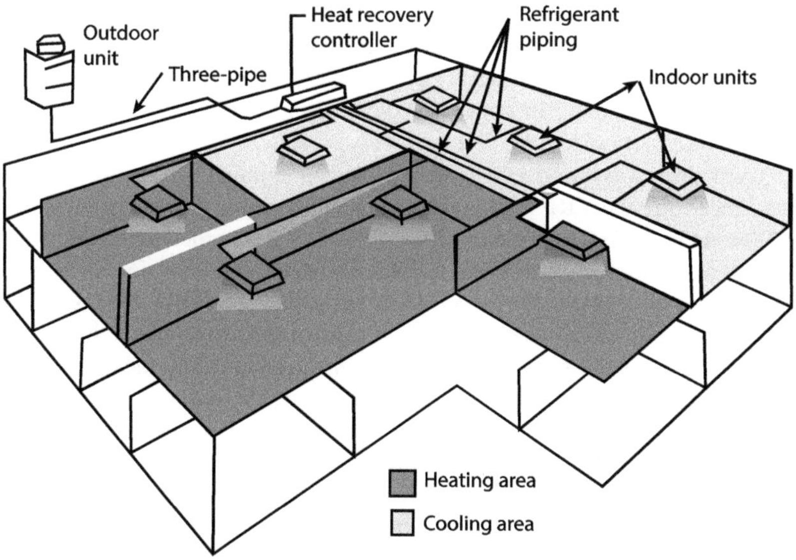

FIGURE 5.2 Variable refrigerant flow heat pump schematic.

FIGURE 5.3 Variable refrigerant flow heat pump units. (*Source: Courtesy of Carrier.*)

building spaces (see Figures 5.2 and 5.3). They are able to serve multiple zones in a building, each with different heating and cooling requirements and have the ability to modulate the amount of refrigerant sent to each zone in accordance with space conditioning requirements. In contrast, conventional HVAC systems operate on a 100%-on or 100%-off schedule.

VRF technology was introduced to the commercial building market in the early 1980s in Japan. Since then, the VRF market has expanded its global presence to Europe in the early 1990s and to the U.S. in the early 2000s. Today, it is regarded as a rapidly growing, cutting-edge technology offering considerable opportunity in the HVAC industry.

VRF systems are essentially enhanced direct-expansion (DX) ductless multi-split heat pump systems making use of advanced control techniques including a variable speed fan, variable speed scroll compressor with inverter technology, and electronic expansion valve. It can incorporate an air source or water source outdoor unit. While traditional multi-split systems turn off or on in response to a master controller, VRF systems continually modulate refrigerant flow to each indoor fan coil unit (FCU) to meet individual zone conditioning requirements. Refrigerant flow is modulated in response to outdoor air temperature and indoor conditions relying on variable-speed compressors which are rated at significantly higher part load efficiency compared to constant-speed systems. Multiple (as many as 50) indoor FCUs capable of operating with variable-speed control in or close to the conditioned space can be piped to a single outdoor condensing unit. With little or no ductwork in the FCUs, the fan static pressure and therefore energy consumption is considerably less than that of a central air system.

VRF systems can be set up to operate in cooling-only mode, heat pump mode (cooling-only/heating-only), or in heat recovery mode (simultaneous heating and cooling capability). Operation in the heat recovery mode where heat is extracted locally from the indoor units enables "free cooling" and "free heating," which can result in enhanced energy savings, particularly in the spring and fall shoulder months. What this means is that one zone can be in cooling mode while another is in heating mode. This is accomplished with the addition of a branch selector box (different manufacturers have different names for this box), which acts as the traffic cop for the refrigerant. Depending on the call from the thermostat and fan coil, the branch selector box will send either hot gas (heating mode) or low-pressure liquid (cooling mode) to the indoor fan coil.

Savings are improved when this technology is coupled with ceiling fans to enhance air distribution as well as a dedicated outdoor air system (DOAS) or an energy recovery ventilator (ERV) that incorporates heat recovery. VRF technology offers significant energy savings and superior occupant comfort by relying on advanced inverter-driven compressor design and the capability of providing simultaneous heating and cooling to better meet occupant needs. Compared with traditional ASHPs, VRF heat pumps offers energy savings due to better part load efficiencies, heat recovery, smaller zones, and reduced duct losses. The extent of the energy savings will depend on the climate zone where the building is located.

VRF technology has been found to be especially attractive in buildings with the following characteristics:

- No existing air conditioning and in need of an upgrade such as many historic buildings.

- Many different zones, each with different occupancy schedules, activities, uses, and temperature preferences.
- Large footprints, where the perimeter might require heat, while the interior requires cooling.
- Occupant comfort and controllability are top priorities.

As such, the following property types are ideal candidates for this high energy efficiency upgrade:

- Office buildings with many enclosed offices, conference rooms, and other separate spaces.
- Educational facilities such as schools and colleges.
- Lodging such as hotels and motels.
- Multifamily apartment complexes.
- Hospitals.
- Nursing homes and assisted living homes.
- Medical offices.
- Strip malls.
- New construction.

Where VRF has been adopted for energy efficiency, it is not unusual to find it used in conjunction with specialized ventilation systems such as decoupled DOAS (see Figure 7.6 in Chapter 7) to handle ventilation and moisture removal. In decoupled designs, the DOAS handles ventilation through separate ducting than the VRF system. DOAS supplies fresh tempered air directly into the space through its dedicated supply air diffuser, while the VRF system heats or cools the air in the space based on the zone's set-point.

While VRF presents an exciting energy savings opportunity, it is not a panacea. There are important considerations that also need to be evaluated before replacing an existing space conditioning system with a VRF system. These include:

1. *The cost of electricity and natural gas or fuel oil*
   When converting to VRF from natural gas–fired (or fuel oil–fired) boilers or furnaces, a relatively inexpensive heating source typically is being replaced by a more expensive heating source (i.e., natural gas being replaced by electricity). Notwithstanding, VRF is still a more efficient heating source, with COPs in the 3–5 range versus boilers with COPs in the 0.8–0.9 range. In addition, converting from a natural gas (or fuel oil) to a VRF electric heating system will increase electricity demand and utility demand charges. It may also involve a potential building electrical upgrade, which can be expensive. As such, the cost impact of fuel switching should be considered carefully, along with the energy savings associated with VRF systems.
2. *Competitive systems to VRF*
   Prior to selecting a VRF system, it always makes good sense to evaluate all options. For example, WSHPs can also service buildings needing simultaneous heating and cooling. During the winter the exterior portions of a large building may need heat while the core of the building, which may

have numerous internal loads and little heat loss, may only need cooling. With a WSHP system, heat pumps located throughout the building can draw off a single water loop that employs heat pumps capable of operating in either a heating or cooling mode. Thus, the core can cool the space and reject waste heat to the loop, while exterior portions of the building can heat the space, drawing heat from the loop.

Chiller-boiler systems utilizing air handling units (AHUs) may offer advantages in certain building types, particularly larger buildings. Initial equipment cost for the system may also be lower than for a VRF system. Moreover, chillers offer high COPs that ultimately may be more cost-effective. An additional advantage may exist if the chiller system includes an ice-storage system that can take advantage of non-peak electricity costs.

3. *Ventilation requirements and associated ductwork*

   Typically, VRF systems provide space cooling and heating by recirculating air within a zone. Outdoor air (ventilation air), which is required to comply with local building codes (and ASHRAE 62.1 requirements related to indoor air quality), must be provided separately.

   Conventional HVAC systems supply both ventilation airflow and the required airflow to deliver properly conditioned air throughout the building. Required ventilation airflow typically makes up less than 20% of this total airflow. A VRF system with DOAS only needs to meet the required ventilation airflow. As such, the VRF-DOAS system uses significantly less airflow than in a conventional system, which can result in substantial fan energy savings.

   Since VRF systems are not designed to extract large amounts of moisture from air, the DOAS is often equipped with energy recovery ventilation (ERV). Energy recovery allows sensible and latent heat to be exchanged between the entering outside air and the exhaust air.

4. *Lack of an economizer*

   Because there is no ductwork to deliver heating and cooling, VRF systems do not have an air-side economizer. However, if a DOAS is included in the installation, particularly with an HRV or ERV, energy recovery is still possible. Careful evaluation of all VRF system savings (i.e., cooling savings, ventilation savings and heating savings) often can offset the lack of an economizer and still result in meaningful net savings.

The cost-effectiveness of VRF in retrofit applications can be attractive with small and medium-sized buildings (i.e., in the 10,000–100,000 SF range). Below and above this size range, VRF may be less attractive compared with other potentially more cost-effective alternatives. While specifying VRF for new construction has shown to provide a potential cost advantage, retrofitting existing buildings presents more of a challenge as retrofit VRF systems typically are more expensive to purchase and install. It is not unusual in retrofit applications to find the total cost of VRF systems with DOAS to be 20%–50% more expensive than conventional systems. Unfortunately, building owners with tenants under triple net leases (where tenants pay utility costs) all too often replace equipment giving priority to the upfront cost.

This is a major reason why conventional DX systems continue to remain popular. Notwithstanding, each project should be evaluated on its own merits with building owner requirements and site-specific factors taken into consideration.

5.2 ENERGY EFFICIENCY METRICS

The energy efficiency metrics of heat pumps operating on the cooling side are the same as for air conditioners (refer to Chapter 4); that is, EER and IEER. On the heating side, the energy efficiency metric is the coefficient of performance (COP), which represents the kWh output divided by the kWh input or Btu output divided by the Btu input. A heat pump COP of 3.0 means that a heat pump will produce three times as much heating output for every unit of energy input (i.e., the heat pump effectively will be 300% efficient). This can be compared, for example, with a high efficiency condensing boiler having an efficiency of 96% (or a COP = 0.96). Heat pumps have greater than 100% efficiency because they do not generate heat from electricity. They merely use electricity to pump available outdoor heat indoors.

The EERs of new high efficiency ASHPs will be similar to those of high efficiency air conditioning equipment. The COPs of ASHPs are typically in the 3.2–3.6 range at full load depending on size (tons). Table 5.1 provides typical full load efficiencies for existing ASHPs that may be in need of replacement.

WSHPs are more efficient at heating when compared with packaged roof-top air conditioners that incorporate natural gas heating or electric resistance heating. The maximum efficiency for heating by burning natural gas in a condensing boiler is about 96% (or a COP = 0.96). Electric resistance heating is virtually 100% efficient (or a COP = 1.0). With a water source heat pump in heating mode, not only is the thermal energy from the water loop being extracted and used, but also the heat of compression in the refrigerant circuit is captured and used as a source of heating. Due to this capability of extracting heat from a heat source (i.e., the water loop) and using the heat of compression, the WSHP can easily provide 4 to 6 units of heating for every unit of energy consumed (COPs in the 4–6 range). Most water source heat pump energy efficiency metrics today include full load EERs typically ranging between 12.2 and 18.9 and COPs ranging between 4.3 and 6.

Commercial VRF systems have attractive energy efficiency metrics:

TABLE 5.1
Typical EERs and COPs for Heat Pumps as a Function of Year Installed

Year	EER	COP
1977	6.8	2.2
1980	7.5	2.5
1989	8.5	2.8
1999	9.1	3.1
2004	9.3	3.2
2010	10.6	3.3
2013	11	3.3

- EERs typically in the 9.5–15.4 range and IEERs typically in the 16.8–30.1 range, depending on size (tons cooling capacity) and type (e.g., air-cooled vs. water-cooled, heat pump or heat recovery mode, ducted vs. non-ducted).
- COPs typically in the 3.2–5.5 range, depending on size (tons cooling capacity) and type (e.g., air-cooled vs. water-cooled, heat pump or heat recovery mode, ducted vs. non-ducted).

VRF systems can retain better heating capacity at lower temperatures. Traditional heat pumps begin de-rating (i.e., COP reduction) at approximately 30–40°F, and eventually around temperatures approaching 0°F they typically rely on 100% electric resistance heating (COP = 1) or another source of heating such as boiler hot water. VRF systems with inverter-based vapor injection compressors have the capability to deliver up to 100% heating capacity down to outdoor temperatures of 0°F, up to 85% down to −13°F and up to 60% down to −22°F. The result is more heating energy savings compared to conventional heat pump systems.

The efficiency of heat pumps depends on how much heat is available in the outdoor air. That simply means that there is a relationship between the outdoor air temperature and heat pump efficiency (see Table 5.2). In colder climates, a heat pump can extract less heat from the cold outdoor air. Heat pump COPs begin to fall (i.e., heating become less efficient) as outdoor temperatures begin to fall below 30–40°F.

When estimating heating energy savings for a heat pump, rather than using a single COP for the baseline existing unit and a single COP for the new high efficiency unit, if a relationship between COP and outdoor air temperature exists for each piece of equipment, this can be used to get a more representative COP using the bin method.

The "bin method" refers to a procedure where weather data is sorted into discrete groups (bins) of weather conditions. Each bin contains the number of average hours of occurrence during the year of a particular range of weather conditions. At a particular

TABLE 5.2
Sample High Efficiency Heat Pump COP De-rating with Falling Outdoor Air Temperature

Outdoor Temperature, °F	Heating COP
47	3.4
40	3.03
30	2.56
17	2.25
15	2.17
10	1.96
5	1.74
0	1.52
−5	1.3

location, the bins would be specified at a given interval (e.g., 5°F) over the entire temperature range that may exist at that location. For example, one might find one bin covering the temperature interval of 65–69°F and that this occurred 128 hours in the year. The average temperature of this bin would then be 67°F. Bin data are based on long-term weather measurements from the National Weather Service. Bin data can be purchased from the National Climatic Data Center bookstore or from ASHRAE. As such, rather than using a single design condition for an entire year, calculations can be made for each bin condition (see Example Problem 5–1).

While this technique has been used to calculate heating and cooling loads in a building, it can also be used to calculate a more representative COP for energy savings analysis. When seeking to get a better estimate of the COP, the COP in each bin temperature range is calculated at the mid-point temperature of the range using a COP as a function of outdoor air temperature equation. Such an equation can be developed using a manufacturer's COP data at various temperatures. The average annual COP can then be calculated using the hours per year of occurrence as the weighing factor for each bin.

Example Problem 5–1

A small office building in Newark, New Jersey has an air source heat pump with a nominal heating capacity of 97,000 Btu/hr. The COPs and heat pump output as a function of outdoor air temperature have been taken from the heat pump technical specifications (see below). The hours (TMY) in the bin temperature range (see below) were for the specific office building location. The building heat load at the various outdoor air temperatures were estimated and are shown in the table. What is the seasonal COP for the heat pump? If 100% electric resistance is used for heating, how much more electricity would be consumed?

Given						Calculated							
Bin Temp. Range °F	Avg Temp in Bin °F	Hours/ yr in Bin	Office Bldg. Heat Load, Btu/hr	Heat Pump COP	Heat Pump Output, BTUH	Heat Pump Input, BTUH	Auxiliary Power, BTUH	Heating System COP	Heating Energy, MM Btu/yr	Heat Pump Output, MM Btu/yr	Heat Pump Input, MM Btu/yr	Auxiliary Heat Input MM Btu/ yr	Total Input MM Btu/yr
60–64	62	783	5,256	3.97	5,256	1,324	0	3.97	4.12	4.12	1.04	0.00	1.04
55–59	57	731	14,017	3.76	14,017	3,728	0	3.76	10.25	10.25	2.73	0.00	2.73
50–54	52	678	22,778	3.54	22,778	6,434	0	3.54	15.44	15.44	4.36	0.00	4.36
45–49	47	704	31,539	3.29	31,539	9,586	0	3.29	22.20	22.20	6.75	0.00	6.75
40–44	42	692	40,300	3.03	40,300	13,300	0	3.03	27.89	27.89	9.20	0.00	9.20
35–39	37	717	49,061	2.77	49,061	17,712	0	2.77	35.18	35.18	12.70	0.00	12.70
30–34	32	721	57,822	2.56	57,822	22,587	0	2.56	41.69	41.69	16.29	0.00	16.29
25–29	27*	553	66,576	2.41	66,576	27,625	0	2.41	36.82	36.82	15.28	0.00	15.28
20–24	22	359	75,336	2.31	61,900	26,797	13,436	1.87	27.05	22.22	9.62	4.82	14.44
15–19	17	216	84,096	2.17	57,530	26,512	26,566	1.58	18.16	12.43	5.73	5.74	11.46
10–14	12	119	92,856	1.96	51,370	26,209	41,486	1.37	11.05	6.11	3.12	4.94	8.06

Given						Calculated							
Bin Temp. Range °F	Avg Temp in Bin °F	Hours/ yr in Bin	Office Bldg. Heat Load, Btu/hr	Heat Pump COP	Heat Pump Output, BTUH	Heat Pump Input, BTUH	Auxiliary Power, BTUH	Heating System COP	Heating Energy, MM Btu/yr	Heat Pump Output, MM Btu/yr	Heat Pump Input, MM Btu/yr	Auxiliary Heat Input MM Btu/ yr	Total Input MM Btu/yr
5–9	7	78	101,616	1.74	45,200	25,977	56,416	1.23	7.93	3.53	2.03	4.40	6.43
0–4	2	36	110,376	1.52	39,030	25,678	71,346	1.14	3.97	1.41	0.92	2.57	3.49
−5 to −1	−3	22	119,136	1.30	32,870	25,285	86,266	1.07	2.62	0.72	0.56	1.90	2.45
−10 to −6	−8	6	127,896	1.07	26,700	24,953	101,196	1.01	0.77	0.16	0.15	0.61	0.76
								Totals	**265.13**	**240.15**	**90.46**	**24.97**	**115.43**

Notes:

*Heat pump balance point (the balance point is the outdoor temperature at which a heat pump reaches its heating capacity). The heat pump will struggle to increase the indoor temperature, but since the outdoor temperature matches its heating capacity, the heat pump will continue to run but have less and less impact on the indoor temperature. A heat pump's balance point defines the lowest outdoor temperature at which it can adequately warm the space it is serving. A supplementary heating system serves as a backup. Rather than allowing the heat pump to run in excessively cold weather, a supplementary heating system can kick in. The supplementary system kicks in below the balance point. The balance point is determined knowing the heating capacity of the heat pump at 17°F and 47°F (the AHRI rating of the heat pump is determined at these two temperatures) and the Btu/hr heating load (demand) of the building at the average bin temperature (typically estimated using a building model based on $Q = U \times A \times \Delta T$ where Q is the total hourly building heat loss through walls, roof, glass, etc. in Btu/hr; U is the overall heat transfer coefficient (Btu/hr-ft^2 °F); A is the net area of walls, roof, ceiling, floor, or glass in ft^2 and ΔT is the temperature difference (65°F − average bin temperature in °F).

For each average bin temperature, the columns in the table were calculated as follows:

Heat pump input, Btu/hr = (heat pump output, Btu/hr)/COP

Auxiliary power, Btu/hr = (building heat load, Btu/hr) − (heat pump output, Btu/hr)

Heating system COP = [building heat load, Btu/hr]/[(heat pump input, Btu/hr) + (auxiliary power input, Btu/hr)]

Annual heat needed for the building load, MM Btu/yr = (building heat load, Btu/hr) × (hours/yr in bin)/(1,000,000 Btu/MM Btu)

Annual heat pump output energy, MM Btu/yr = (heat pump output, Btu/hr) × (hours/yr in bin)/(1,000,000 Btu/MM Btu)

Annual heat pump input energy, MM Btu/yr = (heat pump output, MM Btu/yr) / heat pump COP

Auxiliary heat input, MM Btu/yr = (auxiliary power, Btu/hr) × (hours/yr in bin)/(1,000,000 Btu/MM Btu)

Total input energy, MM Btu/yr = (heat pump input energy, MM Btu/yr) + (input auxiliary energy, MM Btu/yr)

With the heat pump, the total electricity requirement is 115.43 MM Btu/yr. If pure resistance heating was used, the total electricity requirement would be 265.13 MM Btu/yr.

The seasonal COP for the heat pump is (240.15 MM Btu/yr)/(90.46 MM Btu/yr) = 2.65.

This example demonstrates how the part load characteristics of air source heat pumps are influenced by the outdoor air temperature.

Heat pump input energy/total end-use energy consumption = 90.46/115.43 = 78%. This can be compared to using the estimate in Table 5.3, where the heat pump heating as a percent of total space heating end-use energy consumption = 75%.

5.3 ENERGY SAVINGS

The energy savings associated with heat pumps equals the energy savings on the cooling side plus the energy savings on the heating side. Cooling side energy savings rely on the same equations used for air conditioning equipment (refer to Chapter 4). As such, the focus of this section will be solely on the heating side savings.

When replacing an inefficient heat pump with a new higher efficiency heat pump, the energy savings on the heating side can be estimated using Equation 5–1.

$$ES_{H\text{-}kWh} = \left(CAP_H / 3{,}412\right) * EFLH_H * \left(\left(1/COP_{Baseline}\right) - \left(1/COP_{New}\right)\right) \tag{5–1}$$

where $ES_{H\text{-}kWh}$ = space heating end-use energy consumption savings, kWh/yr

CAP_H = heat pump heating output (delivered), Btu/hr (*Note*: Many heat pumps have a heating capacity [output] about 10% less than their cooling capacity.)

$EFLH_H$ = equivalent full load hours for heating (see Tables 6.1 and 6.2 in Chapter 6)

$COP_{Baseline}$ = existing baseline heat pump coefficient of performance, kWh output/kWh input

COP_{New} = new heat pump coefficient of performance, kWh output/kwh input

3,412 = conversion factor, 3,412 Btu/kWh

Let's do a dimensional unit consistency check on Equation 5–1:

$$\text{kWh input/yr} = \left[\left(\text{Btu/hr output}\right) / \left(\text{Btu/hr/kW}\right)\right] * \left(\text{hr/yr}\right) * \left(\text{kWh input/kWh output}\right)$$

The percent heating energy savings can be estimated by dividing the energy savings from Equation 5–1 by:

$$\left(CAP_H / 3{,}412\right) * EFLH_H * \left(1/COP_{Baseline}\right)$$

The result will be the percent heating energy savings:

$$\%\ ES_{H\text{-}kWh} = \left(1-\left(COP_{Baseline} / COP_{New}\right)\right) * 100 \tag{5–2}$$

Multiplying the percent heating energy savings by the existing (baseline) unit's space heating end-use energy consumption in Btu/yr will give the same result as Equation 5–1.

The peak demand savings in the billing period can be estimated using Equation 5–3:

$$DS_{kW} = \left(CAP_H / 3{,}412\right) * \left(\left(1/COP_{Baseline}\right) - \left(1/COP_{New}\right)\right) * CF_{winter} \tag{5–3}$$

where DS_{kW} = peak heating demand savings, kW
CAP_H = output (delivered) heating capacity, Btu/hr
CF_{winter} = coincidence factor in months when heating is used, e.g., winter season (specified in utility TRMs)

Equation 5–3 provides the peak demand savings. All things being equal, assuming no ratchet provisions and that billed demand equals read demand, the peak heating demand savings would occur in the winter months when heating occurs (i.e., the winter season). If this is the case, then the annual peak demand savings would be the winter peak demand savings multiplied by the number of billing periods constituting the winter season. Hence, if the peak demand savings is 5 kW and the winter season when heating occurs lasts 6 months, the annual demand savings would be 30 kW. It is important to understand how the utility charges for demand and whether ratchet provisions exist before estimating demand cost savings.

Since COP is reduced as outdoor air temperature decreases, any savings calculation will need to consider an average COP over the time period (hours) when the heat pump is not relying on supplemental heat. Supplemental heat often is electric resistance heat with a COP = 1. To be conservative, no winter demand savings should be assumed if the air source heat pump uses electric resistance heating as backup at low temperatures.

Example Problem 5–2

A 10-ton roof-top heat pump with a heating capacity of 106,000 Btu/hr on an office building has outlived its usefulness and will be replaced with a new high efficiency unit. The existing unit has an IEER of 10.9 and a COP of 2.2. The new unit has an IEER of 13.5 and a COP of 3.3. The EFLH cooling (TMY) is 650 hr/yr and the EFLC heating (TMY) is 980 hr/yr. What are the energy consumption savings?

Existing Unit			**New Unit**		**Savings**	**% Savings**
Cooling output =	10	ton	10	ton		
	or 120,000	Btu/hr	or 120,000	Btu/hr		
Total Cooling Btu/yr delivered (output) =	78,000,000	Btu/yr	78,000,000	Btu/yr		
	or 22,860	kWh/yr	22,860	kWh/yr		

Existing Unit			New Unit		Savings		% Savings
Cooling kWh input = Btu/yr output/ (IEER × 1,000) =	7,156	kWh/yr	5,778	kWh/yr	1,378	kWh/yr	19.3%
Heating output =	106,000	Btu/hr	106,000	Btu/hr			
Total heating Btu/yr output (delivered) =	103,880,000	Btu/yr	103,880,000	Btu/yr			
	or 30,445	kWh/yr	30,445	kWh/yr			
Heating energy consumed (input) = kWh/yr output/COP	13,839	kWh/yr	9,226	kWh/yr	4,613	kWh/yr	33.3%
				Total Savings:	5,991	kWh/yr	

Equation 4–7 can be used for cooling energy savings and Equation 5.1 for heating energy savings:

$$\text{Cooling Energy Savings, kWh/yr} = (120{,}000)*650*\left[(1/10.9)-(1/13.5)\right]/1000$$

$$= 1{,}378 \text{ kWh/yr}$$

$$\text{Heating Energy Savings, kWh/yr} = (106{,}000*980/3412)*\left((1/2.2)-(1/3.3)\right)$$

$$= 4{,}613 \text{ kWh/yr}$$

Total Savings: 5,991 kWh/yr

The percent energy savings and total heating and cooling energy consumption savings can also be estimated as follows:

Cooling % Energy Savings: $= \left(1-\left(IEER_{baseline}/IEER_{new}\right)\right)$
$= (1 - (10.9/13.5))$
$= 19.3\%$ (kWh/yr savings = 0.193 × 7,156 kWh/yr = 1,378 kWh/yr)

Heating % Energy Savings: $= [1 - (COP_{baseline}/COP_{new})]$
$= [1 - (2.2/3.3)] = 33.3\%$ (kWh/yr savings = 0.333 × 13,839 kWh/yr = 4,613 kWh/yr)

Total energy consumption savings = 5,991 kWh/yr

Example Problem 5–3

An office building in Atlanta, Georgia, is currently served by a natural gas–fired heating system. The building owner would like to convert this to electric heat pump heating as part of an electrification program. The following data is available:

Boiler input capacity =	2,000,000 Btu/hr
Space heating end-use energy consumption (input) =	3,500 MM Btu/yr in the last 12 months
Boiler thermal efficiency =	70%
Heat pump average COP =	2.7
Electricity generation cost =	$0.08 per kWh
Natural gas cost =	$12.00 per MM Btu

What first year heating energy savings might there be with this conversion ignoring the impact on electric demand?

Natural gas cost in last 12 months =	$42,000 (3,500 MM Btu × $12/MM Btu)
Required building space heating load = 0.7 × 3,500 MM Btu	2,450 MM Btu/yr or 717,850 kWh/yr
ASHP average COP =	2.7 kWh output/kWh input
ASHP kWh consumed (input) to satisfy space heating load = 717,850 / 2.7	265,870 kWh/yr input or 907 MM Btu/yr input
Electricity cost for ASHP = (265,870 kWh/yr) × $0.08/kWh =	$21,270
Energy consumption savings = 3,500 MM Btu/yr – 907 MM Btu/yr =	2,593 MM Btu/yr
First year net energy cost savings impact = $42,000 – $21,270	$20,730

The $20,730 net energy savings would be reduced by any increase in electric demand cost and increase assuming the new ASHP has a higher EER/IEER than the existing air conditioning unit.

A Word About $EFLH_H$ . . .

$EFLH_H$ is calculated from the following formula:

$$EFLH_H = E_{Out\text{-}H} / CAP_H \tag{5–4}$$

where $EFLH_H$ = equivalent full load hours for space heating, hr/yr
$E_{Out\text{-}H}$ = output (delivered) space heating end-use energy consumption, Btu/yr
CAP_H = output (delivered) space heating capacity, Btu/hr

When the total heating Btu/yr output, $E_{Out\text{-}H}$, is not available (remember, this is not the input space heating end-use energy consumption estimated from whole building energy use as determined from utility bills), it can be determined using Equation 5–5.

$$E_{Out\text{-}H} = E_{in\text{-}H} * COP_H \qquad (5\text{–}5)$$

where $E_{in\text{-}H}$ = input (consumed) space heating end-use energy consumption, Btu/yr
COP_H = coefficient of performance for space heating, Btu output/Btu input

The output heating capacity is typically the rated output capacity for the heat pump (which is usually known or available in manufacturer specifications).

As indicated in Chapter 4 for the $EFLH_C$, the $EFLH_H$ can be estimated in a number of ways. For example, Table 6.1 may be used, as can the state technical resource manuals (refer to Table 6.2). When using state TRMs, if not developed using typical (TMY) data, be particularly careful to note the year the data was developed as the $EFLH_H$ will be influenced by the weather in that particular year. Notwithstanding, using Table 6.2 to obtain a preliminary estimate is not unreasonable as year-to-year drastic weather changes are uncommon.

As with the theoretical $EFLH_C$ calculation, the theoretical $EFLH_H$ is based upon outdoor air conditions and the building HVAC design set-point, and may be estimated using Equation 5–6.

$$EFLH_H = \left[HDD * 24\right] / \left[65 - T_{Design}\right] \qquad (5\text{–}6)$$

where HDD = heating degree days
T_{Design} = outside air design temperature (low side), °F
24 = 24 hr/day
65 = reference temperature, °F

The design temperatures for select cities in the U.S. are included in Table 4.5, and HDD data are available at www.degreedays.net (where you can obtain the heating degree days at your building's location for the full 12-month period).

Unfortunately, Equation 5–6 ignores the property type and specific HVAC system characteristics, both of which can have a significant impact on the building's heating load. TRM manuals typically provide the $EFLH_H$ by building type, location, and select HVAC characteristics. ASHRAE found that building operating hours may also be a key determinant in defining EFLH for differing building types.[1]

For a preliminary estimate of the percent of total space heating end-use energy consumption attributed to heat pump heating versus that attributed to supplemental heating—electric resistance (COP = 1) or heating from a gas-fired boiler (COP = 0.8 conventional boiler or COP = 0.95 condensing boiler—the author has found that Table 5.3 can provide a reasonable first estimate.

TABLE 5.3

Heat Pump Heating as a Percent of Total Space Heating End-Use Energy Consumption ($E_{In\text{-}H}$)

Geographic Location	States	% of $E_{In\text{-}H}$
New England	CT, RI, MA, NH, VT, ME	60%
Middle Atlantic	NY, PA, NJ	75%
East North Central	IL, MI, WI, IN, OH	75%
West North Central	MO, IA, MN, ND, SD, KS, NE	75%
South Atlantic	VA, DC, DE, MD, WV, NC	90%
South Atlantic	SC, GA	100%
South Atlantic	FL	100%
East South Central	TN, AL, KY, MS	90%
West South Central	TX, AR, LA, OK	90%
Mountain North	CO, ID, MT, UT, WY	60%
Mountain South	AZ, NM, NV	90%
Pacific	HI, S. CA	100%
Pacific	OR, WA	85%

TABLE 5.4

VAV RTU versus VRF-Heat Pump Comparison*(3)

Building Characteristics	RTU-VAV System	VRF HP System
Heating	Gas inside packaged RTU	VRF DX heating coil
	Gas efficiency = 80%	COP = 3.2
Cooling	Unitary DX in RTU	VRF DX cooling coil
	EER = 11.5	EER = 11
Supply Fan Type	VAV	On/Off
Economizer	Yes	N/A

Climate Zone**	Total HVAC Energy VRF Energy Savings, %	Heating VRF Energy Savings, %	Cooling VRF Energy Savings, %	HVAC Fan VRF Energy Savings, %
1A	25	88	21	48
2A	33	64	24	41
2B	14	67	1	49
3A	34	46	26	36
3B-Coast	28	58	23	32
3B	16	46	–2	50
3C	22	50	–4	30
4A	31	40	20	33
4B	14	32	–8	40
4C	33	44	–5	26
5A	35	41	22	33
5B	29	40	–3	37
6A	39	44	21	34
6B	27	37	–26	32
7	39	43	9	32

TABLE 5.4 ***(Continued)***
VAV RTU versus VRF-Heat Pump Comparison*[(3)]

Climate Zone**	Fan Energy as a % of Total HVAC Energy	
	RTU-VAV System	VRF-HP System
1A	10.9	7.8
2A	11.2	10.1
2B	14.8	8.8
3A	11.5	11.4
3B-Coast	26.3	24.1
3B	17.1	10.2
3C	32.3	29.4
4A	10.9	10.5
4B	21.3	14.7
4C	15.4	17
5A	7.7	8.8
5B	13.8	12.2
6A	6.8	7.3
6B	12	11.1
7	6.7	7.6

* Comparison of VAV RTU with electric reheat with VRF-Heat Pump Mode (no heat recovery) with supplementary electric heaters at low outdoor air temperature in a 53,600 ft^2 medium-sized office building located in various climate zones. The building was assumed to comply with the minimum energy code requirements in ASHRAE 90.1-2010 and *only* HVAC energy consumptions are compared. EnergyPlus v.8.4 model was used on a DOE reference medium office building.

** Climate zones are identified below.

Climate Zone	Representative City	Condition
1A	Miami, FL	Very hot, humid
2A	Houston, TX	Hot, humid
2B	Phoenix, AZ	Hot, dry
3A	Atlanta, GA	Warm, humid
3B-Coast	Los Angeles, CA	Warm, dry
3B	Las Vegas, NV	Warm, dry
3C	San Francisco, CA	Warm, marine
4A	Baltimore, MD	Mixed, humid
4B	Albuquerque, NM	Mixed, dry
4C	Seattle, WA	Mixed, marine
5A	Chicago, IL	Cool, humid
5B	Boulder, CO	Cool, dry
6A	Minneapolis, MN	Cold, humid
6B	Helena, MT	Cold, dry
7	Duluth, MN	Very cold

For example, assume a New York Middle Atlantic building location. From Table 5.3 the default amount of heat pump heating would be approximately 75% of $E_{In\text{-}H}$, and this would be produced at the heat pump's average COP. It would also mean that approximately 25% of the space heating end-use energy consumption (supplemental) would be provided by a conventional gas-fired boiler with perhaps a COP = 0.80, or electric resistance heating with a COP approaching 1. If an inefficient ASHP is being replaced by a more efficient ASHP, it would not be unreasonable to assume the backup supplemental heat source would not change.

It would not be unusual to find the energy consumption savings for high efficiency heat pumps in the 20%–30% range for space cooling end-use energy consumption and 10%–15% for space heating end-use energy consumption.

It is not uncommon for a VRF system to save 25%–40% energy over a baseline VAV DX system or a central chilled water-boiler system, depending upon climate zone (see Table 5.4[(2)]). VRF systems also perform better at part load conditions. An Oak Ridge National Laboratory (ORNL) report[(3)] showed the VRF-DOAS system outperformed the RTU-VAV system in the cooling season by 35%, 30%, and 48% at 100%, 75%, and 50% load conditions, respectively.

As climates with more extreme weather often offer a better opportunity for VRF with heat recovery (simultaneous heating and cooling), buildings located in such climates typically generate the most energy savings.

REFERENCES

1. Carlson, S., ASHRAE research project report RP-1120: Development of equivalent full load heating and cooling hours for GCHPs applied in various building types and locations, March 2001, https://www.techstreet.com/ashrae/searches/39741364
2. Kim, D., Cox, S.J., Cho, H. and Im, P., Evaluation of energy savings potential of variable refrigerant flow (VRF) from variable air volume (VAV) in the U.S. climate locations, *Energy Reports Journal*, Vol. 3, pp. 85–93, November 2017.
3. Im, P., Malhotra, M. and Munk, J.D., Evaluation of variable refrigerant flow systems performance on ORNL's flexible research platform: Part 1: Cooling season analysis, ORNL/TM-2016/364 Report, August 2016, https://info.ornl.gov/sites/publications/files/Pub69040.pdf

6 Heating Systems

Most commercial buildings are heated with either boilers or furnaces. Furnaces heat air and distribute the heated air through ductwork. Boilers heat water and provide either hot water or steam for heating. Hot water is typically distributed to coils used to heat air or can be distributed via baseboard radiators.

6.1 COMBUSTION PROCESS

The three elements essential to combustion control are air, fuel, and heat. Combustion is the rapid chemical reaction between oxygen and a combustible material that releases heat. The combustible material is the fuel such as natural gas. Air supplies the oxygen (O_2) for combustion, and heat is the byproduct of this combination. Perfect combustion occurs when the proper mixture of fuel and oxygen under precise conditions are consumed completely in the process. This precise amount is called the stoichiometric point of the fuel. With hydrocarbon fuels such as natural gas or fuel oil, the perfect mix of air and fuel results in two main byproducts: carbon dioxide (CO_2) and water (H_2O). Carbon combines to form CO_2, and hydrogen creates H_2O. Combustion that is not complete leaves partially unburned fuel in the form of waste. Carbon monoxide (CO) and hydrogen (in the form of water) are two common waste products.

Natural gas, the most common combustible fuel, is predominantly methane (CH_4). Its byproducts are CO_2 and H_2O. The balanced equation for this process is given by:

$$CH_4 + 2O_2 -> CO_2 + 2H_2O$$

Air, the other main ingredient of the equation, consists of approximately 78% nitrogen (N_2), 21% oxygen (O_2), and 1% argon (Ar). The resulting heat that is released is known as the heat of combustion.

Boilers are fired with excess air. Excess air is supplied to the burner beyond the required air for complete combustion of the fuel because a boiler firing without sufficient air, or "fuel rich," is operating in a potentially dangerous condition. Therefore, excess air is used to provide a safety factor above the theoretical air required for combustion. Because excess air is heated by the flame, it takes energy away from combustion, thus taking away potential energy for transfer to the water in the boiler. In this way, excess air reduces boiler efficiency.

The heating value or heat content of a fuel is the amount of heat released when combusted. When evaluating fuel heating value, the terms HHV and LHV come into play. HHV is the fuel's higher heating value, which represents the full energy content of the fuel including all products of combustion (i.e., including the latent heat content of water vaporization). The LHV is the fuel's lower heating value, which does not include the latent heat content of water vaporization. For example, the HHV for natural gas is 22,446 Btu/lb (or 1,090 Btu/ft^3), and the LHV is 20,262 Btu/lb (or 983 Btu/ft^3).

DOI: 10.1201/9781032692777-6

6.2 EQUIPMENT DESCRIPTION

Boilers used in commercial buildings can generally be classified as either a fire-tube boiler (see Figure 6.1) or a water-tube boiler (see Figure 6.2) and have four main components: the burner, the combustion chamber, the heat exchanger, and the controls.

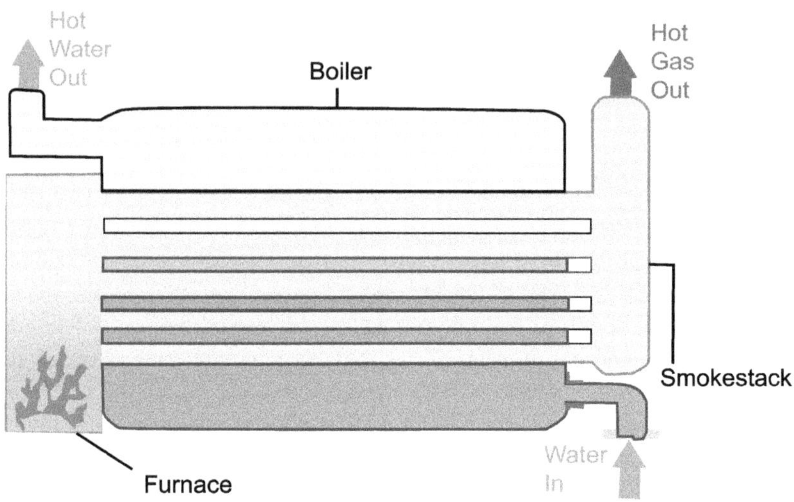

FIGURE 6.1 Fire-tube boiler schematic.

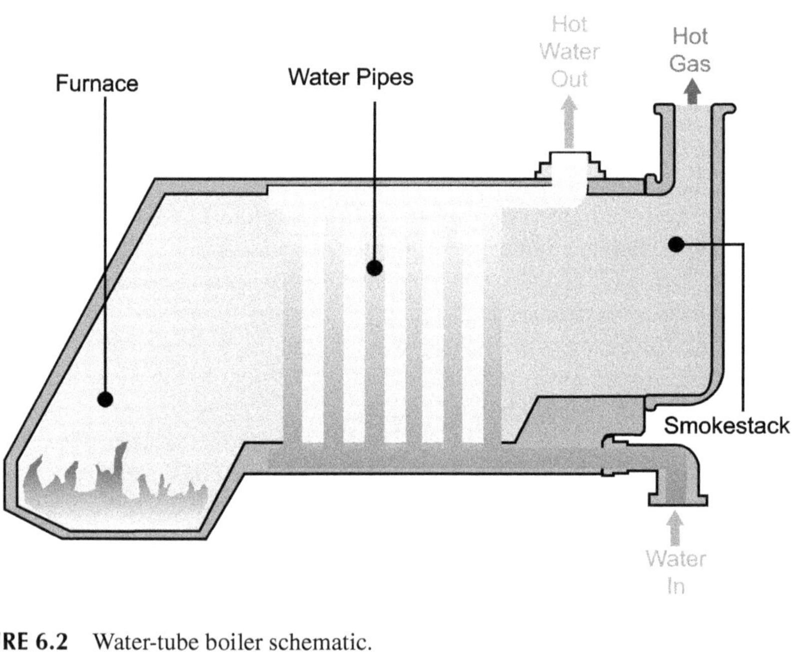

FIGURE 6.2 Water-tube boiler schematic.

Burner: The function of the burners is to mix air and fuel in optimal proportions for complete and efficient combustion and provide an ignition source for the mixture. Oil burners typically use a forced draft burner with fuel atomization using a pressure nozzle. Gas burners can use either a forced draft fan or natural draft with the gas delivery through a series of orifices. The principal fuel sources used in boilers in commercial buildings are natural gas, fuel oil, and electricity. Electricity is regaining some popularity as hot water generated by an electric boiler produces no greenhouse gases at the site. If electricity can be generated from a 100% renewable source such as solar, an electric boiler becomes a zero-emission (no greenhouse gases) heat generator. Notwithstanding, the most popular fuel source today in the U.S. is natural gas.

Combustion chamber: The rate of fuel being supplied to the combustion chamber is controlled to match the heating requirements of the space to maintain the set-point temperature. The rate of oxygen, as a component of air, is controlled to provide complete stoichiometric combustion of the fuel. Too little air and not all the fuel is burned, resulting in pollution, including hazardous carbon monoxide and wasted fuel. Too much air will cool the boiler or furnace and carry away useful heat. Some level of excess air is needed to insure complete combustion. In efficient natural gas and oil-fired units, typical excess air levels are in the 10%–20% range. Excess air above these levels wastes energy and reduces efficiency.

Heat exchanger: Once past the end of the flame zone, the hot products of combustion and unconsumed portions of the air pass through narrower heat exchange channels in the heat exchanger and heat is transferred via convection and conduction.

Controls: Control of a boiler is crucial to ensuring that it operates in the most economical and efficient way. Almost all boilers have modulating burners. The burners are usually controlled by an embedded system with built-in logic to control the output of the burner to match the load and give the best performance. The highest efficiency boilers combine accurate temperature control with high burner turndown to match output to heating demand precisely and cost-effectively. Many process parameters are monitored and controlled, typically managed by a programmable logic controller. These may include, for example, fuel and airflow (combustion air/oxygen), system temperature and pressure, the temperature of the exiting hot water or hot air, flue gas temperature, and temperature and pressure set-points. Today, there is a growing preference to have boiler operation controlled by a building HVAC energy management system.

Conventional boilers and furnaces are designed to prevent water vapor from condensing as it may combine with trace contaminants such as sulfur in the fuel and form an acidic and corrosive condensate that can be harmful to the boiler. With a conventional boiler, the exhaust gases are very hot, typically around 350–400°F. These hot exhaust gases are immediately sent up a stack (i.e., a flue or chimney) and take a lot of thermal energy with them, which limits their energy efficiency. At these exit temperatures water remains in vapor form. Typical efficiencies for conventional boilers today range between 80% and 84%.

Water vapor is produced in the combustion process arising from burning the hydrogen content of the fuel. A condensing boiler (see Figures 6.3 and 6.4) extracts additional heat from the waste gases by condensing this water vapor to liquid water, thus recovering its latent heat of vaporization. Condensing boilers can achieve

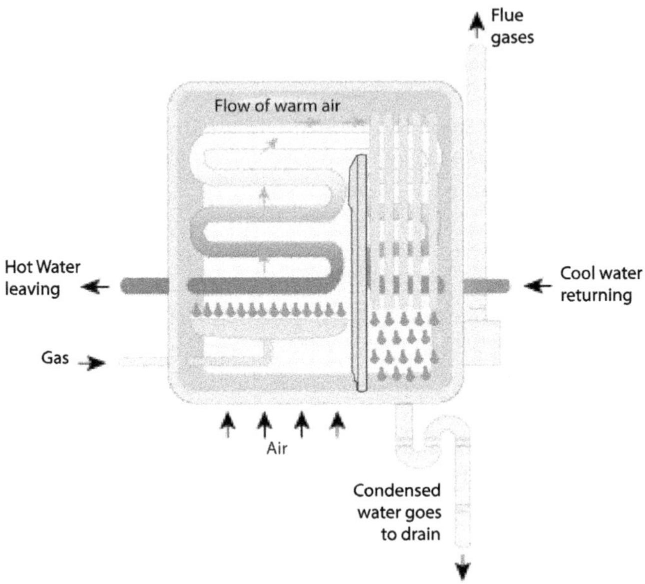

FIGURE 6.3 Condensing boiler schematic.

FIGURE 6.4 Commercial fire-tube gas-fired condensing unit installation. (*Source: Courtesy of Laars.*)

high efficiency by condensing water vapor in the exhaust gases. This increases the boiler efficiency from a maximum of about 84% in non-condensing (conventional) boilers and furnaces to as much as 95%–98%. However, this increase in efficiency comes at an increased equipment cost for corrosion-resistant materials to handle the acidic condensate. When latent energy is extracted from water vapor, acidic condensate will be left behind on the surface of heat exchangers. The condensate has a pH of 3–5, so suitable materials must be used in areas where liquid is present. The heat exchangers are made of high-quality materials and designed to drain freely, which allows them to withstand the condensing operation with no significant corrosion. Aluminum alloys and stainless steel are the materials commonly used. In condensing applications, contaminants in the fuel, particularly sulfur, need to be minimized. Therefore, natural gas is the predominant fuel source in condensing boilers.

6.2.1 Fire-Tube Boilers

In fire-tube boilers, tubes are located in a drum. The combustion gases pass through (inside) these tubes, heating the water in the drum. Fuel and air are combined in a burner with combustion gases going through the tubes, heating the water, and then exiting through a flue. The water to be heated is brought into the boiler, surrounds the tubes containing the fire from the burner, rises to the top under pressure, and produces hot water (or steam) for use in the building. A typical fire-tube boiler installation is shown in Figure 6.5.

FIGURE 6.5 Fire-tube boiler installation. (*Source: Courtesy of Laars.*)

6.2.2 Water-Tube Boilers

A water-tube boiler is one in which the water to be heated passes inside a series of tubes with the hot combustion gases outside the tubes. The water to be heated starts in a cylindrical drum at the bottom of the boiler chamber, then travels upward by convection through the tubes in the combustion chamber and is converted to hot water or steam as it is heated by the flame in the boiler. A water-tube boiler installation is shown in Figure 6.6.

6.2.3 Electric Boilers

Electric resistance may also provide the heat source for a boiler or furnace. Electric boilers produce heat by electricity and are virtually 100% efficient (although this does not reflect the efficiency at which electricity is produced at the power plant and distributed by the utility generating the electricity (refer to site vs. source energy in Chapter 2, Section 2.2.2). Electric boilers are compact, clean, and quiet. They have replaceable heating elements, either electrode or resistance-coil type. With the electrode-type boiler, the heat is generated by electric current flowing from one electrode to another electrode through the boiler water. Resistance-coil type electric boilers have the electricity flowing through a coiled conductor similar to an electric space heater. Resistance created by the coiled conductor generates heat. Resistance-coil electric boilers are not as common as electrode electric boilers.

FIGURE 6.6 Commercial water-tube gas-fired condensing unit installation. (*Source: Courtesy of Laars.*)

FIGURE 6.7 Electric boiler installation. (*Source: Courtesy of Laars.*)

Electric boilers are an alternative to oil or gas boilers where the latter boilers are restricted by regulations (such as electrification regulations or greenhouse gas emission limitations) and in areas where the cost of electric power is minimal. Electric boilers can be fire-tube or water-tube and supply low- or high-pressure steam or hot water. Figure 6.7 shows an electric boiler installation.

Electric resistance heating is commonly used in packaged roof-top air conditioners to supply space heating and in heat pumps to supply supplemental heating in cold weather.

Electric resistance heating has a number of advantages:

- Relatively low first cost.
- Efficiency approaching 100%.
- Near-zero part load penalty.

- No greenhouse gas emissions at the building site (but if the utility uses fossil fuels to generate electricity and deliver it to the building, the greenhouse gas emissions to produce this electricity must be considered).
- A flue gas stack or chimney is not needed.

A high electricity cost (both for energy consumed and energy demand charges) can reduce the apparent advantage of electric boilers and furnaces.

6.2.4 Furnaces

There is a substantial difference between boiler heat compared to furnace heating. A boiler uses hot water or steam from hot water to heat the building. The steam or hot water travels through pipes into heating coils located in the building, such as in air handling units, packaged units, or fan coil units. A furnace uses warm air to heat the building. A strong fan pulls air into the furnace, across the heat exchanger or heating element, and then distributes the warm air throughout the building through the building's duct system and vents or registers. Figure 6.8 presents a schematic of forced air furnace operation, and Figure 6.9 shows an installed system.

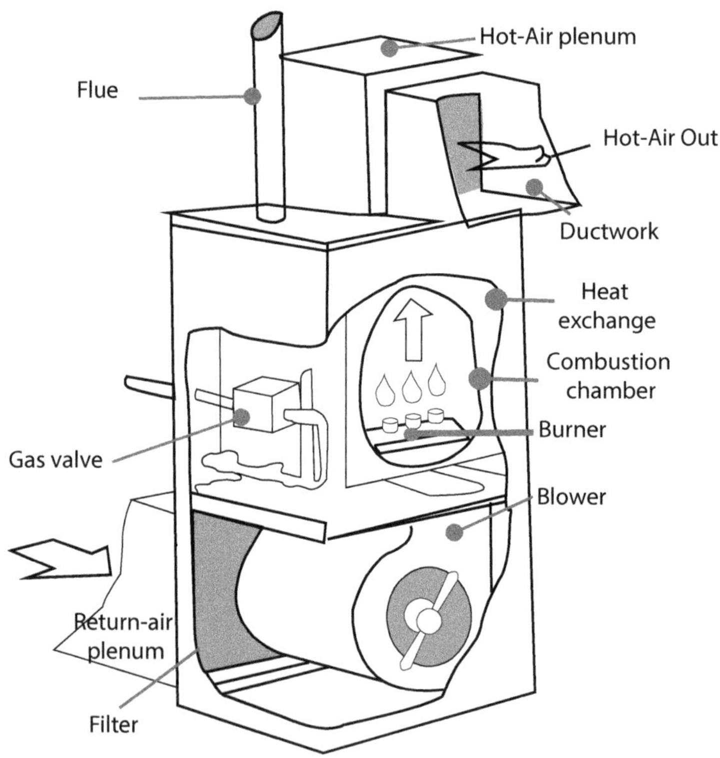

FIGURE 6.8 Forced air furnace schematic.

FIGURE 6.9 Forced air furnace installation.

6.3 SIZING THE HEATING SYSTEM

Specifying the size boiler or furnace needed to satisfy the space heating load of a building is typically determined by careful analysis of building heat losses throughout the year (frequently using the bin method or building simulation modeling). However, a rule of thumb for boiler sizing (capacity) in a building is 35–40 Btu/hr delivered (output)/building ft^2.[(1)] Unfortunately, this rule of thumb does not take into consideration either building type or climate zone. From the author's experience, values can range between 25 to 75 Btu/hr delivered/ft^2 depending on building type and geographic location (and excluding climate zones requiring little or no heating, such as may exist in southern California or southern Florida).

If the capacity of the boiler system, which typically is its output capacity (but not always), is not known, utility fuel bill data can be used to obtain an estimate. Assuming fuel utility bill data is available (which is likely) that indicates the input energy consumed for space heating ($E_{In\text{-}H}$) in Btu/year for the building, this energy input can be converted to energy output ($E_{Out\text{-}H}$) delivered to the building to satisfy its space heating load by multiplying by the estimated boiler efficiency (η_B). The approximate boiler size in Btu/hr can then be estimated using the $EFLH_H$, which does consider building type and location.

$$CAP_H = E_{Out\text{-}H} / EFLH_H \tag{6–1}$$

where CAP_H = boiler output capacity, Btu/hr

$E_{Out\text{-}H}$ = space heating end-use energy delivered (Btu/yr output), which can be calculated from $E_{In\text{-}H} \times \eta_B$

$EFLH_H$ = equivalent full load hours for space heating, hr/yr (see Tables 6.1 and 6.2)

Boiler model numbers typically reflect either a boiler's maximum input capacity or maximum output (delivered) capacity. The maximum output capacity may also

TABLE 6.1
Equivalent Full Load Hours for Heating by Location and Building Type[2]

Location	School	Office	Retail	Hospital
Atlanta, GA	200–290	480–690	380–600	160–430
Baltimore, MD	320–460	720–890	570–770	300–590
Bismarck, ND	460–500	950–990	810–900	530–730
Boston, MA	450–520	960–1,000	760–870	420–680
Charleston, WV	310–440	770–840	620–730	320–550
Charlotte, NC	200–320	530–780	420–670	180–490
Chicago, IL	390–470	820–920	670–810	400–640
Dallas, TX	120–200	340–520	280–440	100–310
Detroit, MI	400–480	970–1,020	790–900	460–710
Fairbanks, AK	560–630	1,050–1,170	930–1,090	690–930
Great Falls, MT	360–430	820–890	680–800	420–640
Houston, TX	90–130	250–350	190–300	70–200
Indianapolis, IN	400–480	840–920	690–820	390–640
Los Angeles, CA	80–160	370–580	250–440	20–180
Louisville, KY	290–430	710–830	570–720	300–550
Madison, WI	390–470	840–900	700–800	440–640
Memphis, TN	170–240	420–600	330–510	140–370
Miami, FL	6–12	34–46	25–37	1–12
Minneapolis, MN	420–500	860–950	720–860	470–700
Montgomery, AL	120–180	330–470	250–400	90–260
Nashville, TN	250–320	590–680	470–590	240–450
New Orleans, LA	67–110	230–320	160–260	46–160
New York City, NY	350–440	790–870	630–760	330–590
Omaha, NE	330–400	720–800	600–720	360–570
Phoenix, AZ	65–110	210–290	170–250	34–140
Pittsburgh, PA	470–500	910–950	750–840	420–650
Portland, ME	400–480	880–980	710–870	420–690
Richmond, VA	270–410	660–820	520–710	250–530
Sacramento, CA	220–360	640–990	430–830	120–540
Salt Lake City, UT	520–540	1,040–1,080	830–930	440–720
Seattle, WA	460–650	1,270–1,370	960–1,170	360–850
St. Louis, MO	280–400	710–800	570–700	320–550
Tampa, FL	36–58	140–190	100–160	22–90
Tulsa, OK	240–300	560–620	450–540	220–410

TABLE 6.2
Equivalent Full Load Hours for Heating in Select State TRMs[3]

Facility Type	CT Statewide	IL Zone 2 Chicago	IL Zone 3 Springfield	Mid-Atlantic Wilmington	Mid-Atlantic Baltimore	Mid-Atlantic Washington, DC	MN Minneapolis–St. Paul (Zone 3)	NY–Small Commercial New York City	NY–Small Commercial Albany	TX Climate Zone 2 Fort Worth	TX Climate Zone 3 Houston
Banks, Financial Centers	1,248										
Religious Worship/Church	1,694	1,657	1,730	754	737	723		722	954		
College/University		404	376				1,616				
College—Cafeteria	594										
College—Classes/Administrative	1,537										
College—Dormitory	1,418										
Commercial Condos	1,172										
Convenience Stores	594	550	585				1,546				
Convention Center	1,695										
Dining—Cafeteria/Fast Food	574										
Dining—Family	1,140										
Entertainment	1,695										
Fast Food Restaurant	594			983	866	766		813	1,426	631	372
Food Stores	1,172			298	241	192		191	554		
Gym	1,537										

(Continued)

TABLE 6.2 *(Continued)*
Equivalent Full Load Hours for Heating in Select State TRMs

Facility Type	CT Statewide	IL Zone 2 Chicago	IL Zone 3 Springfield	Mid-Atlantic Wilmington	Mid-Atlantic Baltimore	Mid-Atlantic Washington, DC	MN Minneapolis–St. Paul (Zone 3)	NY–Small Commercial New York City	NY–Small Commercial Albany	TX Climate Zone 2 Fort Worth	TX Climate Zone 3 Houston
Healthcare Clinic		1,036	1,029				1,830			310	196
Hospital	1,519	1,929	1,863				2,054				
Industrial—1 shift	1,470										
Industrial—2 shifts	1,003										
Industrial—3 shifts	530										
Library	1,248										
Lodging	1,418	2,050	1,780	667	632	601	1,934	619	1,037		
Small										249	147
Large										464	250
Medical Offices	1,248										
Multifamily		422	335				1,904				
Nursing Homes	727										
Office	1,306										
Small Office											
High-Rise (10+ floors)		2,768	2,656				1,760				
Mid-Rise (5–9 floors)		879	824				1,793			262	149
Low-Rise (1–4 floors)		666	647	524	466	416	1,610	431	747	153	69
Office/Retail	1,248										
Parking Garage/ Parking Lot	1,094										

Performing Arts Theaters	1,537										
Refrigerated Warehouse	1,533										
Residential	1,418										
Restaurants	1,170	1,379	1,291	1,016	891	784	1,530	821	1,502	688	429
Retail	1,170							545	984	352	204
Small				674	595	528					
Strip Mall		1,147	1,151				1,393			376	218
Department Store		927	900				1,444				
Big Box				298	241	192		191	554		
Large Retail											
Schools	1,637										
Elementary		840	927	993	926	868	1,961	840	1,300	397	239
High School		807	999				2,098			489	293
Warehouse—Non-Refrigerated	1,533	443	427	576	487	411	1,533	452	916		
Warehouse—Refrigerated	1,533			576	487	411	1,533	452	916		

be identified in boiler specifications as the DOE heating capacity, which represents the gross output of a boiler after accounting for its efficiency and assuming the heat loss through the jacket is wasted. In this case, the energy input to the boiler would be the boiler's output capacity divided by the boiler efficiency. If the boiler model number reflects the boiler's maximum input capacity (firing rate), then to obtain the energy output, the input capacity is multiplied by the boiler efficiency. If you want to obtain a cost estimate for the equipment, cost can be based on the input capacity (firing rate) of the boiler or the output capacity (heat delivered for space heating). As such, be sure to check whether the size is based on maximum input or maximum output capacities.

ASHRAE Report RP-1120(2) and state TRMs(3) may be used to approximate the $EFLH_H$. Tables 6.1 and 6.2 include the ASHRAE RP-1120 and select state TRM data.

The Institute of Boiler and Radiator (IBR) Manufacturers (formerly the Hydronics Institute and currently part of the Gas Appliance Manufacturers Association) is a boiler certification organization that is responsible for two terms you will sometimes see in a boiler specification: "IBR Gross Output" and "IBR Net Output." IBR Gross Output is the amount of heat energy transferred to the water during steady state operation. The assumption is that any heat lost through the boiler jacket is wasted and does not contribute to the building heating load. As such, this rating assumes the boiler is in an unheated space. The IBR Gross Output is calculated excluding the boiler jacket losses, but not the piping and pickup losses (pickup losses refer to the heat energy it takes to bring the system up to the operating temperature). Therefore, this output rating would be used when the boiler is located in an unheated space, but the piping is located within a heated space. The IBR Net Output is a further arbitrary deduction of 15% from the IBR Gross Output that takes into account losses from the distribution piping and the necessary Btus it takes to bring the system up to operating temperature.

The DOE Heating (output) Capacity is the highest rating. The losses associated with the jacket, piping and pickup are assumed to contribute to the building heating load. The DOE Heating Capacity should be used if the boiler and piping is in a heated space. The DOE Heating Capacity rating will always be higher than the IBR Gross Output, but the difference will be relatively insignificant. Since boilers in commercial buildings are typically located in mechanical rooms and the distribution piping located within the heated space, the DOE Heating Capacity is typically used for boilers.

6.4 ENERGY EFFICIENCY METRICS

Various efficiency terms have been used to describe heating system efficiency.

Combustion efficiency is an indication of the burner's ability to burn fuel. The amount of unburned fuel and excess air in the exhaust are used to assess a burner's combustion efficiency. Burners resulting in low levels of unburned fuel while operating at low excess air levels are considered efficient. Well-designed conventional burners firing gaseous and liquid fuels operate at excess air levels of approximately 15% and result in negligible unburned fuel. By operating at the minimum excess air requirement, less heat from the combustion process is being used to heat excess combustion air, which increases the energy available to satisfy the building space heating requirements (heat load).

The Annual Fuel Utilization Efficiency (AFUE) is a dimensionless ratio of useful energy output to energy input expressed as a percentage. The AFUE or fuel-to-energy output efficiency, which does account for boiler radiation and convection losses, is a true indication of overall boiler efficiency. It takes into account cycle losses and standby losses and is averaged over the course of the heating season.

The AFUE is determined by one of two methods: the input-output method or the heat loss method. The input-output efficiency measurement method is based on the ratio of the output-to-input of the boiler. It is calculated by dividing the boiler's total annual heat output (in Btu) by the boiler's total annual energy input (in Btu) and multiplying by 100 to get the percent. The heat loss method is based on accounting for all the heat losses of the boiler. The actual measurement method consists of subtracting from 100% the total percent associated with stack, radiation, and convection losses. The resulting value is the boiler's fuel-to-energy output efficiency.

AFUE applies to boilers that have less than 300,000 Btu/hr input capacity. Thermal efficiency (see below) applies to boilers with an input capacity of 300,000 Btu/hr or greater.

Thermal efficiency is a measure of the effectiveness of the heat exchanger of the boiler and is the commercial equivalent of the AFUE. It measures the ability of the heat exchanger to transfer heat from the combustion process to the water or steam in the boiler, exclusive of radiation and convective losses. Radiation and convective losses are typically negligible, less than 1%–2% in most cases.

The term "boiler efficiency" is often substituted for thermal efficiency or fuel-to-energy output efficiency. It is the heat used in producing hot water divided by the heat liberated in the boiler. When the term boiler efficiency is used, it is important to know which type of efficiency is being represented. Thermal efficiency, for example, does not account for radiation and convection losses. These losses occur from the external surfaces of an operating boiler. For any boiler at operating temperature, the loss is constant. Expressed as a percentage of the boiler's heat output, the loss increases as boiler output is reduced. Hence, operating a boiler at full load lowers the loss percentage. Since the boiler's surface area relates to its bulk, the relative loss is lower for a larger boiler and higher for a smaller boiler.

6.5 ENERGY SAVINGS

Energy savings can be estimated using the difference between the space heating end-use energy consumption (input) in Btu/yr at the existing boiler or furnace efficiency and at the new unit's higher efficiency. This is shown in Equation 6–2.

$$ES_{\text{H-Btu}} = E_{\text{In-H-Baseline}} - E_{\text{In-H-New}} \tag{6–2}$$

where $ES_{\text{H-Btu}}$ = space heating end-use energy consumption savings, Btu/yr

$E_{\text{In-H-Baseline}}$ = baseline input (consumed) space heating end-use energy consumption of existing equipment, Btu/yr

$E_{In\text{-}H\text{-}New}$ = baseline input (consumed) space heating end-use energy consumption of new equipment, Btu/yr

In Equation 6–2, $E_{In\text{-}H\text{-}Baseline}$, if it is not known, can be estimated knowing the whole building energy consumption (from a review of utility bill data) and the percent of this consumption that can be attributable to the space heating end-use (see Chapter 2). $E_{in\text{-}H\text{-}New}$ is estimated using Equation 6–3.

$$E_{In\text{-}H\text{-}New} = E_{Out\text{-}H\text{-}Baseline} / \eta_{New} \qquad (6\text{–}3)$$

where $E_{Out\text{-}H\text{-}Baseline}$ = baseline output space heating end-use energy consumption, Btu/yr

η_{New} = AFUE or thermal efficiency of new high efficiency unit, fraction

$E_{Out\text{-}H\text{-}Baseline}$ is calculated using Equation 6–4. $E_{Out\text{-}H\text{-}Baseline}$ is the amount of heating needed to satisfy the building demand (load), which is assumed to remain unchanged regardless of the boiler used (i.e., existing or new boiler).

$$E_{Out\text{-}H\text{-}Baseline} = E_{In\text{-}H\text{-}Baseline} * \eta_{Baseline} \qquad (6\text{–}4)$$

where $\eta_{Baseline}$ = AFUE or thermal efficiency of baseline existing unit, fraction Equation 6–2 becomes:

$$ES_{H\text{-}Btu} = E_{in\text{-}H\text{-}Baseline} - [\,(E_{In\text{-}H\text{-}Baseline} * \eta_{Baseline}) / \eta_{New}\,]$$

$$ES_{H\text{-}Btu} = E_{In\text{-}H\text{-}Baseline} * [1 - (\eta_{Baseline} / \eta_{New})] \qquad (6\text{–}5)$$

Example Problem 6–1

A small commercial building is replacing its existing 30-year-old conventional boiler having a design efficiency of 78% with a high efficiency condensing boiler of the same size.

The existing boiler's output capacity is 250,000 Btu/hr and the $EFLH_{Heating}$ is 840 hr/yr (TMY).

What is the space heating end-use energy consumption savings? Assume the new condensing boiler has an efficiency of 95%.

From Equations 6–2 to 6–4:

$$ES_{H\text{-}Btu} = E_{In\text{-}H\text{-}Baseline} - E_{In\text{-}H\text{-}New}$$

$$E_{Out\text{-}H\text{-}Baseline} = E_{In\text{-}H\text{-}Baseline} * \eta_{Baseline}$$

$$E_{In\text{-}H\text{-}New} = E_{Out\text{-}H\text{-}Baseline} / \eta_{New}$$

$$E_{Out\text{-}H\text{-}Baseline} = (250{,}000 \text{ Btu/hr}) * (840 \text{ hr/yr})$$

$$= 210{,}000{,}000 \text{ Btu/yr}$$

$$E_{In\text{-}H\text{-}Baseline} = (210{,}000{,}000 \text{ Btu/yr}) \,/\, 0.78 = 269{,}231{,}000 \text{ Btu/yr}$$

$$E_{In\text{-}H\text{-}New} = (210{,}000{,}000 \text{ Btu/yr})/0.95 = 221{,}100{,}000 \text{ Btu/yr}$$

$$ES_{H\text{-}Btu} = 269{,}231{,}000 - 221{,}100{,}000 = 48{,}131{,}000 \text{ Btu/yr}$$

$$\% \text{ savings} = (48{,}131{,}000 \,/\, 269{,}231{,}000) \times 100 = 17.9\%$$

Or Equation 6–5 can be used:

$$ES_{H\text{-}Btu} = E_{In\text{-}H\text{-}Baseline} {}^{*} \left[1 - (\eta_{Baseline})/\eta_{New})\right]$$

$$= 269{,}231{,}000 \times [(1 - (0.78/0.95)] = 48{,}131{,}000 \text{ Btu/yr}$$

The percent energy savings is:

$$\left[1 - (\eta_{Baseline})/\, \eta_{New})\right]{}^{*}100 = 17.9\%$$

You can also calculate the energy savings:

$$E_{Out\text{-}H\text{-}Baseline} {}^{*}\left[(1/\eta_{Baseline}) - (1/\eta_{New})\right]$$

$$= (210{,}000{,}000) \times 0.229$$

$$= 48{,}100{,}000 \text{ Btu/yr}$$

This example demonstrates the importance of noting whether the capacity of the boiler or furnace is the output capacity (how much energy goes to satisfying building space demand) or the input capacity (how much energy is consumed as reflected in fuel bills).

The above calculations assume that the 30-year-old boiler has not degraded in performance over the years (due to age) and maintained its design efficiency of 78%. The author's experience has shown that commercial boiler performance does degrade slightly over time, even if properly maintained. A factor used in the author's experience is an average performance degradation of 0.067% per year. This would change the baseline efficiency to 0.067%/year × 30 years = 2% further reduction in efficiency or 78% – 2% = 76%. This would result in a slightly greater percent space heating end-use energy consumption savings (i.e., 20%).

Example Problems 6–2 to 6–6

Example Problems 6–2 to 6–6 involve fuel switching. You are asked to provide a quick estimate of first year energy savings resulting from fuel switching using the following assumptions:

Oil at 138,690 Btu/gal and $3.50/gal
Gas at 102,900 Btu/CCF and $12/MM Btu or $1.24/CCF
Electric pricing at $0.15/kWh (for electric firing, there will also be associated demand charges, but these will be ignored in the examples)

While these examples are only evaluating first year energy consumption and cost savings, actual savings should be evaluated over the life of the equipment using the methodology presented in Chapter 16.

Example Problem 6–2

What are the first-year energy consumption and energy cost savings associated with fuel switching from oil to natural gas? Assume the following:

$$\text{Energy input to the oil-fired boiler } (\text{TMY}) = 2{,}500 \text{ MM Btu/yr}$$

$$\text{Oil-fired boiler efficiency} = 78\%$$

$$\text{Gas-fired boiler efficiency} = 82\%$$

The energy and cost savings in switching from oil to natural gas can be estimated as follows:

1. Obtain the annual space heating energy consumption (Btu/yr) for oil firing—this is the energy input into the boiler, e.g., 2,500 MM Btu/yr (or 18,026 gal/yr at $63,100/yr)
2. Calculate the energy output from the boiler, e.g., $(2{,}500 \text{ MM Btu/yr}) * 0.78 = 1{,}950 \text{ MM Btu/yr}$
3. Calculate the new energy input to the boiler under gas firing, e.g., $(1{,}950 \text{ MM Btu/yr}) / 0.82 = 2{,}378 \text{ MM Btu/yr}$ $(\text{or } 23{,}110 \text{ CCF/yr at } \$28{,}656\text{/yr})$
4. Calculate the energy consumption savings, e.g., $(2{,}500 \text{ MM Btu/yr}) - (2{,}378 \text{ MM Btu/yr}) = 122 \text{ MM Btu/yr}$
   Note: The energy consumption savings can also be obtained using Equation 6–5: $ES_{H\text{-}Btu} = 2{,}500 \text{ MM Btu/yr} * \left(1-(0.78/0.82)\right) = 122 \text{ MM Btu/yr}$
5. Calculate the energy cost savings by the switch from oil to natural gas, e.g., $63,100/yr — $28,656/yr = $34,444/yr

Example Problem 6–3

What are the first-year energy consumption and energy cost savings associated with fuel switching from electric to natural gas? Assume the following:

$$\text{Energy input to the electric boiler } (\text{TMY}) = 2{,}000 \text{ MM Btu/yr } (586{,}000 \text{ kWh/yr})$$

$$\text{Electric boiler efficiency} = 99\%$$

$$\text{Gas-fired boiler efficiency} = 82\%$$

The energy and cost savings in switching from electricity to natural gas firing are estimated as follows:

1. Obtain the annual space heating end-use energy consumption for electric heating—this is the energy input into the boiler, e.g., 2,000 MM Btu/yr or 586,000 kWh/yr at $87,900/yr. Demand charges should also be included but will be ignored.
2. Calculate the energy output from the boiler using electric heating, e.g., 2,000 MM Btu/yr) * 0.99 = 1,980 MM Btu/yr or 580,140 kWh/yr
3. Calculate the new energy input to the boiler under gas firing, e.g., $(1{,}980 \text{ MM Btu/yr}) / 0.82 = 2{,}415 \text{ MM Btu/yr}$

   $(\text{or } 23{,}470 \text{ CCF/yr at } \$29{,}103/\text{yr})$
4. Calculate the energy consumption savings, e.g., $(2000 \text{ MM Btu/yr}) - (2{,}415 \text{ MM Btu/yr}) = -415 \text{ MM Btu/yr}$
   Note: The energy consumption savings can also be obtained using Equation 6–5: $ES_{H\text{-}Btu} = 2{,}000 \text{ MM Btu/yr} * \left(1 - (0.99/0.82)\right) = -415 \text{ MM Btu/yr}$
5. Calculate the energy cost savings by the switch from electricity to natural gas, e.g., $87,900/yr-$29,103/yr = $58,797/yr

The switch from electric to gas will incur an energy consumption penalty because an electric boiler is much more efficient than a conventional gas-fired boiler; however, there is an energy cost savings because of the considerable difference in prices. The energy cost savings would be even greater if demand charge savings are also factored in.

Example Problem 6–4

What are the first-year energy consumption and energy cost savings associated with fuel switching from oil to electricity? Assume the following:

$$\text{Energy input to the oil-fired boiler } (\text{TMY}) = 2{,}500 \text{ MM Btu/yr}$$

$$\text{Electric boiler efficiency} = 99\%$$

$$\text{Oil-fired boiler efficiency} = 78\%$$

The energy and cost savings in switching from oil-firing to electricity are estimated as follows:

1. Obtain the annual space heating energy consumption (Btu/yr) for oil firing—this is the energy input into the boiler, e.g., 2,500 MM Btu/yr (or 18,026 gal/yr at $63,100/yr)

2. Calculate the energy output from the boiler, e.g., (2,500 MM Btu/yr) * 0.78 = 1,950 MM Btu/yr
3. Calculate the new energy input to the electric boiler, e.g., (1,950 MM Btu/yr) / 0.99 = 1,970 MM Btu/yr

   (or 577,210 kWh/yr at $86,582/yr)
4. Calculate the energy consumption savings, e.g., (2,500 MM Btu/yr) − (1,970MM Btu/yr) = 530 MM Btu/yr

   Note: The energy consumption savings can also be obtained using Equation 6–5: $ES_{H\text{-}Btu}$ = 2,500 MM Btu/yr * (1−(0.78/0.99)) = 530 MM Btu/yr
5. Calculate the energy cost savings (penalty) by switch from oil to electricity, e.g., $63,100/yr — $86,582/yr = -$23,482/yr

The switch from oil to electric will save energy because an electric boiler is much more efficient than a conventional gas boiler; however, there is an energy cost savings penalty because of the considerable difference in prices. The energy cost savings penalty will be even greater when electricity demand charges are factored in.

Example Problem 6–5

What are the first-year energy consumption and energy cost savings associated with fuel switching from oil to electricity using a heat pump? Assume the following:

Energy input to the oil-fired boiler (TMY) = 2,500 MM Btu/yr

Electric heat pump average COP = 2.5

Oil-fired boiler efficiency = 78%

The energy and cost savings in switching from oil-firing to electricity via a heat pump are estimated as follows:

1. Obtain the annual space heating energy consumption (Btu/yr) for oil firing—this is the energy input into the boiler, e.g., 2,500 MM Btu/yr (or 18,026 gal/yr at $63,100/yr)
2. The average heat pump COP of 2.5 is equivalent to 250% efficiency. The units of COP are Btu output/Btu input.
3. Calculate the energy output from the boiler, e.g., (2,500 MM Btu/yr) * 0.78 = 1,950 MM Btu/yr
4. Calculate the energy input supplied by the heat pump, e.g., (1,950 MM Btu/yr output)/(2.5 Btu output/Btu input)

   = 780 MM Btu/yr input (or 228,540 kWh/yr at $34,281/yr)

5. Calculate the energy consumption savings, e.g., $(2{,}500\text{ MM Btu/yr}) - (780\text{ MM Btu/yr}) = 1{,}720\text{ MM Btu/yr}$
6. Calculate the energy cost savings by the switch from oil firing to heat pump heating, e.g., \$63,100/yr-\$34,281/yr = \$28,819/yr

The switch from oil (or gas) to an electric heat pump will save significant energy because a heat pump is much more efficient than a conventional oil (or gas) boiler. As such, there will be significant energy cost savings. However, the energy cost savings will be reduced when additional electricity demand charges are factored in when converting to electricity. Also, heat pump COP decreases as the outdoor temperature decreases and at low temperatures supplemental electric heating may be needed. In this example, it was assumed that the COP was the average seasonal COP. In areas of the country where low winter temperatures are encountered, it is necessary to obtain temperature data and COP data as a function of temperature to determine the average (seasonal) COP (refer to Chapter 5, Section 5.2).

Example Problem 6–6

What would the estimated first-year energy consumption and cost savings be if switching from high efficiency gas firing using a condensing boiler to a heat pump? Assume the following:

$$\text{Energy input to the gas-fired boiler } (\text{TMY}) = 2{,}500\text{ MM Btu/yr}$$

$$\text{Electric heat pump average COP} = 2.5$$

$$\text{Condensing gas-fired boiler efficiency} = 96\%$$

The energy and cost savings in switching from high efficiency gas-firing to electricity via a heat pump are estimated as follows:

1. Obtain the annual space heating energy consumption (Btu/yr) for gas firing—this is the energy input into the boiler, e.g., 2,500 MM Btu/yr (or 24,300 CCF/yr) at \$30,130/yr.
2. The heat pump average COP of 2.5 is effectively equivalent to 250% efficiency. The units of COP are Btu output/Btu input.
3. Calculate the energy output from the boiler, e.g., $(2{,}500\text{ MM Btu/yr}) * 0.96 = 2{,}400\text{ MM Btu/yr}$
4. Calculate the energy input supplied by the heat pump, e.g., $(2{,}400\text{ MM Btu/yr output})/(2.5\text{ Btu output/Btu input})$
   $= 960\text{ MM Btu/yr input } (\text{or } 281{,}280\text{ kWh/yr at \$42,192/yr})$

5. Calculate the energy consumption savings, e.g., (2,500 MM Btu/yr) – (960 MM Btu/yr) = 1,540 MM Btu/yr
6. Calculate the energy cost savings by switch from gas firing to electric heat pump heating, e.g., \$30,130/yr - \$42,192/yr = -\$12,062/yr

While replacing a high efficiency gas-fired boiler with an electric heat pump can save considerable energy per year, the cost penalty may not make economic sense. In this case, the cost penalty is \$12,062 in the first year. The penalty would be even greater if the electricity demand charge impact associated with the switch to electricity is taken into consideration.

REFERENCES

1. Biada, B. and Schultz, C., HVAC basics, *Consulting Specifying Engineer*, pp. 26–30, October 2009.
2. Carlson, S., ASHRAE research project report RP-1120: Development of equivalent full load heating and cooling hours for GCHPs applied in various building types and locations, March 2001, https://www.techstreet.com/ashrae/searches/39741364
3. State Technical Resource Manuals:

   CT: Program Savings Documentation, p. 44 and Appendix 5, 2017, https://www.puc.nh.gov/EESE%20Board/EERS_WG/ct_trm.pdf

   IL: Statewide Technical Reference Manual for Energy Efficiency, V. 5, Vol. 2, pp. 131–132, 2016, https://icc.illinois.gov/api/web-management/documents/downloads/public/IL-TRM%20Version%205.0%20dated%20February%2011,%202016%20Final%20-%20Compiled%20Volumes%201-4.pdf

   Mid-Atlantic: Technical Reference Manual, Version 7.0, p. 136, May 2017, https://neep.org/sites/default/files/resources/Mid_Atlantic_TRM_V7_FINAL.pdf

   MN: Technical Reference Manual for Energy Conservation Improvement Programs, V. 2.1, Commercial Heat Pump Systems, Table 2, p. 42, December 2016, https://mn.gov/commerce-stat/pdfs/mn-trm-v2.1.pdf

   NY: New York Standard Approach for Estimating Energy Savings from Energy Efficiency Programs, Vol. 7, Appendix G, 2019, https://dps.ny.gov/system/files/documents/2022/11/technical-resource-manual-version-7-filed-april-15-2019-effective-january-1-2020.pdf

   TX: Technical Resource Manual, Version 4.0, Vol. 3, Tables 2–22 and 2–23, pp. 2–47 and 2–48, 2017, https://texasefficiency.com/wp-content/uploads/2023/11/trm4v3.pdf

7 Air Handling and Ventilation

7.1 BUILDING VENTILATION

Buildings are ventilated with outdoor air for the comfort and health of the occupants. Depending on the weather conditions, outdoor air may be heated or cooled, humidified or dehumidified, which consumes energy and costs money. Building designers, owners, and operators are faced with the balancing act of bringing in enough outdoor air for the occupants and controlling energy costs. To balance indoor air quality and energy, the building operator needs to understand how the building is being ventilated, including both how the ventilation system is designed to operate and how it is performing.

The design of HVAC systems in commercial buildings is a complex process. Buildings contain many different spaces that exchange air with each other, with the outdoor environment, and with the ventilation system. A system design will specify airflow rates at various points in the ventilation system and how these airflow rates should change in response to weather conditions, internal loads, and time of day. These specifications are based on the activities in the building zones, the thermal loads generated in these zones, the number of occupants, and recommended or minimum outdoor air ventilation rates from appropriate building codes, ventilation standards, and guidelines.(1) The specification of system airflow rates is often limited by uncertainties in how space in the building will be used and in the number of occupants in the zones.

The analysis of ventilation requirements begins with an accounting of the activities in a building. The building is typically divided into zones based on the activities taking place in each of the zones, their location in the building, their floor area, and the number of occupants expected to occupy each zone. The outdoor air requirements for each space are determined based on appropriate ventilation standards, guidelines, or building codes (such as ASHRAE Standard 62.1). Depending on the space type, ventilation rates will be in units of total cubic feet per minute (cfm), cfm per square foot of floor area, or cfm per person.

7.2 BUILDING VENTILATION END-USE ENERGY CONSUMPTION

The 2018 CBECS(2) provides insight into commercial building ventilation end-use energy consumption. Across all climate zones in 2018, ventilation end-use energy consumption for individual building types is shown in Table 7.1. Ventilation end-use energy consumption for all buildings in different areas of the country is shown in Table 7.2 and by climate zone in Table 7.3.

DOI: 10.1201/9781032692777-7

TABLE 7.1
Building Ventilation End-Use Energy Consumption by Property Type

Building Type	CBECS 2018 Ventilation kBtu/ft^2
Education	4.8
Food Sales	13.1
Food Services	16.6
Health Care	24.4
Inpatient	28.6
Outpatient	19.1
Lodging	12.5
Mercantile	10.8
Retail (other than mall)	11.7
Enclosed and Strip Malls	10.1
Office	12.9
Public Assembly	4.0
Public Order and Safety	6.3
Religious Worship	2.9
Service	3.0
Warehouse and Storage	1.7

TABLE 7.2
Building Ventilation End-Use Energy Consumption by Location

Location	CBECS 2018 Ventilation kBtu/ft^2
Northeast	8.4
New England	7.9
Middle Atlantic	8.6
Mid-West	7.6
East North Central	7.9
West North Central	7.0
South	7.9
South Atlantic	8.4
East South Central	7.2
West South Central	7.5
West	8.1
Mountain	8.5
Pacific	7.8

Notes: Locations in the Northeast group include New England (CT, RI, MA, NH, VT, ME) and Mid-Atlantic (NY, PA, NJ).

Locations in the Mid-West include East North Central (IL, MI, WI, IN, OH) and West North Central (MO, IA, MN, ND, SD, KS, NE).

Locations in the South includes South, Atlantic (VA, GA, FL, DC, DE, MD, WV, NC, SC); East South Central (TN, AL, KY, MS); and West South Central (TX, AR, LA, OK).

Locations in the West include Mountain North (AZ, CO, ID, MT, NM, NV, UT, WY) and Pacific (CA, AK, HI, OR, WA).

TABLE 7.3
Building Ventilation End-Use Energy Consumption by Climate Zone*

Climate Zone*	CBECS 2018 Ventilation kBtu/ft²
Cold or Very Cold	7.4
Cool	7.7
Mixed Mild	8.7
Warm	7.8
Hot or Very Hot	7.6

*Climate zones are based on ASHRAE Standard 169–2021.[3] Also refer to www.eia.gov/consumption/commercial/maps.php#defined

It is interesting to note that for all commercial buildings in 2018 in all climate zones, the average building ventilation end-use was approximately 7.8 kBtu/ft². Notwithstanding, the impact of building type is evident. Appendix VII provides building end-use energy consumption modeling data results for various areas of the country and building types. The first source in the appendix includes data for office buildings and schools in four cities throughout the country. For office buildings in Portland, Phoenix, Chicago, and Atlanta, the percent of whole building EUI attributed to ventilation fans ranges between approximately 12% and 20% or between 10.6 and 14.2 kBtu/ft²-yr, which can be compared to 12.9 kBtu/ft²-yr for all office buildings nationally in the 2018 CBECS.

Air change rate (see Appendix IV) is often used as a rule of thumb in ventilation design to estimate make-up airflow rates. Air changes per hour (ACH) refers to the number of times that the total air volume in a room or space is completely removed and replaced in an hour. The units of ACH are 1/hr. If the air in the space is either uniform or perfectly mixed, ACH is a measure of how many times the air within a defined space is replaced each hour. Perfectly mixed air refers to a theoretical condition where supply air is instantly and uniformly mixed with the air already present in a space.

Example Problem 7–1

Estimate the cubic feet per minute (cfm) of fresh air to be supplied to a library room with dimensions of 100 ft wide, 35 ft long, and 10 ft high.

$$\text{Volume of Room} = L * W * H = 35\text{ ft} * 100\text{ ft} * 10\text{ ft} = 35{,}000\text{ft}^3$$

$$\text{ACH} = 4\ (\text{from Appendix IV})$$

$$\text{cfm} = (\text{ACH, 1/hr}) * (\text{Volume of Room, ft}^3)/(60\text{ min/hr})$$

$$= (4) * (35{,}000)\ /\ 60$$

$$= 2{,}333\text{ cfm}$$

TABLE 7.4
Rule-of-Thumb Airflow Rates for Various Types of Properties

Property Type	Ventilation Rate, cfm/ft^2
Apartment (5 floors+)	0.8–1.2
Auditorium, Church, Theater	1.0–3.0
Elementary, High School, College	1.2–2.0
Hospital	1.0–1.5
Hotel, Motel, Dorm	1.0–1.5
Library, Museum	1.1–1.6
Office	1.0–1.6
Shops, Shopping Centers	1.0–1.6

Rules of thumb for airflow rates in cfm per building ft^2 for various types of properties are presented in Table 7.4.[(4)]

7.3 EQUIPMENT DESCRIPTION

There are three general categories of ventilation systems: central, perimeter, and unitary. Central systems are large, centrally located air-handling units that serve many zones through air distribution ductwork. Perimeter systems serve zones located along the perimeter of a building where the thermal loads are most influenced by outside weather. Unitary systems are small, factory-assembled air handling systems as typically incorporated in packaged HVAC roof-top units.

Central Systems

Central systems, sometimes referred to as "all-air" systems, supply conditioned air through a ductwork system. There are many types of central systems.

Single Zone and Multi-Zone Systems

Single zone systems serve one temperature control zone through a single-duct air distribution system. The supply air is heated or cooled depending on the thermal loads within the zone. The supply air temperature is controlled by varying the temperature of the heating or cooling fluid, varying the flow rate of the fluid, and using face and bypass dampers at the coils. Multi-zone systems have two supply airstreams at the air handler, a hot deck and a cold deck, where flow for the building is split into two streams, with one stream being heated (i.e., the hot deck) and one stream being cooled (i.e., the cold deck). These two streams are then mixed to create the right amount of heating and cooling for each zone. The thermal loads of each zone are met with a mixture of the two airstreams carried to the zone by a single duct. The mixing of the hot and cold airstreams for each zone is accomplished with dampers at the air handling unit. Multi-zone systems are best suited for offices, schools, and other similar buildings where a relatively small space requires independent zone thermostatic control.

Single Duct and Dual Duct Systems

In single duct systems, a single distribution duct system carries the supply air to the zones. Dual duct systems have two conditioned airstreams, a hot deck and a cold deck.

Two separate systems of ductwork distribute these two airstreams to mixing boxes in the zones where they are mixed to meet the thermal loads of the zones. Dual-duct systems are commonly used in larger buildings.

Constant Volume and Variable Air Volume

In constant air volume (CAV) systems, the supply airflow rate is constant, and the supply air temperature varies to meet the thermal loads. The supply air temperature is controlled by varying the temperature of the heating or cooling fluid, varying the flow rate of the fluid, changing the proportions of outdoor and recirculation air, and using face and bypass dampers at the coils. Sometimes additional heat is provided in the zones (reheat). In variable air volume (VAV) systems, the supply air temperature is constant, and the supply airflow rate to the zones is varied using VAV boxes of various designs. Each VAV box serves a certain zone and is connected via cable to a thermostat located within the zone. The thermostat indicates to the VAV box if it needs to increase or decrease the flow rate to meet the demand.

One of the more common replacements in building upgrade projects involves replacing the original CAV packaged roof-top units with VAV units that have higher cooling and heating efficiency. While CAV units deliver a constant volume of air whenever the units are on, VAV units (terminals) modulate the airflow to meet the needs of the zones within the building. That is ideal for most situations because every space in a building will experience a slightly different heating/cooling load. VAV systems allow each space to control the amount of cooling and heating, individually. Significant fan energy savings can be realized by using VAV units with VFD-controlled supply fans. They are most suited for zones that have varying cooling and heating loads.

Outdoor Air Intake Control

Outdoor air can be brought into the building through a dedicated fan. In systems without an outdoor air fan, the low pressure on the upstream side of the supply fan can pull outdoor air into the system. Several strategies are used to control the outdoor air intake rate:

- 100% outdoor air intake at all times with no recirculation.
- A fixed outdoor air intake damper position.
- Modulation of the outdoor air intake damper based on the outdoor and return air temperatures (or enthalpy). In economizer systems, the outdoor air intake rate is increased during mild weather to use the outdoor air for cooling the building.

Perimeter Systems

Perimeter systems condition zones along a building's exterior where thermal loads are affected by outdoor weather. Sometimes the central system serves the perimeter, but in other cases there is a dedicated perimeter system. Perimeter heating can also be provided by radiators and electric baseboard heaters, which do not involve outdoor air ventilation.

Unitary Systems

Unitary Systems are factory-assembled systems that provide heating, cooling, and outdoor air. There are many types of unitary systems including packaged roof-top units, through-the-wall air conditioner systems (PTAC), and heat pumps. Both rooftop units and PTAC provide outdoor air, while heat pumps may or may not be connected to a source of outdoor air.

7.3.1 Ventilation System Components

Components of typical ventilation systems include:

- Fans.
- Filters.
- Coils (for heating and cooling).
- Dampers (to control outdoor air intake and to return air recirculation and exhaust to the outdoors).
- Controls.
- Ductwork (principally to distribute supply air throughout the building).
- Terminal units (connect supply air ductwork to supply outlets in the conditioned space).
- Air outlets and inlets (to deliver conditioned air to the occupied space and exhaust air from the space).

7.3.2 Fans and Fan Performance

Fans are the key component of any ventilation system (see Figure 7.1). The performance of a fan is described by a fan curve that relates the static pressure increase across a fan to the airflow rate (cfm) through the fan at a constant fan speed in revolutions per minute (rpm). The static pressure change is generally in units of "inches of water column" or "inches of water gauge." Figure 7.2 shows an example of a set of

FIGURE 7.1 Roof-top HVAC exhaust fan.

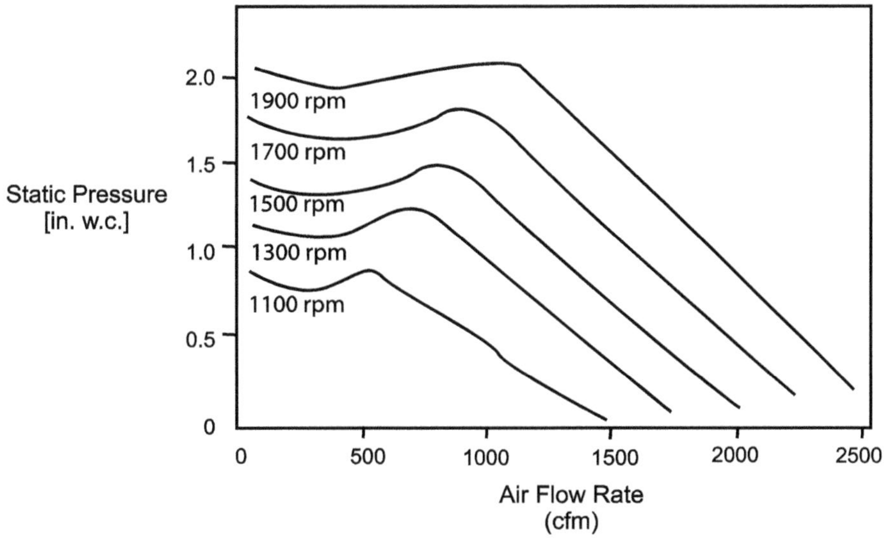

FIGURE 7.2 Example of a set of fan curves.

fan curves at different fan speeds. The term "system resistance" is used when referring to static pressure. The system resistance is the sum of the static pressure losses in the system. It is a function of ductwork configuration, and the pressure drops across equipment. The system resistance varies with the square of the volume of air flowing through the system. Conversely, resistance decreases as flow decreases.

To determine what volume the fan will produce, it is necessary to know the system resistance characteristics. In existing systems, the system resistance can be measured. In systems that have been designed but not built, the system resistance must be calculated. Air pressure decreases through the ventilation system, and this pressure drop is equal to the total airflow resistance of the system components and the ductwork. This pressure drop depends on the airflow rate and is described by a system curve; that is, the pressure loss through the system versus the airflow rate. Figure 7.3 is an example of a system curve. The system curve is constant so long as there are no changes in the system components. The system curve can be affected by changes in damper position, dirty filters, condensation on coils, holes in the ductwork, and obstruction in outlets or inlets.

When a ventilation system fan is operating, it raises the air pressure to overcome the airflow resistance of the system. The intersection of the system curve and the fan performance curve defines the point at which the pressure across the fan and through the system are equal, and thereby defines the airflow rate. If the airflow resistance of the system is accurately estimated during design and the fan is properly selected and installed, then the point of intersection will be at the design airflow rate of the system. Figure 7.4 shows the intersection of a system curve S and a fan performance curve F. If the system resistance increases, for example as filters become dirty, then a new system curve S′ replaces the original system curve S. The fan (F) and system

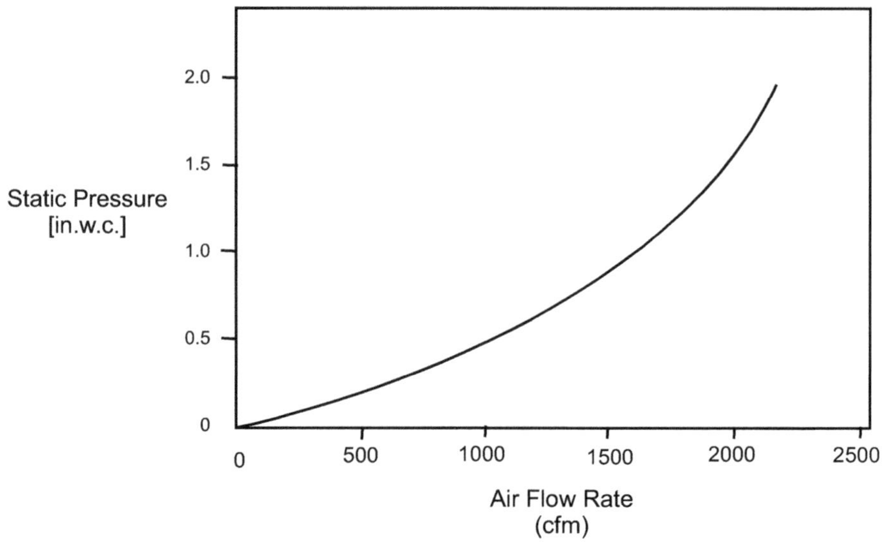

FIGURE 7.3 Example of a system curve.

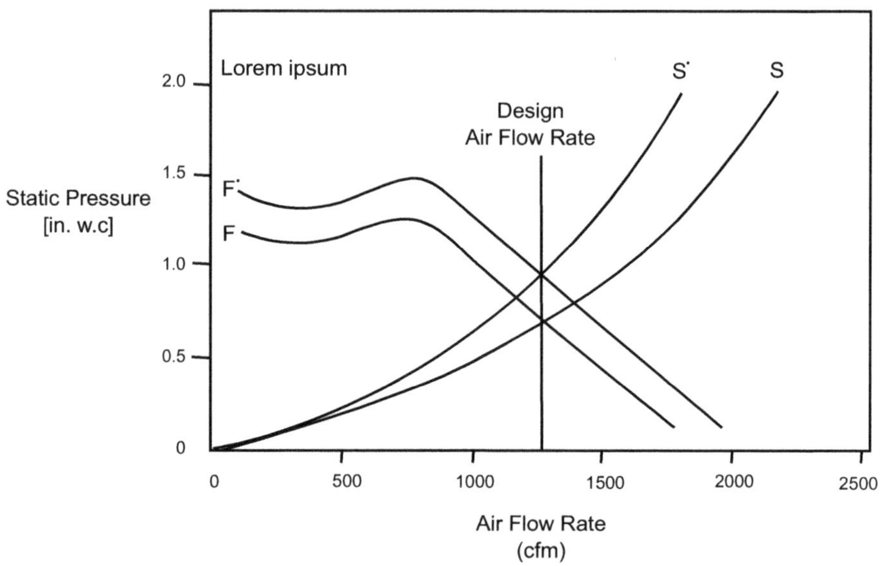

FIGURE 7.4 Example of the interaction between system and fan curves.

(S) curves will intersect at the higher pressure difference and lower airflow rate. The airflow rate can be returned to its design value by increasing the fan speed, such that a new fan curve F′ is in effect.

7.3.3 Air Handling Units

Air handling units (AHUs) are typically found in commercial and industrial buildings. AHUs condition and distribute air within a building (see Figure 7.5). They take fresh ambient air from outside, clean it, heat it or cool it, perhaps humidify it, and then force it through some ductwork to specific areas within a building. Most units will have an additional duct run to then pull the exhaust air out of the rooms and back to the AHU, where a fan will discharge it outside the building. Some of this exhaust air may be recirculated back into the fresh air supply to save energy. AHUs are usually located in the basement, on the roof, or on the floors of the building. AHUs will serve a specified area or zone within a building. Therefore, it is very common to find multiple AHUs around a building.

In the most basic AHU configuration, there are two housings, one for inlet flow and one for return air. At the inlet of the fresh air housing and the discharge of the return air housing there commonly are dampers that can close to prevent air from entering or exiting. The dampers can open to fully allow air in or out, and they can also vary their position somewhere in between to restrict the amount of air that can enter or exit.

After the dampers, there typically are filters to prevent dirt and dust from entering the building. As the filters trap the dirt, the amount of air that can flow through is restricted, thus causing a pressure drop across the filters. Across each bank of filters, there is a pressure sensor that will warn when it is time to replace the filters.

The next components are the cooling and heating coils to heat or cool the air. The air temperature of the supply air is measured as it leaves the AHU and enters the ductwork. This needs to be at the design set-point temperature to keep building occupants comfortable.

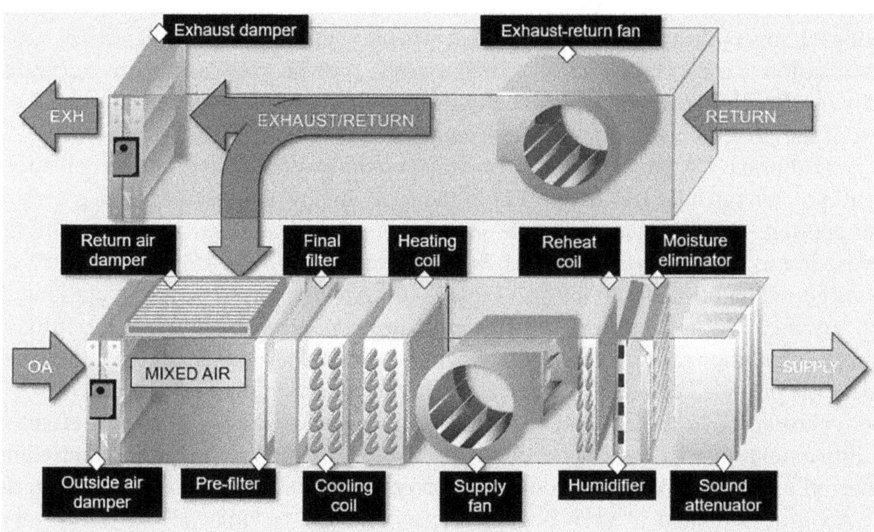

FIGURE 7.5 AHU system schematic. (*Source: Image courtesy of Bill Kosik.*)

AHUs have a fan that pulls the air in from outside and then through dampers, filters, and coils, and distributes it into the ductwork throughout the building. In cold climates, a pre-heater, usually an electric heater, may be found at the inlet of the fresh air intake. The heater will come on and heat the air to protect the components inside the AHU from frost. Otherwise, this could freeze the heating and cooling coils inside the AHU and burst them. After the fan there may be a pressure sensor that will read the static pressure in the duct. In some AHUs the speed of the fan is controlled by the pressure in the duct. A variable speed drive can be connected to the fan for variable volume systems.

The return AHU housing in its simplest form has just a fan and a damper inside. The fan is pulling the air from around the building and pushing it out of the building. The damper is located at the exit of the AHU housing and will close when the AHU turns off.

Some buildings need to control the humidity of the air supplied to the building, such as buildings in colder climates where continuous heating can make the air drier and result in uncomfortable air quality for occupants. A humidity sensor at the outlet of the supply AHU housing can measure the moisture in the supply air. If the moisture content is below a set-point value, then moisture needs to be introduced using a humidifier. If the air is too humid, this situation can be reduced by the cooling coil. As the air hits the cooling coil, the cold surface will cause the moisture within the air to condense.

7.3.4 DOAS

A dedicated outdoor air system (DOAS) unit will use similar equipment to air conditioning and space heating equipment discussed in Chapters 4 and 6, such as DX or chilled-water cooling, gas-fired or hot-water heating coils, filtration sections, variable-speed supply and exhaust fans, motorized dampers and so forth to separately and fully treat outside air independently, prior to it entering the building's systems and mixing with other supply and return airstreams. Other components commonly found in a DOAS unit could be air-to-air energy-recovery devices (i.e., heat exchangers, heat pipes, sensible heat wheels), total enthalpy or desiccant dehumidification wheels, heat-recovery reheat coils, preheat coils and frost-prevention coils in cold climates, and occasionally even humidifiers depending upon the location of the project, code requirements, and the facility type.

In traditional DX packaged roof-top units, outdoor ventilation air makes up only a small portion (usually less than 50%) of the total airflow rate used for space heating and cooling. A DOAS can decouple ventilation air (the amount of which will only be needed to meet the ventilation code requirement in ASHRAE 62.1) from the air required to condition the space (see Figure 7.6). Any increased intake of outdoor air can significantly impact the cost of energy through increased cooling, humidification, and heating requirements. With a typical decoupled DOAS system, 100% outside air is provided directly to a space, while the conditioning units operate in 100% recirculation mode to heat and cool the space. It also makes it more cost-effective to implement air-to-air energy recovery between exiting and incoming airstreams, since all of the outside air is brought in at a central location. Typically, air from the DOAS unit is ducted to each occupied space, while the other HVAC units serve only their specific spaces. DOAS is most effective in hot, humid climates. However, when coupled with air-to-air energy recovery, it can be cost-effective in virtually all climates. Figure 4.23 shows a DOAS roof-top installation.

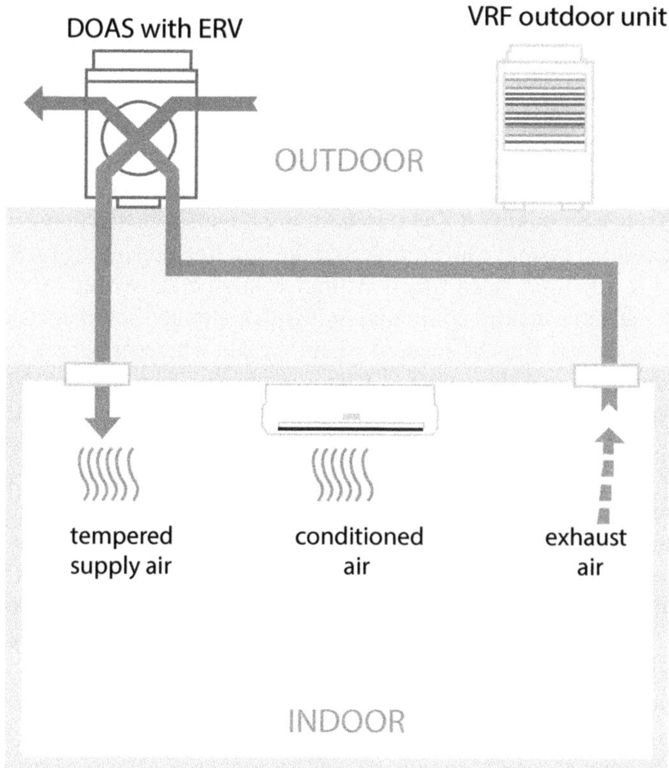

FIGURE 7.6 DOAS with energy recovery in a decoupled design with a VRF system.

While an ERV can handle approximately 50% of the building load (see Chapter 4, Section 4.3.4) to bring outside air to the desired room temperature, the addition of mechanical heating and cooling allows the DOAS to fully condition outside air. Pairing a DOAS with a high-performance heating and cooling system, such as a high efficiency heat pump or decoupled variable refrigerant flow system, with a high efficiency heat recovery ventilator or energy recovery ventilator will maximize the energy savings. In decoupled designs, where DOAS handles ventilation and a VRF system handles heating and cooling, ventilation air is delivered directly to conditioned spaces at the desired room temperature. Adding demand control ventilation (DCV) will further improve energy efficiency in these decoupled systems.

DOAS achieves energy savings by optimizing the use of ventilation air and pre-cooling/heating the outdoor air using energy recovery. Energy savings in the range of 8%–12% for space heating end-use energy consumption and 15%–20% for space cooling end-use energy consumption are readily achievable. In addition, since VAV systems commonly use more outside air than DOAS systems to assure proper ventilation air distribution, DOAS systems can use 20% to 50% less outside air than VAV systems, which reduces ventilation fan energy consumption.

DOAS units are available as packaged equipment or as split systems to support decoupled designs.

7.3.5 Demand-Controlled Ventilation

Buildings often provide at least two to three times more air than the design ventilation rate. All that additional ventilation means more air needing to be heated and cooled, resulting in higher energy costs. In demand-controlled ventilation (DCV) each room has a CO_2 gas sensor that allows ventilation to be activated before air quality in any room is reduced beyond administratively set limits. Ventilation can then be reduced or turned off when air quality (CO_2) standards are satisfied. In this manner, the capacity of the ventilation system can vary around the building depending on which rooms are occupied at any time. Such functions may easily be automated by communicating sensor signals to a central command unit with computer-based decision routines. Quality indoor air may thus be secured whenever and wherever people are present.

When ventilation is only activated where required, energy is saved by reducing the amount of heating and cooling needed, as well as by reducing the energy associated with operating the fans in the ventilation system. This contrasts with conventional systems, where ventilation most often is continuously operating regardless of occupant presence, although possibly with night reductions.

The normal CO_2 level in fresh air is approximately 400 ppmv (parts per million by volume) or 0.04% CO_2 in air by volume. Typical CO_2 levels found indoors are 400–1,000 ppmv. Indoor CO_2 levels above 1,000 ppmv usually cause poor indoor air quality (e.g., sick building syndrome). DCV adjusts the amount of outside air ventilation to keep measured CO_2 levels below 1,000 ppmv.

Energy savings are based on many factors, including the hours of operation, return air dry bulb temperature, return air enthalpy, system total airflow, percent outside airflow used, estimated average outside airflow reduction, and cooling and heating efficiencies. In simulation studies on an Atlanta office building, for example, with a chiller and natural gas-fired boiler and a multi-zone VAV system, DCV resulted in cooling end-use energy consumption savings up to 11.8%, a heating end-use energy consumption savings up to 27%, and a fan (ventilation) end-use energy consumption savings up to 8.6%.[(5)] In another DCV simulation study investigating space heating, cooling, fan, and pump energy savings for office buildings located in Atlanta, Chicago, Miami, and Oakland, most of the energy savings associated with the buildings in Atlanta, Chicago, and Oakland was in the space heating end-use.[(6)] In Miami, as would be expected, the majority of the energy savings was in the cooling end-use.

7.4 ENERGY EFFICIENCY METRICS

Fans are at the heart of ventilation energy use. Fan energy is defined by its horsepower, which is calculated from the fan power equation:

$$\text{BHP} = (\text{cfm} * \Delta\text{SP})/(6356 * \eta_{\text{Fan}}) \quad (7\text{–}1)$$

where BHP = fan brake horsepower, hp (*Note*: Brake horsepower is the amount of real horsepower going to the fan, not the horsepower used by the motor.)
cfm = airflow rate, actual ft^3/min.
ΔSP = change in static pressure across the fan, in. W.C.
η_{Fan} = fan efficiency, fraction (refer to Table 7.6)

TABLE 7.5
Select State TRM Default HVAC AHU Ventilation Fan Motor Hours

Property Type	CT	NY	DE	Mid-Atlantic	Hawaii	PA	TX	IL	CO	Avg
Convention Center	1,954	1,954	1,954	1,954						1,954
Courthouse	3,748	3,748	3,748	3,748						3,748
Entertainment	1,952	1,952	1,952	1,952						1,952
Entertainment—Performing Arts	2,586	2,586	2,586	2,586						2,586
Entertainment—Movie	1,954	1,954	1,954	1,954						1,954
Entertainment—Sports Arena	1,954	1,954	1,954	1,954						1,954
Fire Station (unmanned)	1,953	1,953	1,953	1,953						1,953
Fitness Center	5,836	5,836	5,836	5,836						5,836
Food Stores/Grocery Stores	4,055	4,055	4,055	4,055	6,389	6,718		5,840	6,389	5,195
Hospitals	7,674	7,674	7,674	7,674		8,760		6,871	8,374	7,814
Hospitals/Healthcare	7,666	7,666	7,666	7,666	5,187	8,760	8,760			7,624
Library	3,748	3,748	3,748	3,748						3,748
Lodging	3,064	3,064	3,064	3,064	3,719	8,760		6,872	3,719	4,416
Multifamily Condos	4,055	4,055	4,055	4,055						4,055
Multifamily Common Area	7,665	7,665	7,665	7,665						7,665
Museum	3,748	3,748	3,748	3,748						3,748
Nursing Homes	5,840	5,840	5,840	5,840		8,760				6,424
Office	3,748	3,748	3,748	3,748	6,192			1,766	6,192	4,163
Office—Large						4,503	4,424			4,464
Office—Small						3,976	4,006			3,991
Office—Banks/Financial	3,748	3,748	3,748	3,748						3,748
Office—Medical	3,748	3,748	3,748	3,748						3,748
Office—Retail	3,748	3,748	3,748	3,748						3,748
Parking Garage	4,368	4,368	4,368	4,368						4,368
Penitentiary	5,477	5,477	5,477	5,477						5,477
Police/Fire Station	7,665	7,665	7,665	7,665						7,665
Post Office	3,748	3,748	3,748	3,748						3,748

(Continued)

TABLE 7.5 *(Continued)*
Select State TRM Default HVAC AHU Ventilation Fan Motor Hours

Property Type	CT	NY	DE	Mid-Atlantic	Hawaii	PA	TX	IL	CO	Avg
Religious—Church	1,955	1,955	1,955	1,955						1,955
Restaurants	4,182	4,182	4,182	4,182	4,155			4,654	4,155	4,242
Restaurants: Bar Lounge	4,182	4,182	4,182	4,182						4,182
Restaurants: Cafeteria/Fast Food	6,456	6,456	6,456	6,456		7,313				6,627
Restaurants: Family Dining	4,182	4,182	4,182	4,182		5,286	5,256			4,545
Restaurants: Fast Food	6,376	6,376	6,376	6,376			6,716			6,444
Retail	4,057	4,057	4,057	4,057	3,261		5,548	3,438	3,261	3,967
Retail—Large						4,907				4,907
Retail—Small						5,030				5,030
Retail—Auto	4,056	4,056	4,056	4,056						4,056
Retail—Bakery	2,854	2,854	2,854	2,854						2,854
Retail—Convenience Store	6,376	6,376	6,376	6,376						6,376
Retail—Laundromat	4,056	4,056	4,056	4,056						4,056
Retail—Mall Concourse	4,833	4,833	4,833	4,833						4,833
Schools—Education General				3,665						3,665
Schools—University	2,187	2,187	2,187	2,187		6,053	4,590	4,216	3,631	3,405
Schools—College Cafeteria	6,376	6,376	6,376	6,376						6,376
Schools—College Classes/Admin.	2,586	2,586	2,586	2,586						2,586
Schools—College Dormitory	3,066	3,066	3,066	3,066						3,066
Schools—Gymnasium	2,586	2,586	2,586	2,586						2,586
Schools—Jr/Sr HS	2,187	2,187	2,187	2,187		3,900	4,173	2,203		2,718
Schools—Pre-school/Elem.	2,187	2,187	2,187	2,187		3,790	4,173	2,203	3,899	2,852
Schools—Technical/Vocat.	2,187	2,187	2,187	2,187						2,187
Town Hall	3,748	3,748	3,748	3,748						1,874
Warehouse—Non-refrigerated	2,602	2,602	2,602	2,602				3,222	6,389	3,337
Warehouse—Refrigerated	2,602	2,602	2,602	2,602		4,041		3,222	6,389	3,437

N.B. New York, Delaware, and Mid-Atlantic TRMs rely on Connecticut TRM data.

6,356 = conversion factor using 33,000 ft-lb_F/min per HP and 5.192 lb_F/ft^2 per in. W.C. = 6,356 (ft^3/min) (in. W.C.)/HP

The fan energy equation is the fan power equation multiplied by the hours of fan operation with the appropriate terms for motor efficiency, load factor, and the conversion of horsepower to kilowatts.

$$E_{In\text{-}Fan} = 0.7457 * BHP * H_{Fan} * LF/\eta_{motor} \quad (7\text{–}2)$$

where $E_{In\text{-}Fan}$ = energy use of fan, kWh
BHP = brake horsepower, hp
H_{Fan} = fan operating hours (see Table 7.5)
LF = load factor (ratio of avg. operating load to peak load [nameplate rating of the motor], default at 80%)
η_{motor} = motor (drive) efficiency, fraction (typically 80%–95%)
0.7457 = conversion factor: 0.7457 kw/hp

Table 7.5 presents default fan motor hours, H_{Fan}, from select state TRMs. In this table, NY, DE, and Mid-Atlantic TRMs rely on CT TRM data.[7]

7.5 ENERGY SAVINGS

Making changes to fans are governed by the fan affinity laws. These laws can provide insight into potential energy savings.

$$cfm_2/cfm_1 = rpm_2/rpm_1 \quad (7\text{–}3)$$

$$\Delta SP_2/\Delta SP_1 = (rpm_2/rpm_1)^2 \quad (7\text{–}4)$$

$$BHP_2/BHP_1 = (rpm_2/rpm_1)^3 = (cfm_2/cfm_1)^3 \quad (7\text{–}5)$$

where cfm_1 = airflow rate at original operating speed, actual ft^3/min.
cfm_2 = airflow rate at new operating speed, actual ft^3/min.
rpm_1 = original operating speed, rev./min.
rpm_2 = new operating speed, rev./min.
ΔSP_1 = static pressure diff. at original operating condition, in. W.C.
ΔSP_2 = static pressure diff. at new operating condition, in. W.C.
BHP_1 = brake horsepower at original operating speed, hp
BHP_2 = brake horsepower at new operating speed, hp

Variable speed is the most efficient means of fan capacity control. Adjusting the flow rate (cfm) of the fan by controlling the frequency of the power to the fan motor reduces horsepower according to Equation 7–5. Reducing the speed of the fan by 50% reduces the flow rate by 50% but requires only 12.5% of the power required at the full flow rate (full capacity). This is a power savings of 87.5%. Reducing fan speed to reduce flow is the most efficient method for reducing power consumption. (Refer to Chapter 14 for additional information on the energy savings associated with variable speed drives.)

Example Problem 7–2

A building ventilation system is designed to handle 30,000 cfm at a total pressure drop of 7.5 in. W.C. What is the fan horsepower if the fan is 70% efficient? What would the horsepower be if the flow rate is reduced 20% by reducing the fan speed? What would the energy consumption (input) of the fan operating at the reduced flow rate be assuming a motor efficiency of 90% and fan motor operating hours equal to 3,750 hours?

Using Equation 7–1 and assuming an LF = 0.8, the BHP is:

$$BHP = (cfm * \Delta SP)/(6356 * \eta_{Fan})$$

$$- 30{,}000 * 7.5 / (6356 * 0.7)$$

$$= 50.6 \text{ hp}$$

If the flow rate is reduced 20% by reducing the fan speed, and applying Equation 7–5 from the fan affinity laws:

BHP_2/BHP_1	$=(cfm_2/cfm_1)^3$
BHP_2	$= 50.6 \times (24{,}000/30{,}000)^3$
	$= 50.6 \times (0.8)^3$
	$= 25.9$ hp

To get the energy consumed (input) by the fan, use Equation 7–2:

$$E_{In\text{-}Fan} = 0.7457 * BHP * H_{Fan} * LF/\eta_{motor}$$

$$= 0.7457 * 25.9 * 3{,}750 * 0.8 / 0.9$$

$$= 64{,}379 \text{ kWh}$$

There are many different types of fans used in HVAC systems, including forward-curved/airfoil centrifugal supply air fans, centrifugal/plug/axial return air fans or axial relief fans, air-cooled axial condenser fans, centrifugal condenser fans, propeller cooling tower fans, boiler draft fans and exhaust fans. Some central systems have return air fans to pull air from the conditioned space back to the air handler. Other systems have no return air fan, and the supply air fan is used to pull return air from the conditioned space. The various fans have different efficiencies as shown in Table 7.6.

Centrifugal fans are the most common fans used in ducted air conditioning systems. Forward-curved fans are typically the lowest in cost and are often the most forgiving. As such, they are very popular in packaged units.

TABLE 7.6
Typical Efficiencies for Different Types of Fans

Fan Type	Typical Efficiency	Category
Backward-Curved Fans	70%–80%	Centrifugal
Forward-Curved Fans	60%–75%	Centrifugal
Air Foil Fans	70%–86%	Centrifugal
Radial-Blade Fans	60%–70%	Centrifugal
Propeller	45%–50%	Axial
Tube/Vane Axial Fans	70%–85%	Axial

REFERENCES

1. ASHRAE 62.1–2022 Standard, *Ventilation for acceptable indoor air quality*, Atlanta, GA: American Society of Heating, Refrigerating and Air-Conditioning Engineers, Inc., 2022.
2. 2018 Commercial Building Energy Consumption Survey (CBECS). Released December 21, 2022, see www.eia.gov/consumption/commercial/data/2018/index.php?view=consumption#el-el1.
3. ASHRAE Standard 169–2021, Climatic data for building design standards, 2021, https://www.ashrae.org/technical-resources/bookstore/weather-data-center.
4. HVAC rule of thumb calculator, see www.engproguides.com/hvac-rule-of-thumb-calculator.html
5. Lu, X., Yang, T., O'Neill, Z. and Zhou, X.H., Energy and ventilation performance analysis for CO_2-based demand controlled ventilation in multiple zone VAV systems with multiple recirculation paths. In Building Performance Analysis Conference and SimBuild, ASHRAE and IBPSA-USA, pp. 306–316, 2020, https://www.ashrae.org/file%20library/conferences/specialty%20conferences/2020%20building%20performance/papers/d-bsc20-c037.pdf
6. O'Neill, Z., Li, Y., Cheng, H., Zhou, X.H. and Taylor, S., Energy savings and ventilation performance from CO_2-based demand controlled ventilation: Simulation results from ASHRAE RP-1747, *Science and Technology for the Built Environment*, Vol. 26, pp. 257–281, 2020.
7. UI/Eversource C&LM Program Savings Document, 12th Edition, Appendix 5, 2017, https://www.puc.nh.gov/EESE%20Board/EERS_WG/ct_trm.pdf

8 DHW Heating

8.1 EQUIPMENT DESCRIPTION

Domestic hot water (DHW) heating systems typically utilize natural gas or electricity directly in the water heater, or the water may be heated indirectly off the existing space heating system. They are generally characterized as:

- Conventional storage tank water heaters that typically use either gas in a condensing or non-condensing mode or electricity as the heating source.
- Tankless (on-demand/instantaneous) water heaters that can use either electricity or gas in a condensing or non-condensing mode.
- Heat pump electric water heaters (which includes storage).
- Indirect water heaters (operating off the main boiler and using only a storage tank).

The highest efficiency system is the heat pump water heater, followed by the tankless types using natural gas in a condensing mode or electricity.

8.1.1 Conventional Storage Tank Water Heaters

Storage water heaters (see Figure 8.1) are the most common type of water heater. They are insulated storage tanks that typically hold from 30 to more than 100 gallons

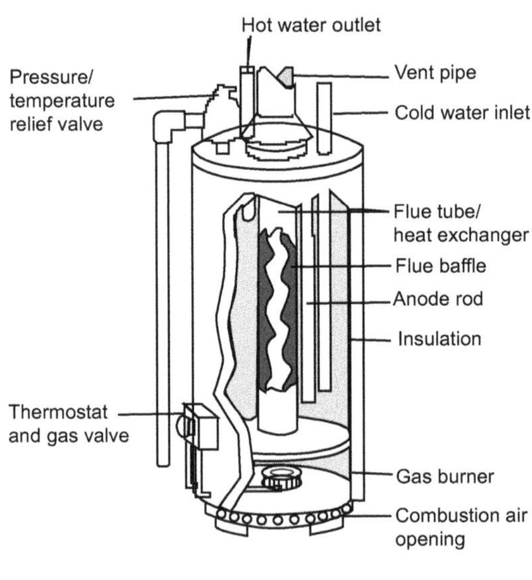

FIGURE 8.1 Storage water heater schematic.

DOI: 10.1201/9781032692777-8

of hot water. The heating source can be natural gas, propane, fuel oil, or electricity. The water always stays heated, so it is not necessary to wait for hot water. A gauge inside the tank reads the temperature and the unit comes on when the temperature drops below the set-point. These water heaters are relatively inexpensive, but they use more energy than other types of water heaters.

Conventional storage water heaters typically have an electric or natural gas heat source. When natural gas-fired, they can be of the non-condensing type (see Figure 8.2) or the condensing type (see Figure 8.3). High efficiency condensing gas tank water heaters capture excess exhaust heat (heat of condensation associated with condensing water in the flue gas) and transfer it back into the water to save on energy costs.

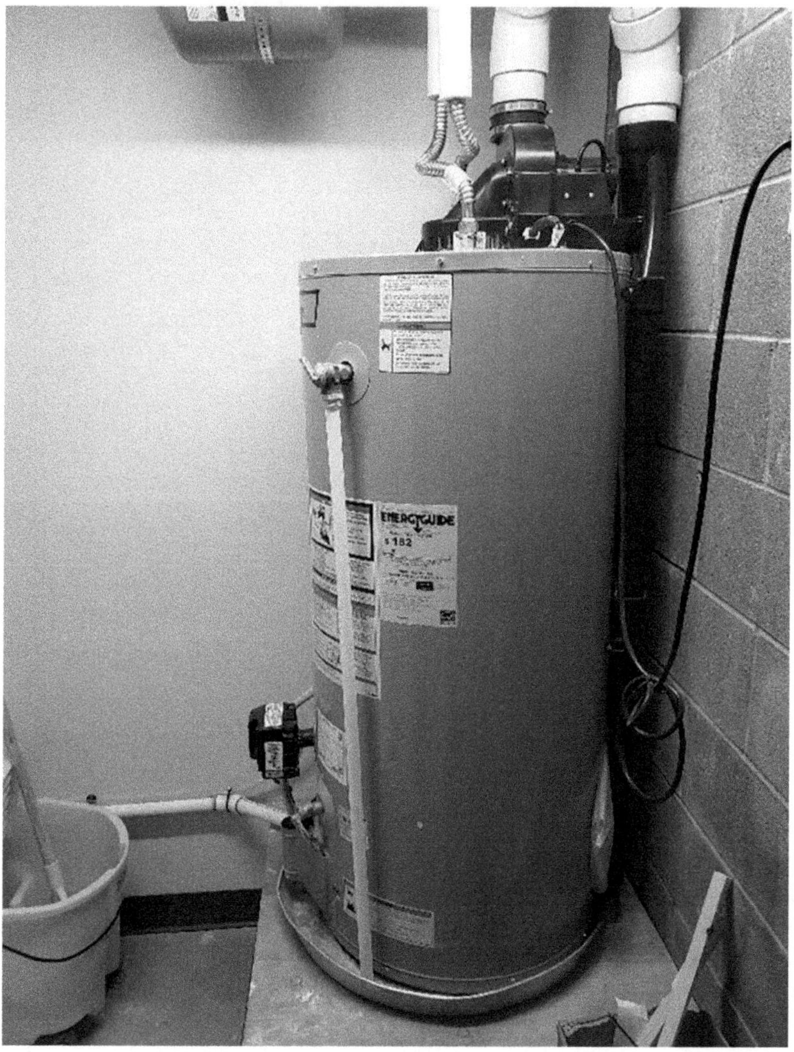

FIGURE 8.2 Gas-fired, non-condensing, storage water heater installation.

FIGURE 8.3 Gas-fired, condensing storage water heater installation.

8.1.2 Tankless Water Heaters

Tankless water heaters, also called on-demand or instantaneous water heaters, heat water only when needed, saving money and energy. There is no water storage tank. Instead, there are super-heated coils that fill with water and heat the water as needed. As such, they are smaller than conventional storage water heaters and take up considerably less storage space.

Tankless water heaters typically have an electric or natural gas heat source. The natural gas system can be of the conventional (non-condensing) type or the condensing type. A gas-fired, condensing tankless water heater installation is shown in Figure 8.4, and electric tankless water heater installation in Figure 8.5. Figure 8.6 shows three sizes of commercial tankless electric water heater units.

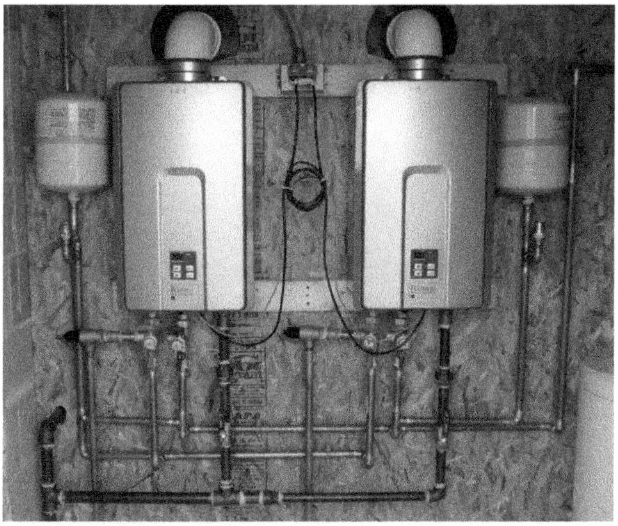

FIGURE 8.4 Gas-fired, condensing tankless water heater installation.

FIGURE 8.5 Tankless electric water heater installation.

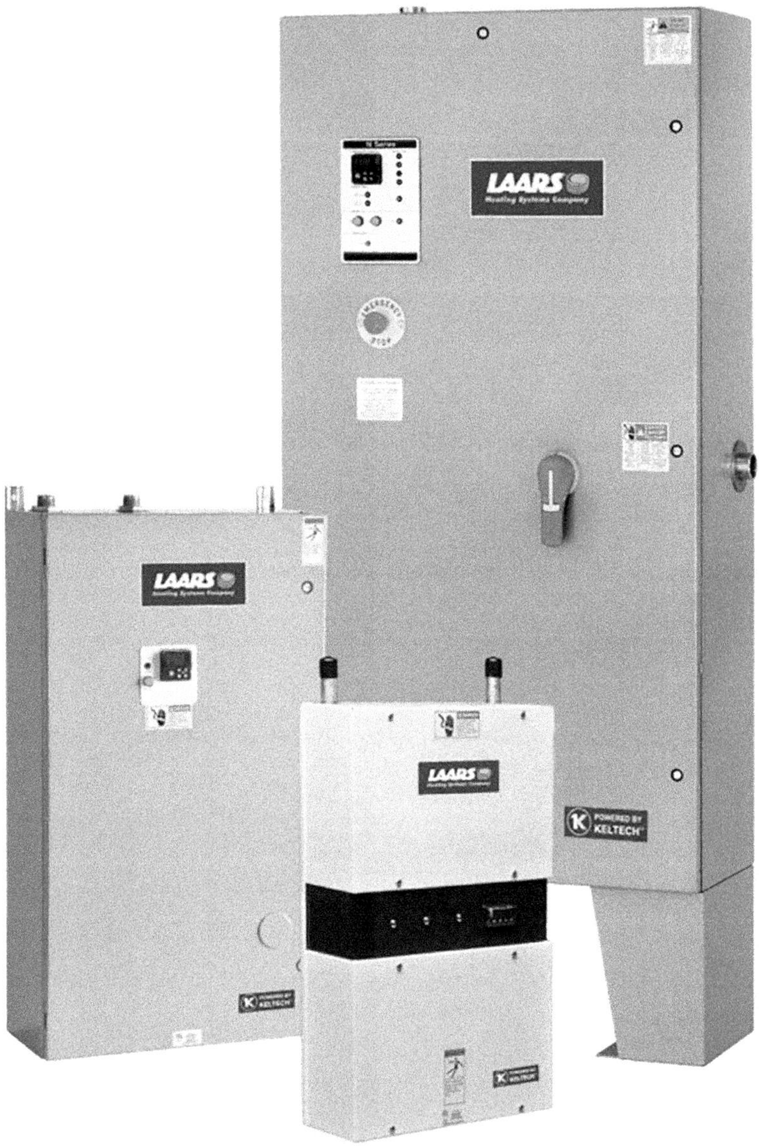

FIGURE 8.6 Tankless electric water heaters: 5–25 kW, 18–50 kW, and 36–144 kW. (*Source: Courtesy of Laars.*)

8.1.3 Heat Pump Water Heaters

Heat pump water heaters have a small heat pump on top of the storage water heater (see Figure 8.7). These water heaters are unique in that they use heat in the surrounding air. This means that electricity is only used to move heat from the surrounding air

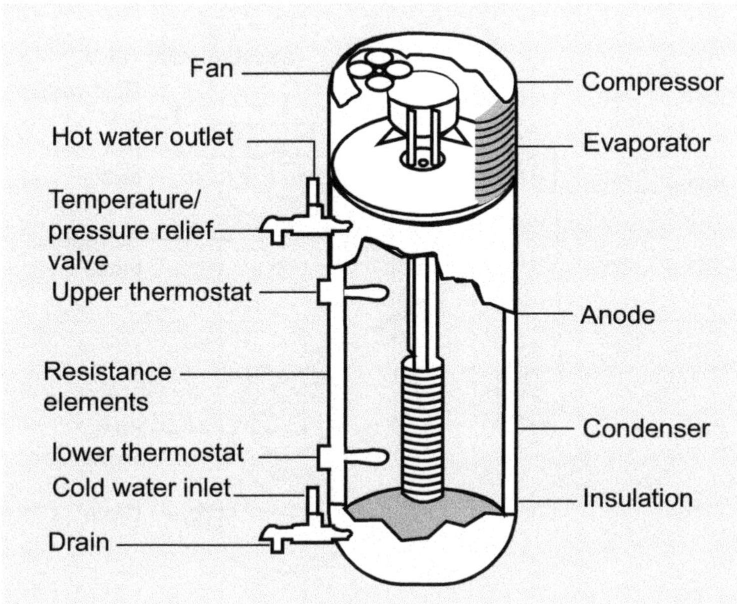

FIGURE 8.7 Heat pump water heater schematic. (*Source: Credit: U.S. DOE, Office of Energy Efficiency and Renewable Energy, Heat Pump Water Heaters.*)

to the water, instead of the alternative where electricity is used to generate heat. Heat pump water heaters will not operate efficiently in a cold space since they tend to cool the space they are in (unless ducted to the outside). Installing them in a space with excess heat, such as a furnace room, will increase their efficiency. Output capacities for commercial air source heat pumps typically range between 17 and 70 kW with integral water storage tank capacities between 50 and 120 gallons. Heat pump water heaters can use up to 60% less electricity than traditional styles of water heaters. Because the heat pump is on the top of the unit, additional headroom may be needed for this type water heater—as much as 8 feet of vertical clearance. An unconditioned surrounding area is also needed where the heat pump can absorb heat from the surrounding air. The most energy efficient mode of operation is in the heat pump mode, but in times of high demand electric resistance it may be used to heat the water (this is the least energy efficient mode). There is also a hybrid mode that adjusts between these two modes depending on heating requirements.

8.1.4 Indirect Water Heaters

An indirect water heater (see Figure 8.8) uses hot water off the main boiler and circulates it through a coiled heat exchanger located in the storage tank. An indirect water heater is, in essence, little more than a well-insulated storage tank that holds a heat exchanger. The indirect unit doesn't produce its own heat; instead, it relies on the building's existing heating plant.

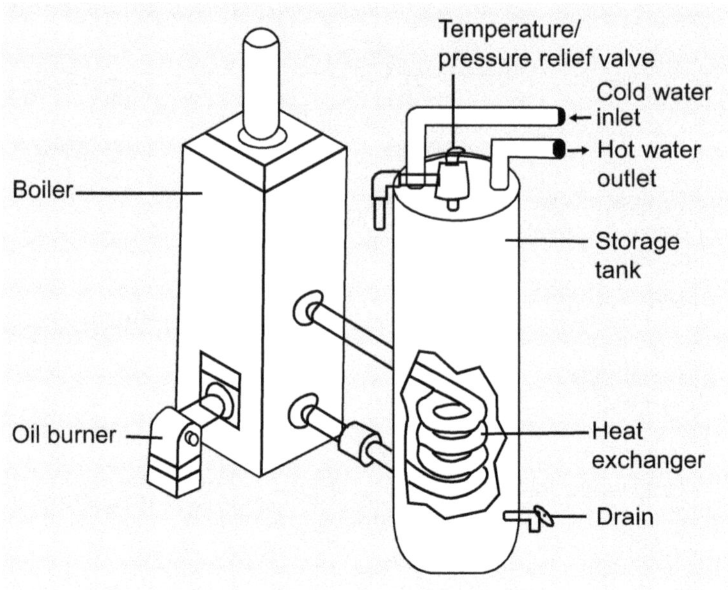

FIGURE 8.8 Indirect water heater schematic. (*Source: Credit: U.S. DOE, Office of Energy Efficiency and Renewable Energy, Tankless Coil and Indirect Water Heaters.*)

8.2 ENERGY EFFICIENCY METRICS

The energy efficiency metric for DHW heaters is the energy factor (EF). EF is basically a measure of the portion of input energy (natural gas or electricity) that is transferred to the hot water produced. The units of EF are Btu in the hot water delivered per period of time divided by the Btu in the fuel input over that same period of time. The higher the EF, the more efficient the water heater. As such, EF is mainly a function of the efficiency of the heating element and heat losses (such as standby losses from the water heater storage tank). EF considers both combustion and standby losses. For example, if the water heater has an EF of 0.60, this means that 60% of the energy input is effectively being used to produce hot water. Gas-fired water heaters generally have a lower EF than electric models because a significant amount of energy is lost through flue exhaust. Since tankless water heaters eliminate storage tank standby losses, the EF of these systems is higher than conventional storage water heaters. (*Note*: To enable greater consistency in equipment energy efficiency comparisons across brands, EF ratings were replaced in June 2017 with the uniform energy factor [UEF] developed using a modified test method. UEF ratings are determined by assigning water heaters into one of four different categories of hot water usage and then evaluating their performance based on that usage. These categories are called bins. A water heater is assigned a UEF within its bin based upon its first hour rating. A higher UEF means a water heater is more energy efficient and will cost less to operate compared to other water heaters in the same bin. A water heater's

TABLE 8.1
Typical Hot Water Heating System Energy Factors

Gas-fired DHW Heaters

Baseline

- Storage tank, conventional gas-firing, vented outside, older units: EF = 0.53–0.59

High Efficiency

- Storage tank, conventional non-condensing gas firing (75%–80% efficiency), vented outside, double-wall (well insulated): EF = 0.67–0.77
- Storage tank, condensing high efficiency (90%–95%) gas heating type: EF = 0.86
- Tankless, instantaneous, non-condensing high efficiency heating: EF = 0.82
- Tankless, instantaneous, condensing high efficiency heating: EF = 0.95

Oil-fired DHW Heaters

Baseline

- Storage tank, conventional oil-fired, vented outside, EF = 0.59–0.64

High Efficiency

- Storage tank, conventional non-condensing oil firing (75%–80% efficiency), vented outside, double-wall (well insulated): EF = 0.67–0.77

Electric DHW Heaters

Baseline

- Storage Tank: EF = 0.90–0.94

High Efficiency

- Tankless: EF = 0.94–0.98

Heat Pump Water Heaters

- EF = 2.0–3.5

UEF should only be compared with water heaters within the same bin. However, the author has found that if you have a UEF rating for a commercial heating system, its EF will usually be within a few percent of its UEF.)

Typical EFs for different DHW heating systems are presented in Table 8.1.

8.3 ENERGY SAVINGS

The energy savings associated with a replacement hot water heater is the savings associated with more efficient hot water heating plus any savings in standby losses. On demand/instantaneous DHW systems have no standby losses.

$$\text{Energy savings, Btu/yr} = \text{Savings in Hot Water Heating} + \text{Savings in standby losses} \qquad (8\text{–}1)$$

The energy savings associated with hot water heating is given by Equation 8–2.

$$ES_{DHW\text{-}Btu} = GPY * \left(8.34\ \text{lb/gal}\right) * (1\ \text{Btu/lb}^{\circ}\text{F}) * \Delta T * [(1/EF_{Baseline}) - (1/EF_{New})] \qquad (8\text{–}2)$$

where $ES_{DHW\text{-}Btu}$ = energy savings in DHW heating, Btu/yr
GPY = gallons/yr of hot water (output)
8.34 = density of water, lb/gal
1.0 = heat capacity of water, Btu/lb °F
ΔT = inlet/outlet water temperature difference, °F
$EF_{Baseline}$ = energy factor for the existing baseline system
EF_{New} = energy factor for the new high efficiency system

In Equation 8–2, the dimensional unit consistency check results in:

$$\text{Btu/yr} = \text{Gal/yr} * \text{lb/gal} * \text{Btu/lb}^{o}\text{F} *^{o}\text{F} * \left(\text{Btu in fuel input/Btu in hot water delivered}\right)$$

The percent energy saved will be the ratio of the energy saved to the water heater input energy:

$$\%\ \text{DHW energy saved} = [(ES_{DHW\text{-}Btu})/(\text{GPY} *8.34 *1 *\Delta T/EF_{Baseline})]*100$$

or

$$\%\ \text{DHW heating energy savings} = \left[1-\left(EF_{Baseline}/EF_{New}\right)\right] \text{ x } 100\% \qquad (8\text{–}3)$$

The percentage energy savings in Equation 8–3 can be multiplied by the water heater input energy (i.e., the DHW heater end-use energy consumption) to get the Btu/yr energy saved.

Example Problem 8–1

Given the following:

ΔT= 75 °F
$EF_{Baseline}$= 0.67 gas-fired, non-condensing
EF_{New} = 0.86 gas-fired, condensing
250,000 GPY = gal/hot water delivered (output)/year

What is the energy consumption savings?

$$\text{Btu/yr input to deliver hot water} = (250{,}000 \text{ gal/yr}) * (8.34 \text{ lb/gal}) * \left(1 \text{ Btu/lb}^{o}\text{F}\right) * \left(75^{o}\text{F}\right)/\left(0.67 \text{ Btu out/Btu in}\right)$$

$$= 233.4 \text{ MM Btu/yr}$$

The energy consumption savings is determined using Equation 8–2:

$$ES_{DHW\text{-}Btu} = GPY * (8.34\ lb/gal) * (1\ Btu/lb^oF) * \Delta T * [(1/EF_{Baseline}) - (1/EF_{New})]$$

$$= 250{,}000 * 8.34 * 1 * 75 * [(1/0.67) - (1/0.86)]$$

$$= 51.557\ MM\ Btu/yr\ saved$$

The percent savings is obtained directly using Equation 8–3:

$$\%\ DHW\ heating\ energy\ savings = [1-(EF_{Baseline}/EF_{New})] \times 100\%$$

$$= [1-(0.67/0.86)] \times 100\%$$

$$= 22.10\%$$

or % DHW energy saved = (51.557/233.4) × 100 = 22.1%

Example Problem 8–2

Assuming the DHW end-use energy consumption (which represents the fuel input Btu) is available (i.e., 200 MM Btu/yr), what is the energy consumption savings in replacing a gas-fired non-condensing storage water heater with a condensing storage heater?

Assume gas-fired non-condensing storage water heater has an $EF_{Baseline} = 0.55$

Assume the gas-fired condensing storage water heater has an $EF_{New} = 0.86$

Baseline energy in HW produced (output)
= 200 MM Btu/yr *0.55 = 110 MM Btu/yr

$$\text{Gas input for new unit} = (110\ MM\ Btu/yr) / 0.86 = 128\ MM\ Btu/yr$$

$$\text{Energy savings} = (200\ MM\ Btu/yr) - (128\ MM\ Btu/yr) = 72\ MM\ Btu/yr$$

$$\%\ \text{energy savings} = 72 / 200 = 36\%$$

or alternately:

$$\%\ \text{savings} = 1-(EF_{Baseline}/EF_{New}) = 1-(0.55/0.86) = 1-0.64 = 0.36\ \text{or}\ 36\%$$

$$ES_{DHW\text{-}Btu} = 0.36 * 200 = 72\ MM\ Btu/yr$$

Example Problem 8–3

Assume the existing unit in Example Problem 8–2 is instead being replaced with a gas-fired on-demand condensing unit. What is the energy consumption savings?

Assume the gas-fired, on-demand condensing water heater has an $EF_{New} = 0.95$

$$\% \text{ savings} = 1 - (EF_{baseline}/EF_{New}) = 1 - (0.55/0.95) = 0.42 \text{ or } 42\%$$

$$\text{Energy savings} = (200 \text{ MM Btu/yr}) * 0.42 = 84 \text{ MM Btu/yr}$$

Example Problem 8–4

Assume the existing unit in Example Problem 8–2 is instead being replaced with an on-demand electric water heater. What is the energy consumption savings?

Assume the on-demand electric water heater has an $EF_{New} = 0.98$

$$\% \text{ savings} = 1 - (EF_{baseline}/EF_{New}) = 1 - (0.55/0.98) = 0.439 \text{ or } 43.9\%$$

$$\text{Energy savings} = (200 \text{ MM Btu/yr}) * 0.439 = 88 \text{ MM Btu/yr} = 25{,}800 \text{ kWh/yr}$$

(*Note*: Electricity for water heating is sold on a demand rate basis. This means, in addition to the energy charge (kWh), the utility can charge for the peak demand impact (kW) that a customer switching to electric heating imposes on the electrical service. This would be an additional cost (penalty) that would be subtracted from the energy cost savings. Moreover, it may be necessary to increase the electricity service to the building, which would increase the installation cost.)

Example Problem 8–5

Assume the existing unit in Example Problem 8–2 is instead being replaced with a storage heat pump water heater. What is the energy consumption savings?

Assume the storage heat pump water heater has an $EF_{New} = 2.0$

$$\% \text{ savings} = 1 - (EF_{baseline}/EF_{New}) = 1 - (0.55/2.0) = 0.725 \text{ or } 72.5\%$$

$$\text{Energy savings} = (200 \text{ MM Btu/yr}) * 0.725 = 145 \text{ MM Btu/yr}$$

TABLE 8.2
Average Daily Gallons, Maximum Gallons/Hour, and Storage Tank Capacity by Type Facility

		Average gal/day	Maximum gal/hr	Rule of Thumb Storage Capacity, gal/1,000 SF	Comments
Dormitories		17.2 GPD/ student			General
	Men's Dormitories	13.1 GPD/ student	3.8 GPH/student		
	Women's Dormitories	12.3 GPD/ student	5 GPH/student		
Food Service					
	Full-Service Restaurant	2.4 gal/avg. no. meals served/ day	1.5 gal/max meals/hr	36	340 meals/ operating day/1,000 SF
	Fast Food	0.7 gal/avg no. meals served/ day	0.7 gal/max meals/hr	38.9	784.6 meals/ operating day/1,000 SF
Hospital		90 GPD/patient		21.4	3.8 patients/1,000 SF
Hotels		14 GPD/room		8.8	
Motels				23.9	
	20 or fewer units	20 GPD/room	6 gal/unit		
	21–99 units	14 GPD/room			
	60 units	14 GPD/room	5 gal/unit		
	100+ units	10 GPD/room	4 gal/unit		
Multifamily*		46 GPD/unit			
	20 or fewer units	42 GPD/unit	12 gal/unit		
	50 units	40 GPD/unit	10 gal/unit		
	75 units	38 GPD/unit	8.5 gal/unit		
	100 units	37 GPD/unit	7 gal/unit		
	200+ units	35 GPD/unit	5 gal/unit		
Nursing Homes		18.4 GPD/bed	4.5 gal/bed		
Office		1.0 GPD/person	0.4 gal/person	0.7	2.3 people/1,000 SF, typ. 250 days/yr
Religious/ Church		7.02 GPD/1,000 SF			
Retail		2.0 GPD/ employee		0.6	1 employee/1,000 SF
	Shopping Centers	300 GPD/1,000 SF			

(Continued)

TABLE 8.2 (*Continued*)
Average Daily Gallons, Maximum Gallons/Hour, and Storage Tank Capacity by Type Facility

		Average gal/day	Maximum gal/hr	Rule of Thumb Storage Capacity, gal/1,000 SF	Comments
	Grocery	2 GPD/employee		0.7	1.1 employees/ 1,000 SF
Schools					
	Elementary	0.6 GPD/person	0.6 gal/person	4.6	9.5 persons/1,000 SF, typ. 200 days/yr
	Jr/Sr High School	1.8 GPD/person	1.0 gal/person	7.6	9.5 persons/1,000 SF, typ. 200 days/yr
	University	0.5 GPD/student			
	Community College	1.9 GPD/student			
Supermarket		3.43 GPD/1,000 SF			
Warehouse		2.0 GPD/ employee		0.3	0.5 employee/1,000 SF

*Multifamily minimum storage capacities (assuming 2.5 persons/unit):

No. of Units	Min. Storage Capacity, Gallons
1–3	50
4	60
5–6	72
7–8	85
9–10	100
11–15	113
16–20	130
21–25	148
26–30	162
31–35	175
36–40	188
41–45	200
46–50	210
51–75	255
76–100	300
101–125	325
126–150	360

No. of Units	Min. Storage Capacity, Gallons
151–175	395
176–200	410
201–250	500
251–300	600
301–350	700

Source: A.O. Smith, Hot Water Requirements, Apartments, AOSSG88080, March 2010.

Example Problem 8–6

A 125,000 SF office building in San Diego with 625 occupants operates 5 days/week for 50 weeks per year. No DHW end-use energy consumption data is available. Estimate the annual energy (in MM Btu) of DHW that needs to be produced (delivered). Also estimate the energy consumption savings by replacing an existing non-condensing gas storage water heater with a new on-demand condensing storage water heater. Use a calculation approach[1–3] based on water usage per person in the building, consideration of water and air temperatures, and storage tank standby losses to estimate the energy savings. Compare the detailed calculation results with the simplified EF approach to estimate energy savings.

N.B. The major determination in sizing and the basis of all computations is establishing the probable demand for hot water. In addition, any unusual conditions which might be related to hot water consumption must also be recognized and planned for. Table 8.2[4–6] can provide guidance on developing water demand requirements.

Assume non-condensing, gas-fired storage DHW heater $EF_{Baseline} = 0.67$.
Assume condensing, on-demand gas-fired storage DHW heater $EF_{New} = 0.96$.
Assume the following ΔTs:

1. ΔT between water heater set-point and supply water temp. = 90°F. (Assumes 145°F water heater set-point and 55°F water inlet temp. Water heater set-point temperatures are generally in the range 120–160°F. The coldest water inlet temperature experienced during the year should be the base from which the maximum system temperature rise is established.)
2. Ambient ΔT between water heater set-point and surrounding air temp. = 85°F. (Assumes the ambient room temperature where heater storage tank is located is at an average temperature of 60°F.)

From Equation 8–1:

Gas energy savings, Btu/yr
= Savings in hot water heating + savings due to elimination of standby losses

Baseline input energy

$$= \left[\left(\text{GPD} * 250\ \text{day/yr} * 8.34\ \text{lb/gal} * \Delta T^{o}\text{F} * 1\ \text{Btu/lb}^{o}\text{F}\right) / EF_{Baseline}\right]$$

+ standby losses

$$GPD = (1\ GPD/person\ from\ Table\ 8.2) * 625\ persons = 625\ GPD$$

$$\text{Baseline input energy} = [(625*250*8.34*90*1)/0.67] + \text{standby losses}$$

$$= 175\ \text{MM Btu/yr} + \text{standby losses}$$

To estimate standby losses, it is first necessary to estimate the storage tank size: Estimated tank size from Table 8.2: 0.7 gal storage/1,000 SF.

$$\text{Estimated storage tank size} = 0.7\ \text{gal} *125 = 88\ \text{gal, assume } 100\ \text{gal}$$

For estimating the standby heat loss, $SL_{Baseline}$, the overall heat loss coefficient, UA, is needed.

UA values may be estimated for different types of tank insulation using Table 8.3.[7]
Assume the storage tank has 1 in. fiberglass insulation:
UA for a 100 gal 1 ins. storage tank from Table 8.3 (interpolated) = 10.2 Btu/hr °F.

$$SL_{Baseline}\text{standby heat loss} = UA_{baseline} * \text{Ambient } \Delta T = 10.2 * 85 = 867\ \text{Btu/hr}$$

It is now necessary to estimate the number of standby hr/yr when the standby losses exist: assume 60% of the hr/yr or 0.6 × 8,760 = 5,256 hr/yr;

$$\text{Standby losses, Btu/yr} = 867\ \text{Btu/hr} * 5{,}256\ \text{hr/yr} = 4.6\ \text{MM Btu/yr}$$

$$\text{Gas input associated with these losses} = 4.6/0.67 = 6.87\ \text{MM Btu/yr}$$

The energy savings can now be calculated:

$$\text{Energy Savings, MM Btu/yr} = [625*250*8.34*90*1*[(1/0.67)-(1/0.96)]] + 6.87$$

$$= 52.8\ \text{MM Btu/yr} + 6.87\ \text{MM Btu/yr}$$

$$= 59.67\ \text{MM Btu/yr}$$

$$\text{Gas energy input existing} = 175\ \text{MM Btu/yr} + 6.87\ \text{MM Btu}$$

$$= 181.9\ \text{MM Btu/yr}$$

% Energy Savings = 59.67/181.9 = 32.8%

Using Equation 8–3 to estimate the energy savings:

$$\%\ \text{Energy Savings} = (1-(0.67/0.96))*100 = 30.2\%$$

Energy savings, MM Btu/yr = 0.302 * 181.9
= 55 MM Btu/yr (within 10% of the more detailed calculations)

TABLE 8.3
UA Values for Different Size Storage Tanks Using Different Amounts of Fiberglass Insulation

	$UA_{baseline}$ (Btu/hr °F)		
Gal	**Bare Tank**	**1″ Fiberglass**	**2″ Fiberglass**
140	47.9	12	6.2
200	60.9	14.7	7.6
250	68.1	16.8	8.6
350	88.8	21.5	11
400	95.3	23.3	11.9
500	115.9	26.3	13.3
750	146.9	34.9	17.7
1,000	177.9	43.5	22.1

Typical energy savings associated with DHW heating system upgrades might be in the range of 20%–45% of the DHW end-use energy consumption, depending principally upon the water heater being replaced (to establish the baseline EF) and the type of high efficiency water heater selected.

REFERENCES

1. U.S. DOE, Office of Energy Efficiency and Renewable Energy, Sizing a new water heater, www.energy.gov/energysaver/sizing-new-water-heater.
2. UI/Eversource Program Savings Documentation, 12th Edition, 48–51, 2017, https://www.puc.nh.gov/EESE%20Board/EERS_WG/ct_trm.pdf
3. NY Standard Approach for Estimating Energy Savings from Energy Efficiency Programs, Version 7, 296–309, April 2019, https://dps.ny.gov/system/files/documents/2022/11/technical-resource-manual-version-7-filed-april-15-2019-effective-january-1-2020.pdf
4. New York State Technical Resource Manual, Version 7, GPD Table on p. 306, April 15, 2019, https://dps.ny.gov/system/files/documents/2022/11/technical-resource-manual-version-7-filed-april-15-2019-effective-january-1-2020.pdf
5. ASHRAE Pocket Guide for Air Conditioning, Heating, Ventilation, Refrigeration, 8th Edition, Table 6.4, 2014.
6. Sezgen, O. and Koomey, J.G., Technology Data Characterizing Water Heating in Commercial Buildings, J1995, December 1995, https://digital.library.unt.edu/ark:/67531/metadc673111/.
7. New York Standard Approach for Estimating Energy Savings from Energy Efficiency Programs, p. 167. October 15, 2010, https://dps.ny.gov/system/files/documents/2022/11/technical-resource-manual-version-8-filed-august-11-2020-effective-january-1-2021.pdf

9 Lighting

9.1 EQUIPMENT DESCRIPTION

Lighting is a major consumer of electricity in commercial buildings and a target for energy savings through use of energy efficient light sources. Common lighting upgrades are presented in Table 9.1.

A high efficiency lighting upgrade in a commercial building will typically focus on the installation of LED lighting with occupancy sensors as needed. Exterior building lighting (e.g., wall pack lighting), may also be upgraded to LED, as well as parking lot lighting. Interior lighting is often upgraded to LED, typically replacing older T12 or T8 lighting. LED lighting upgrades are usually accompanied by significant utility incentives.

9.1.1 Building Lighting

The major types of lighting (lamps) found in commercial buildings includes:

- Standard fluorescent
- Compact fluorescent (CFL)
- Incandescent
- High-intensity discharge (HID)
- Halogen
- Light-emitting diode (LED).

Each of these lighting types has characteristics that make them best suited to specific lighting needs. Fluorescent lamps and compact fluorescent lamps are more energy efficient and have longer life spans than incandescent bulbs. Halogen lamps are a category of incandescent bulb that offer high-quality light with increased energy efficiency. High-intensity discharge lamps produce very bright light, making them useful for lighting large areas. As such, they are commonly used in high-ceiling spaces with open layouts such as sports arenas, storage areas, and big-box retail stores. LEDs are a highly efficient and durable type of directional lighting that is rapidly evolving into the go-to lighting type for commercial building energy efficiency upgrades.

The terms "efficacy" and "efficiency" of lighting are often confused. Efficacy is the measurement of the amount of useful light output per unit of electricity (or how well a light source turns input power [watts] into the desired output, which is lumens) and is measured in lumens per watt. Efficiency, on the other hand, is calculated by comparing the amount of light produced by the light source to the total amount of light coming out of the fixture. It is measured in percentage and is calculated by dividing the fixture lumens by the lamp lumens. This is important to understand since many types of fixtures trap the emitted light inside the fixture. Many downlights, wall packs, ceiling troffers, and other fixtures have low efficiency,

DOI: 10.1201/9781032692777-9

TABLE 9.1
Common Lighting Upgrades

Existing	Upgrade
Incandescent	CFL, LED
Tungsten	
Halogen	
T12 Linear Fluorescent (1–1/2″ tube)	T8, T5, LED
T8 Linear Fluorescent (1″ tube)	Low Wattage T8, T5, LED
T5 Linear Fluorescent (5/8″ tube)	LED
Compact Fluorescent (CFL)	LED
High Intensity Discharge (HID)	CFL, T8, T5, LED
(common for outdoor lighting)	
Mercury Vapor (blue/purple)	
High Pressure Sodium (yellow)	
Xenon (warm white/blue white)	
Metal Halide	
Incandescent Exit Signs	LED Exit Signs
On/Off Control	Bi-Level Switching
	Occupancy Sensors
	Daylight Sensors (photocells)
Exterior Lighting On/Off	Photocell Control

requiring considerably more lumens from the lamp than makes it outside the fixture to provide useful light output. The distinction between efficacy and efficiency is not always carefully maintained in published sources, so it is not uncommon to see “efficiencies” expressed in lumens per watt or “efficacies” expressed as a percentage. Typical efficacies and lamp lifetimes are presented in Table 9.2 (with upgrade by the author).[(1)]

Lamp selection is based on efficacy (lumens per watt), along with color temperature, color rendering index, life and lumen maintenance, availability, switching, dimming capability, and cost. For example, many T8 and T5 linear fluorescent and compact fluorescent lamps have been excellent choices for buildings because they are relatively energy efficient, have great color rendering properties and long life, and are readily available, easily controlled, and very affordable.

Though each of these lighting types is best suited for particular lighting needs, rising energy costs and new regulations are compelling facility managers to replace energy-intensive lighting with more efficient alternatives. For example, compact fluorescent (CFL) lighting has replaced standard fluorescent lighting in many commercial structures, particularly office buildings.

Today, LEDs are considered to be the most energy-efficient and durable type of directional lighting for most commercial buildings. LEDs provide the same brightness as conventional bulbs but consume 90% less energy than incandescent and CFL lighting. They also last 15 times longer, which means they require less maintenance.

TABLE 9.2
Typical Efficacies and Lifetimes of Lamps

	Efficacy (lumens/watt)	Typical Rated Lifetime (hours)
Standard fluorescent		
Fluorescent—T5	25–55	6,000–7,500
Fluorescent—T8	35–87	7,500–20,000
Fluorescent—T12	35–92	7,500–20,000
Compact fluorescent	40–70	10,000
Incandescent	10–19	750–2,500
High-intensity discharge		
Mercury vapor	25–60	16,000–24,000
Metal halide	50–115	3,000–20,000
High-pressure sodium	50–140	16,000–24,000
Halogen	14–20	2,000–3,500
LED	27–150	40,000–100,000

9.1.2 Parking Area Lighting

Energy consumption associated with building parking areas is excluded from benchmarking analysis. Hopefully, a building's parking area electricity consumption has its own meter separate from the building's meter(s). However, if a single electric meter tracks both building and parking area electricity consumption, then parking area electricity consumption must be separated from building electricity consumption for benchmarking analysis. A methodology to estimate parking lot energy consumption has been developed by EPA.(2)

Energy consumption associated with a parking area in the U.S. typically involves two main factors: the energy required to provide adequate lighting and the energy required for ventilation (if needed). The specific requirements for each of these components will generally depend on the type of parking (e.g., an open parking lot, a completely enclosed garage, or a partially enclosed garage). Table 2.6 in Chapter 2 presents the site energy consumption for the different types of parking areas. The assumptions with respect to lighting and ventilation that EPA made to develop energy consumption estimates are briefly discussed below.

9.1.2.1 Lighting

Lighting requirements are typically described in terms of lighting power density (LPD, in watts per square foot). The following lighting densities were assumed by EPA to be typical:

- LPD for completely enclosed and partially enclosed parking garages is assumed to be 0.30 watts per square foot.
- LPD for open parking lots is assumed to be 0.15 watts per square foot.

In addition to consideration of the LPD, it is also necessary to account for the hours of the day during which lighting will be required. Completely enclosed and partially enclosed parking garages are assumed to require 24 hours of lighting, as no other sources of light are available. Open parking lots, which have direct access to natural light during the daytime, are assumed to require only 16 hours of lighting.

9.1.2.2 Ventilation

Ventilation is required for completely enclosed parking garages that have no access to natural ventilation. It is used to remove tailpipe emissions and ensure safety within the structure. Ventilation requirements are characterized in terms of a flow rate. A ventilation rate of 1.0 cfm/ft^2 of parking area was assumed by EPA for the calculations.[(3)] This rate translates to a fan power of 0.39 horsepower per 1,000 square feet.

In order to compute the ventilation requirement over a year, it is necessary to understand the hours of operation for the ventilation fans. Current best practice in parking garage operation is to automatically adjust fan operation based on demand and to run fans in a setback mode when full power is not necessary. Therefore, EPA's methodology assumes that the fans are operating at full power for 6 hours per day and are in setback mode (0.05 cfm/ft^2) for 18 hours per day. Ventilation is assumed to be unnecessary for partially enclosed parking garages and open parking lots, which have access to natural ventilation.

9.2 ENERGY EFFICIENCY METRICS

Lighting provides a significant opportunity for savings. The key energy efficiency metric is the kWh of energy used for lighting, with lighting specifications written in units of the specific number of watts per square foot (W/ft^2). In addition to lighting energy consumption which will be reduced with a typical lighting upgrade such as switching to LED lighting, there are interactive effects; for example, cooling end-use energy consumption (in the summer) will be reduced, and space heating end-use energy consumption (in the winter) will increase.

9.3 ENERGY SAVINGS

The energy savings associated with a lighting upgrade can be estimated by using a detailed spreadsheet such as is commonly prepared by a lighting contractor, or by use of lighting power densities (LPD).

9.3.1 Lighting Spreadsheet Method

Estimating energy consumption savings is typically provided by performing a lighting audit. Typical spreadsheet (see Table 9.3) headings would include the following for both existing (baseline) lighting and the proposed lighting that replaces the existing lighting:

- Location of each light
- Fixture type for each light

TABLE 9.3
Typical Contractor Lighting Spreadsheet (Excluding New Fixture Cost Column)

Building________________
Address________________
City/State/Zip________________
Utility________________

	Existing Lighting							Post-Retrofit						
Location	Occupancy Sensor (Y/N)	Description of Fixture	Fixture Watts	No. of Fixtures	Operating Hrs/Year	kW	kWh	Occupancy Sensor (Y/N)	Description of Fixture	Fixture Watts	No. of Fixtures	Operating Hrs/Year	kW	kWh
1st Fl. Off.	N	2×4 3L32WT8	89	32	1,825	2.85	5,198	Y	2×4 2L28WT8	42	32	1,396	1.34	1,876
Open Area	N	2×26W CFL	56	7	3,216	0.39	1,261	Y	LED	14	7	2,595	0.1	254
____														
____														
____														
				Total No. Fixtures		Total kW	Total kWh				Total No. Fixtures		Total kW	Total kWh

Notes: $\text{kWh Energy Savings} = \text{Total kWh}_{\text{existing}} - \text{Total kWh}_{\text{post-retrofit}}$

$\text{kW Demand Savings} = \text{Total kW}_{\text{existing}} - \text{Total kW}_{\text{post-retrofit}}$

$\%\ \text{kWh Energy Savings} = \left(\left(\text{Total kWh}_{\text{existing}} - \text{Total kWh}_{\text{post-retrofit}}\right) / \text{Total kWh}_{\text{existing}}\right) * 100$

$\%\ \text{kW Demand Savings} = \left(\left(\text{Total kW}_{\text{existing}} - \text{Total kW}_{\text{post-retrofit}}\right) / \text{Total kW}_{\text{existing}}\right) * 100$

- Whether the location has or does not have occupancy sensor control
- Fixture wattage for each light
- Quantity of fixtures
- Operating hours
- kWh consumption
- kW consumption.

A lighting contractor will also include in the spreadsheet a column that provides the installation cost for what is being proposed.

In the spreadsheet shown in Table 9.3, the total number of fixtures, total kWh, and total kW columns would be summed. The electricity consumption savings would be the baseline kWh minus the kWh of the upgrade or:

$$\text{Electrical energy saved } (\text{kWh}) = \left[\text{Sum of existing fixture kWh} - \text{Sum of post-retrofit kWh}\right] \quad (9\text{–}1)$$

If operating hours are not available, default lighting operating hours presented in Table 9.4[(4)] may be used.

TABLE 9.4
Select State TRM Default Lighting Hours of Operation

	Lighting Hours						
Type Facility	**CT & NY**	**AK**	**CA**	**IL & MN**	**PA**	**TX**	**VT**
Assembly		2,638	2,610		2,610	2,638	3,055
Auto Related	4,056				4,056		
Bakery	2,854						
Banks, Financial Centers	3,748						
Church	1,955						
College—Cafeteria	6,376						
College—Classes/Administrative	2,586		2,400				
College—Dormitory	3,066						
Commercial Condos	4,055						
Convenience Stores	6,376	6,900					
Convention Center	1,954						
Courthouse	3,748						
Daycare					2,590		
Dining: Bar Lounge/Leisure	4,182	4,368	4,830	3,673	2,613	4,368	4,089
Dining: Cafeteria/Fast Food	6,456	6,188	4,840			6,188	
Dining: Family	4,182						
Entertainment	1,952						
Entertainment—Performing Arts	2,586						
Entertainment—Movie	1,954						
Entertainment—Sports Arena	1,954						
Exercise Center	5,836						

(Continued)

TABLE 9.4 (*Continued*)
Select State TRM Default Lighting Hours of Operation

	Lighting Hours						
Type Facility	**CT & NY**	**AK**	**CA**	**IL & MN**	**PA**	**TX**	**VT**
Fast Food Restaurants	6,376						
Fire Station (unmanned)	1,953						
Food Stores/Grocery Stores	4,055	4,706	4,910	5,802	4,660		6,019
Gymnasium	2,586						
Hospitals	7,674	5,730	5,260	5,095	5,182		4,007
Hospitals/Healthcare	7,665						
Laundromats	4,056						
Library	3,748						
Lodging	3,064			5,311	3,356	3,055	4,808
Lodging Common Area		6,630		5,311		6,630	
Mall Concourse	4,833					4,813	
Medical Offices	3,748						
Multifamily Condos	4,055						
Multifamily Common Area	7,665	4,772			5,950		
Museum	3,748						
Nursing Homes	5,840	4,271	4,160		4,160		
Office	3,748	3,737	2,600	4,439	2,567	3,737	3,642
Office/Retail	3,748						
Parking Garage	4,368	7,884			6,552	7,884	
Penitentiary	5,477						
Police/Fire Station	7,665			8,766	5,366		
Post Office	3,748						
Religious Building	1,955	1,824		1,664	1,810	1,824	2,648
Restaurants	4,182						
Retail	4,057						
Retail Department Stores		3,965	3,380	4,719	2,829	3,668	4,103
Schools—University	2,187	3,577		3,540	2,348	3,577	3,416
Schools—Jr/Sr HS	2,187	2,777	2,280	4,311	1,632	2,777	2,456
Schools—Pre-school/Elem.	2,187	2,777	2,140	2,442	1,632	2,777	2,456
Schools—Technical/Vocat.	2,187						
Sports Arena	1,954						
Town Hall	3,748						
Warehouse Non-Refrigerated	2,602	3,501	3,420	4,746	2,316	3,501	4,009
Warehouse Refrigerated	2,602	3,798	4,770			3,798	

The demand (kW) associated with each fixture is calculated from the [(watts/fixture)/(1,000 W/kW)] × [number of these fixtures], or

$$\text{Demand, kW} = \text{Sum of}\left[\left(\text{Watts/fixture}\right) * \text{Number of Fixtures}/\left(1000\ \text{W/kW}\right)\right] \quad (9\text{–}2)$$

The demand savings would be the kW associated with the baseline minus the kW associated with the upgrade, or:

$$\text{Demand Savings, kW} = \text{Existing kW Demand} - \text{Post Retro-fit kW Demand} \quad (9\text{–}3)$$

Equation 9–3 provides the demand savings. All things being equal, assuming no ratchet provisions and that billed demand equals read demand, the peak demand savings would occur throughout the year. Under these assumptions, the annual peak demand savings would be:

$$\text{Annual Demand Savings, kW} = (12 \text{ billing periods/yr}) * \text{Demand Savings, kW} \quad (9\text{–}4)$$

It is important to understand how the utility charges for demand and whether ratchet provisions exist before estimating demand cost savings. Demand savings would also typically be multiplied by the associated utility-specified coincidence factors (i.e., for summer and for winter). The sum of the two would be the total annual demand savings. For example, CF_{summer} can range in Connecticut[5] from 0.702 to 0.904 depending on commercial building use, and from 0.539 to 0.770 for CF_{Winter}.

External building lighting such as nighttime parking lot lighting typically is not included in the demand savings as this demand reduction is likely to be occurring during the utility's off-peak usage period.

Example Problem 9–1

A lighting contractor's spreadsheet shows the following:

Existing						Post-Retrofit					
Fixture Type	Watts/ Fixture	Quantity	Operating Hours	kWh	kW	Fixture Type	Watts/ Fixture	Operating Quantity	Hours	kWh	kW
2x4 T8	55	1,020	4,015	225,242	56.1	LED	22	1,020	4,015	90,097	22.4
2x2 T8	60	65	4,015	15,659	3.9	LED	28	65	4,015	7,307	1.8
Total				240,901	60					97,404	24.2

What are the energy consumption and demand savings?

Energy Consumption Savings:	(240,901 – 97,404) =143,497 kWh or 59.6%
Energy Demand Savings:	(60 – 24.2) = 35.8 kW or 59.7%

9.3.2 Lighting Power Density Method

If a lighting contractor spreadsheet is not available, a rough estimate of the lighting energy consumption savings that may be realized can be obtained using the lighting power densities (LPD in watts/SF) from Tables 9.5 and 9.6. Table 9.5 includes the overall LPD complying with the ASHRAE 90.1 standard in effect in different years for various building types. Table 9.6 has the LPD complying with the 2022 ASHRAE 90.1 standard (and 2004 standard for comparison) for select space types that may be inside the building.

Example Problem 9–2

A 25,000 ft² office building in Connecticut, constructed in 1999, is replacing existing lighting with LED lighting in 2022. A lighting contractor's spreadsheet is not available, but the building owner would like a rough estimate of the energy savings that might be realized assuming electricity costs \$0.18/kWh.

Assuming the office building constructed in 1999 complied with the ASHRAE 90.1–1999 standard in effect at that time, the lighting power density would be 1.3 watts/ft² (see Table 9.5).

Assuming compliance with the 90.1–2022 ASHRAE standard, the LPD would be reduced to 0.62 watts/SF.

$$\text{The watts savings} = (1.3 - 0.62) * 25{,}000 = 17{,}000\text{ W}$$

Assuming the lighting hours in Table 9.4 (3,748 hr/yr):

$$\text{The annual kWh savings would be:}$$
$$(17{,}000\text{ W} * 3{,}748\text{ hr/yr}) / 1{,}000\text{ W/kW} = 63{,}716\text{ kWh/yr}$$

$$\text{The consumption cost savings would be:}$$
$$63{,}716\text{ kWh/yr} * \$0.18\text{ kWh} = \$11{,}469$$

plus a cost savings associated with peak demand reduction (17 kW in each billing period).

By reducing operating hours, occupancy sensors can further reduce the lighting electricity consumption savings by 25%–30%. To estimate the energy savings that might be realized with occupancy sensors, Equation 9–5 may be used.

Occupancy Sensor kWh Energy Savings:

$$\text{Occupancy Sensor Savings, kWh} = \left(0.3 * \left(H_{Lighting}\right)/1000\right) * \sum_{n=1}^{N}(OSF_n)*(WF_n) \quad (9\text{–}5)$$

where $H_{Lighting}$ = lighting hours of use for property type (default to Table 9.4)

N = number of different fixture types with occupancy sensors

n = fixture number

OSF_n = number of fixtures of type n (with same watts/fixture) that have occupancy sensors

WF_n = input watts for fixture type n

0.3 = 30% avg. energy savings due to occupancy sensors

1,000 = conversion factor from W to kW

Comprehensive LED lighting retrofit projects typically can reduce lighting energy consumption by 40%–70% over a non-LED baseline. The author has found a good average to be in the 55%–65% range.

Example Problem 9–3

An office building has a baseline non-LED lighting end-use energy consumption of 50,000 kWh/yr. Estimate the total lighting energy consumption savings assuming conversion to LED lighting with occupancy sensors. Assume the LED conversion will reduce lighting energy consumption by 55%.

Lighting savings will be:

$$\text{kWh/yr savings} = 0.55 * 50{,}000 = 27{,}500 \text{ kWh/yr}$$

Assume occupancy sensors can reduce the remaining electricity used by 30%.

$$\text{Occupancy sensor savings} = (50{,}000 - 27{,}500) * 0.3 = 6{,}750 \text{ kWh/yr}$$

$$\text{Total lighting project savings} = 34{,}250 \text{ kWh/yr}$$

Lighting project energy consumption savings over baseline:

$$\text{Energy savings} = 34{,}250/50{,}000 = 0.685 \text{ or } 68.5\%$$

TABLE 9.5
Lighting Power Density (LPD) Requirements for Buildings per ASHRAE 90.1-2022 Standard

Building Area Type	LPD, W/SF per ASHRAE 90.1*				
	2022	2019	2010	2007	1999
Automotive Facility	0.73	0.75	0.82	0.93	1.50
Convention Center	0.64	0.64	1.08	1.21	1.40
Courthouse	0.75	0.79	1.05	1.21	1.40
Dining—Bar, Lounge	0.74	0.80	0.99	1.30	1.50
Dining—Cafeteria, Fast Food	0.70	0.76	0.90	1.39	1.80
Dining—Family	0.65	0.71	0.89	1.58	1.90
Dormitory	0.52	0.53	0.81	1.02	1.50
Exercise Center	0.72	0.72	0.88	1.02	1.40
Fire Station	0.56	0.56	0.71		1.30
Gymnasium	0.75	0.76	1.00	1.11	1.70
Healthcare Clinic	0.77	0.81	0.87	1.02	1.60
Hospital	0.92	0.96	1.21	1.21	1.60
Hotel/Motel	0.53	0.56	1.00	1.02	1.70

(Continued)

TABLE 9.5 (*Continued*)
Lighting Power Density (LPD) Requirements for Buildings per ASHRAE 90.1-2022 Standard

Building Area	LPD, W/SF per ASHRAE 90.1*				
Type	2022	2019	2010	2007	1999
Motel			0.88	1.02	2.00
Library	0.83	0.83	1.18	1.30	1.50
Manufacturing Facility	0.82	0.82	1.11	1.30	2.20
Motion Picture Theater	0.43	0.44	0.83	1.21	1.60
Multifamily	0.46	0.45	0.60	0.84	1.00
Museum	0.56	0.55	1.06	1.11	1.60
Office	0.62	0.64	0.90	1.02	1.30
Parking Garage	0.17	0.18	0.25	0.28	0.30
Penitentiary	0.65	0.69	0.97	1.02	1.20
Performing Arts Theater	0.82	0.84	1.39	1.58	1.50
Police Station	0.62	0.66	0.96	1.02	1.30
Post Office	0.64	0.65	0.87	1.11	1.60
Religious Facility	0.66	0.67	1.05	1.30	2.20
Retail	0.78	0.84	1.40	1.49	1.90
School/University	0.70	0.72	0.99	1.21	1.50
Sports Arena	0.73	0.76	0.78	1.11	1.50
Town Hall	0.67	0.69	0.92	1.11	1.40
Transportation Facility	0.56	0.50	0.77	1.02	1.20
Warehouse	0.45	0.45	0.66	0.84	1.20
Workshop	0.86	0.91	1.20	1.39	1.70

*Adapted from ASHRAE Standards 90.1–1999, 90.1–2007, 90.1–2010, 90.1–2019, and 90.1–2022.

TABLE 9.6
Lighting Power Density (LPD, W/ft²) by Space per ASHRAE 90.1-2022 Standard for Select Space Types*

Space Type	ASHRAE 90.2-2022	ASHRAE 90.1-2004
Atrium		
<20 ft. height	0.32	*First 3 Floors: 0.6*
≥20 ft. and ≤40 ft. height	0.41	*Each Additional Floor: 0.2*
>40 ft. height	0.51	
Audience Seating Area		0.9
Auditorium	0.57	
Gymnasium	0.23	0.4
Motion Picture Theater	0.27	1.2
Performing Arts Theater	1.10	2.6
Sports Arena	0.27	
All Other	0.23	
Banking Activity	0.56	1.5

TABLE 9.6 (*Continued*)
Lighting Power Density (LPD, W/ft²) by Space per ASHRAE 90.1-2022 Standard for Select Space Types*

Space Type	ASHRAE 90.2-2022	ASHRAE 90.1-2004
Classroom/Lecture Hall/Training Room		1.4
Shop Classroom	1.17	
All Other	0.72	
Computer Room	0.75	
Conference/Meeting/Multipurpose Room	0.88	1.3
Convention Center—Exhibit Space	0.50	1.3
Control/Editing Room or Booth	0.73	
Copy Print Room	0.56	
Corridor	0.44	
Courtroom	1.08	1.9
Dining Areas		
Bar/Lounge/Leisure	0.76	1.4
Cafeteria/Fast Food	0.36	
Family Dining	0.52	2.1
All Other	0.42	
Dormitory—Living Quarters	0.48	1.1
Electrical/Mechanical Room	0.71	1.5
Emergency Vehicle Garage	0.51	
Equipment Room	0.73	
Food Preparation Area	1.19	1.2
Gymnasium/Fitness Center		
Exercise Area	0.82	0.9
Playing Area	0.82	1.4
Guest Room (Hotel/Motel, etc.)	0.41	1.1
Healthcare—Hospital		
Control Room (MRI, CAT Scan, etc.)	0.78	
Exam/Treatment	1.33	1.5
Corridor	0.61	1.0
Imaging Room	0.94	
Lounge	0.77	0.8
Medical Supply Room	0.56	1.4
Nursery	0.87	0.6
Nurse's Station	1.07	1.0
Operating Room	2.31	2.2
Patient Room	0.78	0.7
Physical Therapy Room	0.82	0.9
Recovery Room	1.18	0.8
Telemedicine	1.44	
Laboratory		1.4
In or as a Classroom	1.05	
All Other	1.21	

(*Continued*)

TABLE 9.6 ***(Continued)***
Lighting Power Density (LPD, W/ft²) by Space per ASHRAE 90.1-2022 Standard for Select Space Types*

Space Type	ASHRAE 90.2-2022	ASHRAE 90.1-2004
Laundry/Wash Areas	0.51	
Library		
Reading Area	0.86	1.2
Stacks	1.18	1.7
Loading Dock, Interior	0.87	
Lobby		
Elevator	0.64	
Hotel	0.48	1.1
Motion Picture Theater	0.20	1.1
Performing Arts Theater	1.21	3.3
All Other	0.80	
Locker Room	0.43	
Lounge/Breakroom		
Mother's Wellness Room	0.68	
All Other	0.55	
Museum		
General Exhibition Area	0.31	1.0
Restoration Area	1.24	1.7
Office		1.1
≤150 SF	0.73	
>150 SF and ≤300 SF	0.66	
>300 SF	0.56	
Parking Garage		0.2
Daylight Transition Zone	1.06	
All Other	0.11	
Pharmacy Area	1.59	
Religious Facility		
Seating Area	0.72	1.7
Fellowship Hall	0.50	0.9
Worship/Pulpit/Choir Area	0.75	2.4
Restroom	0.74	0.9
Retail		
Dressing/Fitting Area	0.45	0.6
Hair Care	0.65	
Mall Concourse	0.57	1.7
Massage	0.81	
Nail Care	0.75	
Sales Area	0.85	1.7
Seating Area, General	0.21	0.9
Stairwell	0.47	0.6
Storage Room		
<50 SF	0.49	*Active Storage: 0.8*

TABLE 9.6 (*Continued*)
Lighting Power Density (LPD, W/ft²) by Space per ASHRAE 90.1-2022 Standard for Select Space Types*

Space Type	ASHRAE 90.2-2022	ASHRAE 90.1-2004
≥50 SF	0.35	*Inactive Storage: 0.3*
Warehouse-Storage Area		
Medium-to-Bulky Palletized Items	0.33	0.9
Smaller Items, Packing Areas	0.69	
Workshop (including workshop classrooms)	1.17	1.9

*Adapted from ASHRAE Standards 90.1–2004 and 90.1–2022.

9.4 INTERACTIVE EFFECTS

Lighting systems convert a relatively small fraction of their electrical input into useful light output. Much of the rest is released directly as heat into the conditioned space. An LED upgrade to an existing building lighting system (e.g., fluorescent lighting) will incur a heating penalty (in the winter months when heating is required) since LED lighting releases less heat than the existing fluorescent fixtures. As such, the building's heating system must make up for this "loss" during the heating season. An LED upgrade will also reduce the building's cooling need since LED lighting will release less heat that has to be removed in the summertime, thereby reducing the cooling load. The net effect of this interactivity is an energy reduction (savings) in the summertime when air conditioning is needed and an energy increase (penalty) in the winter months when heating is needed. It should be noted that any energy savings due to an upgrade of building exterior lighting will have no interactive effects.

A number of state technical resource manuals provide estimates for the space heating penalty and cooling energy savings.[6–10] For example, in Connecticut, the interactive effects of a lighting upgrade are given by:

$$\text{Reduced cooling load, kWh} = \left(\text{kWh energy savings due to lighting upgrade}\right) * \text{F/COP} \qquad (9\text{–}6)$$

where F = fraction of kWh savings that must be removed by the cooling system

$F = 0.35$ if economizer is present

If no economizer:

$F = 0.48$ for Bldg. ft² $A < 2{,}000$

$F = 0.48 + \left[\left(0.195 * \left(A - 2000\right)\right)/18000\right]$ for Bldg. ft² $A = 2{,}000$ to $20{,}000$

$F = 0.675$ for Bldg. ft² $A > 20{,}000$

COP = coefficient of performance for the cooling system (if the EER is known for the cooling system, the COP = EER/3.412)

$$\text{Increase in heating load, MM Btu} = 0.0003649 * (\text{kWh energy savings due to lighting upgrade}) \text{ for natural gas heating} \quad (9\text{–}7)$$

$$\text{Increase in heating load, MM Btu} = 0.0007129 * (\text{kWh energy savings due to lighting upgrade}) \text{ for oil heating} \quad (9\text{–}8)$$

$$\text{Increase in heating load, MM Btu} = 0.000291 * (\text{kWh energy savings due to lighting upgrade}) \text{ for electric heating} \quad (9\text{–}9)$$

Using the above equations for all climates in the country would introduce error. For example, in Miami, Florida, there would be no heating penalty due to its very warm year-round climate. A better method (preferred) is to estimate the cooling load reduction and heating energy penalty using the annual hours of heating and cooling, which can be obtained from bin weather data (see Chapter 5, Section 5.2 and Example Problem 5–1). An alternative to using hourly bin weather data is to estimate how long (days) the heating and cooling seasons are and converting the number of days to number of hours. This can be accomplished using National Weather Service climate data as follows:

1. Go to www.weather.gov/wrh/climate.
2. Click on the map where the building is located.
3. You will be sent to another screen where you will:
   - Confirm your building's location.
   - Under Product, select Calendar Day Summaries.
   - Under Options:
     - Under Year range, leave "por – 2024", which will provide data from when it was first collected.
     - Under Variable, select Average Temperature.
     - Under Summary, select Mean.
   - Under View, select Go.
4. You will be sent to a screen containing the table "Mean Average Temperature by Day" for your building's location.
5. Heating would likely include those days where the average temperature was <65°F.
6. Cooling would likely include those days where the average temperature was >65°F.
7. Convert the heating and cooling days to hours. This will provide an estimate of the heating and cooling hours using average daily temperature.

The heating penalty and cooling load reduction would be estimated as follows:

Heating Penalty

$$\text{MM Btu heating increase} = \left[(\text{kWh saved due to lighting upgrade}) \times (3{,}412\ \text{Btu/kWh}) \times (\text{fraction conversion to heat}) * (\text{Heating hr/yr} / 8{,}760)\right] / \left[(1{,}000{,}000\ \text{Btu/MM Btu}) * (\text{heating system efficiency})\right] \quad (9\text{–}10)$$

where the percent conversion to heat can be estimated using Table 9.7 and the heating hours per year can be estimated using average daily temperature weather data.

Cooling Reduction

Using the cooling system efficiency in kW/ton and assuming 100% of the building space is conditioned, the kWh cooling energy saved

$$= \left[(\text{kW/ton}) / (12{,}000\ \text{Btu/ton})\right] * (\text{kWh saved due to lighting upgrade}) * (3{,}412\ \text{Btu/kWh}) * (\text{fraction conversion to heat}) * (\text{Cooling hr/yr} / 8{,}760) \quad (9\text{–}11)$$

TABLE 9.7
Input Power (W) Converted to Heat (Conduction + Convection) for Various Light Sources[11]

Light Source	% Conversion to Heat
Incandescent	19%
Fluorescent (typical linear)	42%
Metal Halide	37%
LED	75%–85%

Example Problem 9–4

A lighting project involving a T8 lighting system upgrade in Windsor, Connecticut, is expected to save 30,117 kWh/yr.

Average daily temperature data suggests that heating occurs 4,848 hr/yr and cooling 2,544 hr/yr. Assume a gas heating system efficiency of 80% and a cooling system efficiency of 1.05 kW/ton. Estimate the cooling energy saved and the heating energy penalty. Also, how does it net out?

Heating penalty is determined using Equation 9–10:

$$\text{MM Btu heating increase} = \left[(\text{kWh saved due to lighting upgrade}) * (3{,}412\ \text{Btu/kWh}) * (\text{fraction conversion to heat}) * (\text{Heating hr/yr} / 8{,}760)\right] / \left[(1{,}000{,}000\ \text{Btu/MM Btu}) * (\text{heating system efficiency})\right]$$

From Table 9.7, 42% is converted to heat:

$$\text{Heating penalty} = (30{,}117 * 3{,}412 * 0.42 * 4{,}848/8{,}760)/(1{,}000{,}000*0.8) = 29.9 \text{ MM Btu/yr.}$$

The cooling energy saved is obtained from Equation 9–11:

$$\text{kWh cooling energy saved} = [(\text{kW/ton})/(12{,}000 \text{ Btu/ton})]*(\text{kWh saved due to lighting upgrade}) * (3{,}412 \text{ Btu/kWh}) * (\text{fraction conversion to heat}) * (\text{Cooling hr/yr} / 8{,}760)$$

$$\text{kWh cooling energy saved} = (1.05/12{,}000)*30{,}117*3{,}412*0.42*(2{,}544/8{,}760) = 1{,}097 \text{ kWh/yr or } 3.7 \text{ MM Btu/yr}$$

The net energy impact would be 3.7 – 29.9 = – 26.2 MM Btu/yr.

Example Problem 9–5

An office building in Hartford, Connecticut, installed an LED lighting upgrade that saved 75,000 kWh/yr. What was the resulting impact on the cooling and heating energy consumption? The building uses natural gas for heating and the air conditioning system (with an economizer) has an EER = 10.1.

Since the building is in Connecticut, we have equations from the CT Technical Resource Manual to estimate cooling energy saved and the heating penalty.

From Equation 9–6, the reduction in cooling is given by: (kWh savings in the lighting upgrade) × F/COP.

Since there is an economizer, F = 0.35

$$COP = EER / 3.412 = 10.1 / 3.412 = 2.96$$

$$\text{Energy savings from reduced cooling} = (75{,}000 \text{ kWh}) * 0.35 / 2.96 = 8{,}868 \text{ kWh}$$

The heating penalty is estimated using Equation 9–7:

$$\text{Heating penalty} = 0.0003649 \text{ MM Btu} * (\text{kWh lighting savings/yr}) = 27.4 \text{ MM Btu}$$

The net impact on building energy use would then be:

$$(8{,}868 \text{ kWh}/(293 \text{ kWh/MM Btu})) - 27.4 \text{ MM Btu} = 30.3 - 27.4 = 2.9 \text{ MM Btu}$$

The interactive effects of energy saved in this building due to the LED upgrade in reducing the cooling load and the energy increase in heating load would almost net out.

REFERENCES

1. 2011 Buildings energy data book, Table 5.6.9, Office of Energy Efficiency and Renewable Energy, U.S. Department of Energy, http://large.stanford.edu/courses/2015/ph240/davidson1/docs/2011_BEDB.pdf
2. ENERGY STAR Portfolio Manager, Technical Reference, Parking and the ENERGY STAR Score, www.energystar.gov/sites/default/files/tools/Parking_August_2018_EN_508.pdf
3. ANSI/ASHRAE Standard 62.1 and the International Mechanical Code (ICC 2009a) allow 0.75 cfm/ft^2, while the National Fire Protection Association (NFPA) Standard 88A recommends a minimum of 1.0 cfm/ft^2.
4. State TRMs: CT (2017) https://www.puc.nh.gov/EESE%20Board/EERS_WG/ct_trm.pdf, NY (2019) https://dps.ny.gov/system/files/documents/2022/11/technical-resource-manual-version-7-filed-april-15-2019-effective-january-1-2020.pdf, AK (2018) https://apsc.arkansas.gov/wp-content/uploads/AR_TRM_V9.1_Volume_1_2_and_3_on_8-31-22.pdf, CA (2016) https://www.cmua.org/files/2016%20CMUA%20POU%20TRM_Final_v692016.pdf, IL (2021) https://www.ilsag.info/wp-content/uploads/IL-TRM_Effective_010121_v9.0_Vol_2_C_and_I_09252020_Final.pdf, MN (2014) https://mn.gov/commerce-stat/trm/releases/1.3.pdf, PA (2015) https://www.puc.pa.gov/docket/M-2012-2313373, TX (2017) https://texasefficiency.com/wp-content/uploads/2023/11/trm5v3.pdf, VT (2018) https://www.puc.nh.gov/EESE%20Board/EERS_WG/vt_trm.pdf, Lighting Hours of Operation Tables.
5. UI/Eversource, 2017 Connecticut program savings document, 12th Edition, Appendix 1, October 31, 2016, https://www.puc.nh.gov/EESE%20Board/EERS_WG/ct_trm.pdf
6. New York Standard Approach for Estimating Energy Savings from Energy Efficiency Programs—Residential, Multi-Family, and Commercial/Industrial Measures, Version 7 2020, pp. 434–439 and Appendix D, April 15, 2019, https://dps.ny.gov/system/files/documents/2022/11/technical-resource-manual-version-7-filed-april-15-2019-effective-january-1-2020.pdf
7. Delaware Technical Resource Manual, pp. 231–234 and Appendix E, April 30, 2012, https://regulations.delaware.gov/register/august2016/proposed/DE%20TRM%20complete%202016.pdf
8. Public Utility Commission of Texas, Texas Technical Reference Manual, Version 5.0, Vol. 3: Nonresidential Measures, 2018 Program Year, pp. 2–6 to 2–11, October 30, 2017, https://hawaiienergy.com/wp-content/uploads/PY19-TRM-v1.pdf
9. Arkansas Technical Reference Manual, Version 7.0, Section 3.6.3, and Appendix I, p. 303 ff, August 30, 2013, https://apsc.arkansas.gov/wp-content/uploads/AR_TRM_V9.1_Volume_1_2_and_3_on_8-31-22.pdf
10. XcelEnergy 2017/2018 Demand-Side Management Plan, Electric and Natural Gas, pp458–460, November 17, 2016 (rev), https://www.xcelenergy.com/staticfiles/xe-responsive/Company/Rates%20&%20Regulations/Regulatory%20Filings/CO-Demand%20side%20management-2017%2018%20DSM%20PLAN_FINAL.pdf
11. U.S. DOE, Energy Efficiency and Renewable Energy, Building Technologies Program, Thermal management of white LEDs, PNNL-SA-51901, February 2007, https://www1.eere.energy.gov/buildings/publications/pdfs/ssl/thermal_led_feb07_2.pdf
12. 2017 Connecticut Program Savings Document, 12th Edition, UI/Eversource, pp. 70–72, and Appendix 5, October 31, 2016, https://www.puc.nh.gov/EESE%20Board/EERS_WG/ct_trm.pdf

10 Energy Management Systems

10.1 EQUIPMENT DESCRIPTION

Energy management is one of the many tools that can be used to operate buildings more efficiently. An energy management system (EMS)—sometimes also referred to as a building energy management information system (BEMIS) or building energy management system (BEMS)—is a control system designed to help monitor, control, measure, and optimize energy consumption needs of a building. It typically collects data and controls energy-consuming devices such as heating and cooling equipment, fans, pumps, dampers, lighting, refrigeration equipment, or other building systems, and can process it in real time.

An EMS typically consists of three elements: (1) sensors that will measure things like temperature, pressure, and light levels which are used to initiate responses by the system; (2) controllers that will compare a signal received from a sensor to a desired set-point and send out a signal to a control device for action; and (3) a computer providing a user interface that pulls all the information together and performs analytics. EMS systems can help building owners and managers know how the building's various systems are performing at any given moment and are able to send an alert on potential problems.

Three types of control hardware are typically used in EMS systems: pneumatic, analog electronic, and direct digital control (DDC). Historically, pneumatic controls were used in most large building applications. They are reasonably well understood by designers and maintenance personnel, which was a compelling reason for their use. A second advantage was that pneumatic valve and damper actuators, besides being inherently modulating, are inexpensive and reliable when compared to the electric motors and gears required to produce the same control force with electric actuation. Most pneumatic control equipment, however, requires a very clean source of supply air. The air must be dry and free of oil. While it may not be difficult to install a system with clean air, one failure or mistake, such as overfilling a compressor with oil, can permanently foul an entire pneumatic system.

Modern analog electronic control equipment has several advantages. Power is provided electrically, and there are no air compressors, driers, or filters to maintain. Signals flow at the speed of light, and controllers essentially have an instantaneous response (although signal processing is instantaneous, temperature sensors and motors still have time constraints). Electronic devices can be extremely accurate and free of drift. Resistance temperature detectors made of platinum, nickel, or other metals do not require field calibration and are drift free. They are also inexpensive, and it is easy to display electronic signals using digital readouts, providing ready diagnostics. In the past, electronic control systems were often more

DOI: 10.1201/9781032692777-10

expensive than pneumatic systems, partly because of the high cost of electric valve and damper motors. However, electronic control systems today are very competitive with pneumatic systems and may even be less expensive when integrated with a building automation system (BAS) because many sensors can perform control and automation functions, whereas additional sensors are required for BAS functions with pneumatic systems.

Today, DDC has become the standard in control systems. Large integrated systems as well as small-scale control units for single-fan systems are available for both new and retrofit applications. DDC accepts analog electronic inputs from appropriate sensors, converts them to digital, uses these digitized signals to produce digital outputs via software control algorithms resident in a local microprocessor, and, finally, converts the digital output signals to analog electronic signals to drive control actuators that may be electrical or pneumatic. Perhaps the greatest single advantage of DDC is the flexibility it provides because control is implemented through software rather than hardware. Time sequencing, reset control, economizer cycles, and a variety of other operational schemes are accomplished by programming a microprocessor, not by modifying or adding hardware. Another benefit of DDC is precision and a complete absence of controller drift. For example, once a temperature set-point is entered into a digital controller, it remains in the microprocessor's memory unambiguously until it is changed by an operator or until the controller fails. The concept of recalibration is simply not part of digital hardware, and maintenance should be required only when the hardware fails. All problems do not vanish, however, with DDC. Analog sensors may still require periodic recalibration.

An unfortunate drawback of DDC systems is the proliferation of programming languages and protocols making system commissioning and maintenance difficult, particularly since the user interface into the nuts and bolts of the control logic usually is not user-friendly. Programming of controls is also an issue in that it is often not transparent to the design team and/or readily available for review by the design and commissioning team. Controls programming typically is in the hands of specialists, thus becoming a less integral part of the design process to the potential detriment of integrated design.

A computerized system, however, has some interesting and valuable side benefits.

- *Optimization*: It will allow for continuous analysis of the performance of the various systems, compare actual performance with a standard, and make adjustments as needed to improve efficiency.
- *Data management*: Because the computer can collect a significant quantity of data about operating conditions, it can be programmed to analyze and summarize the data in ways that are useful for operators, building owners, and tenants.
- *System commissioning*: DDC and building automation systems can be used to enhance commissioning, for example, of a new HVAC system. In fact, significant parts of the commissioning activity can be automated.
- *Maintenance schedules*: Preventive maintenance programs can be stored in computer memory and printed out on schedule to remind the operator that they must be performed. Maintenance sensors also can be made a part of

the monitoring equipment. Such sensors can detect overheating of motors or bearings, excessive vibration, loud and extraneous noises, and numerous other malfunction indicators.

- *Energy consumption*: Where air conditioning and heating systems serve multiple tenants, it may be necessary to meter the services supplied to each tenant. Special meters with outputs integrated in the EMS can provide an automatic measuring and billing service. Further, any sudden deviation from normal in a tenant's energy consumption would be detected and cause a warning message for investigation by operating personnel.

10.2 EMS VERSUS BMS/BAS

Unfortunately, terms like EMS, BMS, and BAS are often used interchangeably. As a result, people fail to realize that while BMS and BAS might be synonymous, EMS is very different.

A BMS is a computer-based system installed in buildings to control and monitor mechanical and electrical equipment, including HVAC equipment, lighting, power systems, fire systems, and security systems. While some vendors tried to differentiate BAS as an advanced version of BMS, BAS is effectively an evolution of a BMS by adding smarter analytics and advanced automated controls. Notwithstanding, building owners and operators use BMS and BAS interchangeably.

An EMS, as the name suggests, is focused primarily on the energy aspects of a building. These systems collect data about the energy usage of a wide range of energy consuming systems within a building, such as HVAC systems, lighting, and power systems. They enable building managers to get an overview of how the consumption of energy within a building takes place, monitor this in real time, and provide a control capability. They can also be programmed to identify and recommend possible actions that can enhance the building's energy efficiency. The data collected by an EMS can enable a facility and building manager to make more informed decisions about energy activities within their buildings.

The direct benefits of EMS systems may include:

- Energy cost savings (in both energy consumption and energy demand).
- Prioritization of no cost and low-cost energy saving opportunities in day-to-day operations.
- Reduced greenhouse gas emissions.
- Greater knowledge of building energy use and consumption, and identification of opportunities for improvement; current building energy consumption can also be compared to historical energy consumption.
- Enable a more informed decision-making process.
- Centralize, normalize, and visualize data—automatically bring data streams together from different sources into a common database to allow for visualization of key performance indicators.
- Utility bill management—track, understand, and process data from utility bills.
- Measurement and verification—quantify and verify the energy savings performance of individual energy conservation measures or efficiency projects.

- Automated fault detection and diagnostics—automatically detect equipment-level or system-level faults and diagnose their causes.
- Supervisory control—perform automated changes to underlying building systems for optimization.
- Operations and maintenance optimization—integrate the above capabilities with operations and maintenance processes to increase efficiency.

EMS and BMS can work well together. While energy management systems focus on micro-level data, building management systems typically focus on macro-level information. This positions them as complementary. Unfortunately, EMS and BMS speak two completely different languages, and as such, need an intermediary to translate between the two. The most common translator is BACnet.

BACnet is a communication protocol[(1)] for building automation and control (BAC) networks. It was designed to allow communication of building automation and control systems for applications such as heating, ventilating, and air conditioning control, lighting control, access control, and fire detection systems and their associated equipment. The BACnet protocol provides mechanisms for computerized building automation devices to exchange information, regardless of the building service they perform.

10.3 SELECTING AN EMS

Selecting an EMS requires consideration of many building attributes including building size, use, occupancy, and energy expenditure.

Building size determines a host of details of an energy management system's usefulness and appropriateness. The size of a building generally can determine the overall energy spend, the types of equipment that are installed to heat and cool the building, the number of occupants, and so on. Generally, the larger the building, the more sophisticated the EMS.

Building use can determine the requirements for an EMS. A warehouse has different requirements than an office building, which must consider things like occupant comfort; data centers have different needs than office buildings, hospitals, or schools.

Occupancy is another factor in a building's energy consumption. Occupant comfort while still generally being managed by the building HVAC system is one of the most important considerations for selecting an EMS. Unoccupied spaces, for example, can create misleading information regarding a building's energy consumption. An EMS can normalize energy expenditure based on occupancy and determine the energy use intensity (EUI) of the building.

The monthly or annual energy expenditure of a building is one of the first considerations in selecting an EMS. Sophisticated and expensive systems are not appropriate for small buildings that use little energy. For such buildings, the savings from the EMS would find it difficult to overcome the initial and ongoing costs, and return on investment (ROI) and payback will more than likely be unacceptable. However, for larger buildings with greater energy expenditures, an EMS can generate savings resulting in a more acceptable ROI and payback period.

The intended EMS use also affects the choice of the system and should be discussed at length prior to making an EMS selection. Some building owners may just want better visibility into their energy consumption and not need a more sophisticated system with much more capability. Other building owners may face situations where a sophisticated EMS makes business sense, such as a laboratory or hospital where the inside environment of the building may be critical for safety reasons.

Although many of the above EMS considerations are deemed important to the decision-making process, the reality is that the selection process ultimately comes down to cost and energy savings. Does this EMS make financial sense to the company? What is the ROI that the EMS will provide, and how long will it take to pay back the initial and ongoing costs related to the system? Generally, the EMS vendor can help with determining this, but caution and due diligence must be performed on these numbers. It is easy to show savings with an EMS if it just turns equipment off.

10.4 EMS AND HVAC CONTROL

In the context of energy efficiency, EMS refers to a system designed to achieve energy efficiency through process optimization by reporting on granular energy use by individual pieces of equipment. Newer, cloud-based energy management systems provide the ability to remotely control HVAC and other energy-consuming equipment; gather detailed, real-time data for each piece of equipment; and generate intelligent, specific, real-time guidance on finding and capturing the most compelling savings opportunities.

The scope of a project may span from a single building to a group of buildings such as university campuses, office building complexes, retail stores or factories. Most of the energy management systems also provide capability for the reading of electricity, gas, and water meters. The data obtained from these meters can then be used to perform self-diagnostic and optimization routines on a frequent basis and to produce trend analysis and annual consumption forecasts.

10.5 ENERGY SAVINGS

Due to the complex nature of EMS systems and their controls, gross energy and demand savings need to be custom calculated based on project-specific details including hours of operation, HVAC system equipment and efficiency, and points controlled.

In general, integrating state-of-the-art sensors and controls into most commercial buildings can save as much as 30% of whole building site energy consumption, depending upon building characteristics including HVAC equipment design and operation. These savings are achieved through high-performance sequencing of operations, optimizing settings based on occupancy patterns, and detecting and diagnosing inadequate equipment operation or installation problems. Furthermore, state-of-the-art sensors and controls can curtail as much as 10%–20% of commercial building peak load.(2)

Estimates provided on whole building energy savings have wide variability (5%–40%) due to differences in building characteristics.(3) Birtles, John, and Smith found that energy consumption savings can range up to 27% compared to a building with no EMS installed.[(4)] Roth, Llana, Detlef, and Brodrick found energy savings up to 20% in space heating energy consumption and 10% for lighting and ventilation energy consumption, combining to a 5% to 20% overall energy consumption savings range.[(5)] Lee and Cheng did an investigation of the energy savings of EMS systems in publications from 1976 through 2014.[(6)] Their statistical analysis showed average annual building energy consumption savings of EMS systems in the 11.39%–16.22% range. For HVAC equipment, average energy consumption savings were in the 14.07%–16.66% range.

Typical energy savings experienced by the author for commercial buildings seeking property assessed clean energy loans for energy efficiency upgrading were typically on the order of 10%–15% of whole building energy consumption. On an end-use basis, it was not unusual to find energy consumption savings in the 10%–15% range for the cooling and ventilation (kWh/yr) end-uses and 15%–20% for the space heating (Btu/yr) end-use.

REFERENCES

1. ANSI/ASHRAE Standard 135-2016, BACnet™: A data communication protocol for building automation and control networks, 2016, Addendums through 2019.
2. U.S. DOE Office of Energy Efficiency and Renewable Energy, Building controls, www.energy.gov/eere/buildings/building-controls
3. UN Climate Technology Center and Network, Building energy management systems, www.ctc-n.org/technologies/building-energy-management-systems-bems
4. Birtles, A.B., John, R.W. and Smith, J.T., Building services engineering research and technology, *Study of the Performance of an Energy Management System*, Vol. 5(4), pp. 155–172, November 1984.
5. Roth, K., Llana, P., Detlef, W. and Brodrick, J., Automated whole building diagnostics, *ASHRAE Journal*, Vol. 47(5), 2005.
6. Lee, D. and Cheng, C., Renewable and sustainable energy reviews, energy savings by energy management systems: A review, Vol. 56, pp. 760–777, April 2016.

11 Cogeneration Systems

11.1 COMBINED HEAT AND POWER

Combined heat and power (CHP) is not a technology but an approach to applying various technologies. This approach generates electricity and useful thermal energy in a single, integrated system. Heat that is normally wasted in conventional power generation is recovered as useful energy, avoiding the energy losses that are incurred when heat and power are generated separately. CHP projects can offer substantial economic, environmental, and energy benefits to their owners, including reduced overall energy costs, improved energy system reliability, and reduced thermal energy consumption.

While the conventional method of producing heat and power separately has a typical combined efficiency of 45%, CHP systems often have total efficiencies of 70% to 80%. This increased efficiency reduces or eliminates the purchase of energy from off-site generating sources such as an electric utility and consequently saves on utility costs. Transmission losses associated with traditional grid-sourced power are also avoided.

A highly efficient CHP system uses less input fuel to produce the same amount of output energy as the conventional generation of heat and power. The conventional system has two parts: the power station that generates electricity and the boiler that generates heat. Most of the energy in power station fuel is lost in the conversion to electricity. Although the boiler may be able to convert up to 85% of the energy in its fuel into steam or another form of thermal energy, this efficiency cannot make up for the inefficiency of the power station generator. In contrast, the CHP approach minimizes overall conversion losses by using a single system to generate both electricity and heat.

CHP is a versatile and flexible energy resource. As a system, it can be powered by engines, turbines, fuel cells, or other prime movers. A variety of fuels can be used to feed the system, including natural gas, biomass, diesel, and waste energy streams. As a scalable power source, a CHP system can be sized to meet energy demand loads ranging from that of a small commercial building to that of a large industrial facility. Such systems can be deployed in either existing or new buildings. They can be good investments and provide great benefit when appropriately designed to meet local needs.

CHP can comprise a number of different equipment types and use a number of different fuels.(1) Each installed project is a combination of many different project elements. These elements are chosen based upon the unique constraints and challenges presented by each potential CHP host site, such as cost limitations, space availability, local emissions profiles, and noise concerns. These systems are most often classified by the type of machine or engine that creates energy from the fuel or fuels (referred to as the prime mover) and generally falls into one of five categories: reciprocating engines, fuel cells, microturbines, steam turbines, and combustion

DOI: 10.1201/9781032692777-11

turbines. However, commercial real estate applications are generally limited to the first three categories, with reciprocating engines being the more popular by far.

1. *Reciprocating (combustion) engines*, which can be sized up to about 7 MW output, are the most common type of prime mover used in commercial building applications. A facility may choose a reciprocating engine for its starting speed and ability to quickly reach the desired power level, as well as for its reliability. A 100 kW unit, for example, operating at capacity will consume (input) approximately 1,150 standard ft^3 per hour (SCFH) of natural gas or 11.5 SCFH/kW output. Heat recovery per kWh of electricity produced typically ranges between 6,000 and 6,500 Btu/hr per kW. Preventive maintenance (e.g., filter, coolant, spark plugs) typically is recommended every 500–2,000 hours. For natural gas-fired engines (the most common input fuel source) and an 1,800 rpm engine speed, a top-end minor overhaul (cylinder head and turbocharger rebuild) is generally recommended between 8,000 and 15,000 hours. A major overhaul is typically needed every 30,000–36,000 hours. The time between overhauls is a function of the engine rpm, with the time increasing as engine speed decreases. Figure 11.1 shows a reciprocating CHP unit, with an operating schematic presented in Figure 11.2. Figure 11.3 provides a typical installation.
2. *Fuel cells* create electricity using an electrochemical reaction process, concurrently producing heat byproducts such as hot water and hot air. They are similar to batteries in that they both produce direct current (DC) through an electrochemical reaction without direct combustion of a fuel source. However, where a battery delivers power from a finite amount of stored energy, fuel cells can operate indefinitely provided that a fuel source is continuously supplied. Fuel cells have recently been used in the hospitality and residential sectors, utilizing heat for space conditioning and for general potable hot water needs in residential units and hotel rooms. Fuel cells can be a relatively expensive investment but are known for their clean operating characteristics, high efficiency, and quiet operation. Fuel input is approximately 0.1 standard cubic feet per minute (SCFM) of natural gas (based on HHV) per rated kW. Useful (recovered) heat is typically in the 1,500–2,600 Btu/hr per kW rated capacity. Figure 11.4 shows a fuel cell plant with heat and electricity production and Figure 11.5 a fuel cell installation.
3. *Microturbines* are small, often modular combustion turbines with availabilities nearing 500 kW. They can run on a variety of fuels and resemble small jet engines. These turbines typically offer lower emissions than other types of CHP equipment and contain fewer moving parts, which can be advantageous in some applications. Fuel input is approximately 0.011–0.013 MM Btu/hr of natural gas (based on HHV) per rated kW. Useful heat is typically in the 4,000–6,000 Btu/hr per kW rated capacity. Figure 11.6 shows a schematic of a microturbine unit, and Figure 11.7 shows a microturbine flow chart with electricity and heat production. Figure 11.8 shows a microturbine installation.

FIGURE 11.1 75 kW CHP unit. (*Source: Courtesy of Tecogen.*)

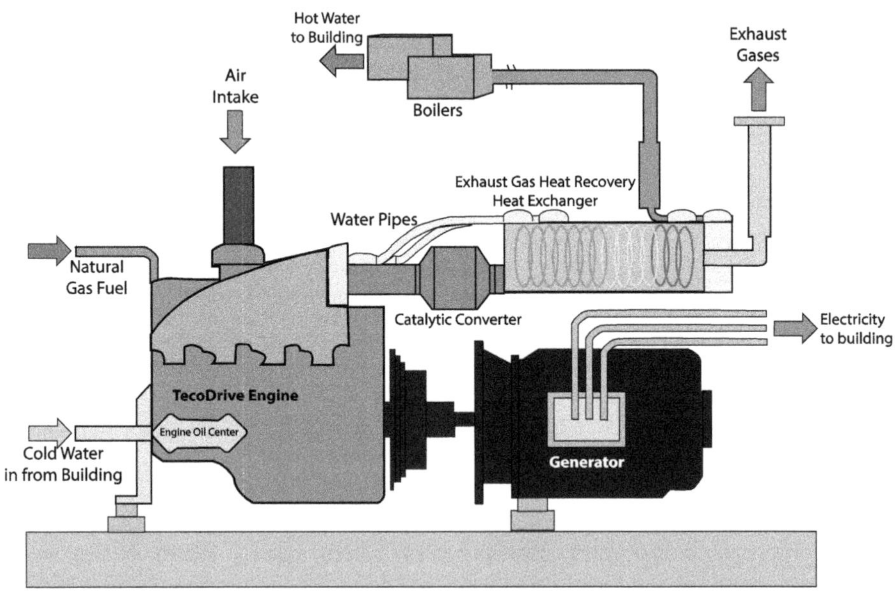

FIGURE 11.2 Operating schematic of CHP unit. (*Source: Courtesy of Tecogen.*)

FIGURE 11.3 Reciprocating engine CHP installation at an office building. (*Source: Courtesy of Tecogen.*)

The other two categories of CHP systems include:

4. *Steam turbines* that use boiler-generated steam to generate electricity or other types of energy. Steam turbines can run on a variety of fuels and can range in size from 50 kW to 250 MW.

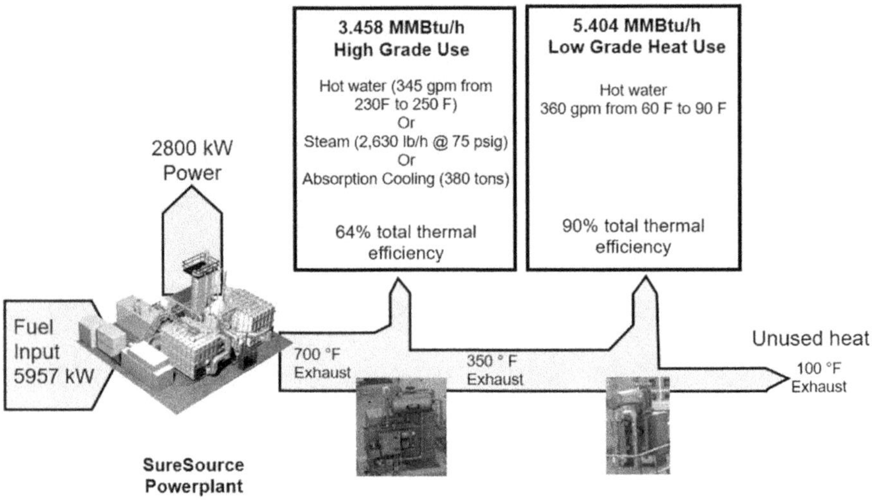

FIGURE 11.4 Fuel cell plant with heat and electricity production schematic. (*Source: Courtesy of Fuel Cell Energy.*)

FIGURE 11.5 Fuel cell installation. (*Source: Courtesy of Fuel Cell Energy.*)

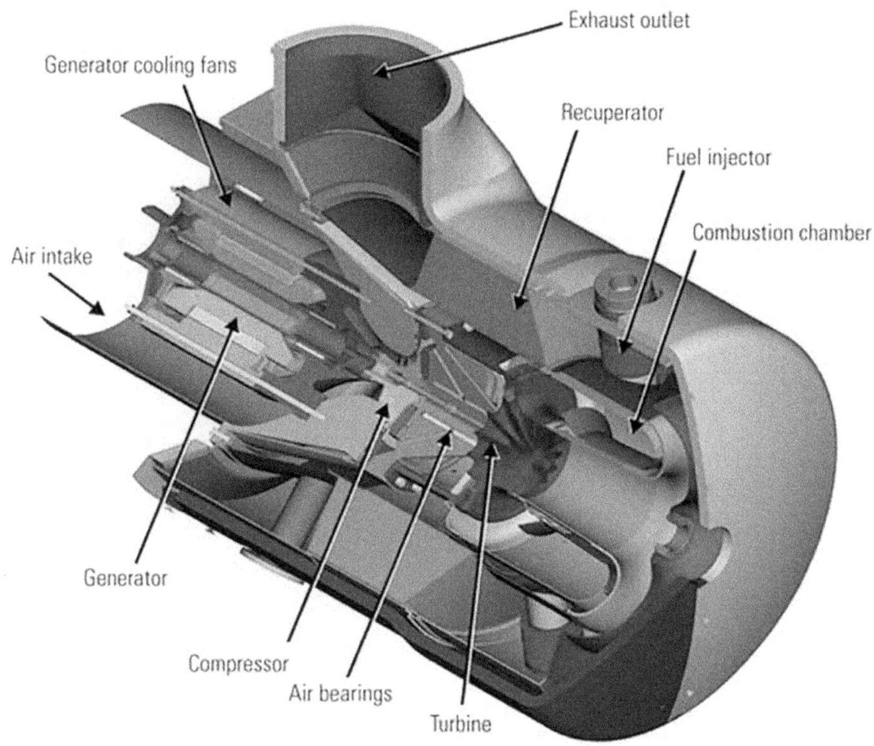

FIGURE 11.6 Microturbine unit schematic. (*Source: Credit: U.S. DOE, CHP Technology Fact Sheet Series, Microturbines, Capstone Turbine Corp. Graphic.*)

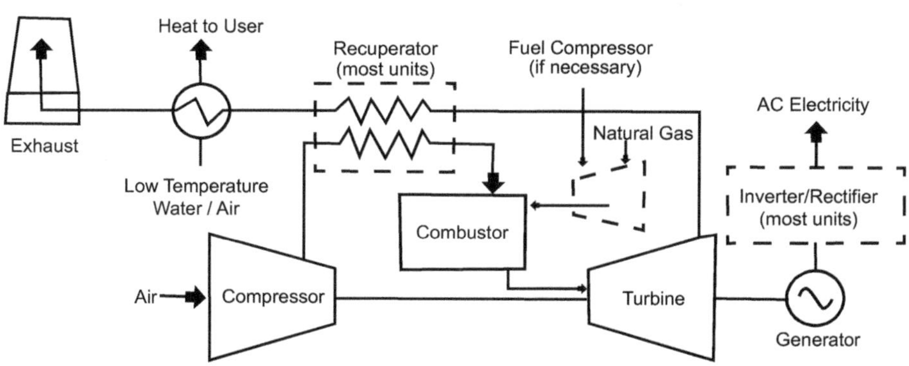

FIGURE 11.7 Microturbine flow diagram with electricity and heat production.

FIGURE 11.8 Microturbine installation.

5. *Combustion turbines*, which can burn a variety of different gases and are typically used in applications that require high-pressure and high-temperature steam. These turbines can be sized up to well over 100 MW and are often used by utilities interested in deploying CHP.

Fuel availability and long-term price forecasts often dictate which prime mover technologies are considered and ultimately chosen at a given site. CHP systems are often operated on fuels such as natural gas purchased from a local utility or purchased as commodities from vendors. Some prime movers can also run on waste heat, waste pressure, or waste fuel streams from industrial processes. Biomass or biogas fuels can also be used when municipal waste management activities, forest management activities, or local agricultural waste streams are located in close proximity. One of the benefits of CHP solutions is that they can take advantage of multiple fuel sources to create useful outputs from what would have otherwise become waste or pollution. These systems offer considerable flexibility and can meet a wide variety of energy needs.

11.2 PERFORMANCE METRICS

CHP systems can satisfy a variety of commercial end-users, providing heat and power to hospitals, schools, university campuses, hotels, nursing homes, office buildings, and apartment complexes. A comparison of those systems that have been installed in commercial buildings is presented in Table 11.1.(2) All performance and cost characteristics are typical values. Costs have been updated to 2022 and will vary depending on site-specific conditions and regional variations.

In Table 11.1, a key metric is the thermal output in Btu of recovered heat per kWh of electricity produced. The kWh produced would be the system capacity (kW) multiplied by the hours per year the system would be operating times the availability factor (which typically ranges between 95% and 98%).

Buildings with CHP systems can generate their own electricity and make use of the waste heat; for example, producing steam or hot water that is typically used for space heating, domestic hot water or absorption (chiller) cooling. With overall efficiencies in the 60%–80% range (i.e., 60%–80% of the fuel input energy is recovered), CHP systems use energy much more efficiently than conventional utility power plants that have an average efficiency of 35%–38%. They can take advantage of less expensive natural gas and low-carbon fuels like biogas, enabling significant efficiency improvement and greenhouse gas reduction when compared to combustion-based power generators.

TABLE 11.1
Comparison of Typical CHP Characteristics for Systems Installed in Commercial Buildings

Characteristic	Reciprocating Engine	Microturbine	Fuel Cell
Typical Size Range	10 kW–10 MW	30 kW–330 kW	5 kW–2.8 MW
Electric Efficiency (HHV), %*	30–42	25–29	38–42
Overall Efficiency (HHV), %*	77–83	64–72	62–75
Total Installed Cost, $/kW	$2,100–$4,400	$3,700–$4,800	$6,800–$8,800
O&M Cost, $/kWh	$0.013–0.036	$0.012–0.024	$0.046–0.055
Thermal Output, Btu/hr/kW	2,900–6,500	4,400–6,400	2,200–2,600

*Efficiencies are based on the higher heating value (HHV) of natural gas, i.e., 1,090 Btu/ft^3.

Example Problem 11–1

A large office building with a data center has a minimum electricity need (24/7, 52 weeks/year) of 640,000 kWh/yr and a minimum year-round heat requirement (hot water) of 3,700 MM Btu/yr. The building owner would like to install a reciprocating engine CHP unit designed to meet these minimum requirements. What size unit would meet this objective? How much natural gas would be required? What would the overall efficiency of the unit be?

Hrs/yr operating (assume 95% availability)	8,322 hr/yr
kW size:	640,000/8,322 = 77 kW
Assuming a recoverable thermal output of 6,500 Btu/kWh:	
Heat recovered:	640,000 × 6,500/1,000,000 = 4,160 MM Btu/yr
Natural gas input, assuming 11.5 SCFH/kW:	= 11.5 × 77 × 8,322 = 7,369,000 SCF/yr
Assuming natural gas at 1,090 Btu/SCF:	8,032 MM Btu/yr
Ratio heat recovered to heat input:	52%
Overall unit efficiency:	
Heat recovered:	4,160 MM Btu/yr
Electricity produced:	2,184 MM Btu/yr
Overall efficiency:	= (4,160 + 2,184)/8,032 = 79%

A 77 KW unit would likely meet the minimum heat and electricity requirements of the building and use approximately 74,000 CCF of natural gas per year.

Example Problem 11–2

Assuming an electricity cost of $0.19/kWh and a natural gas cost of $13/MM Btu, what would the estimated energy cost savings be for the unit in Example Problem 11–1? What would the simple payback time be if the total installed cost of the unit is estimated to be $4,300 per KW capacity? What would be the lifecycle cost?

Utility electricity savings per year:	640,000 kWh/yr × $0.19/kWh = $121,600
Savings associated with not having to produce 4,160 MM Btu/yr of hot water in an 80% efficient gas-fired boiler; fuel input = 4,160/0.8 = 5,200 MM Btu/yr;	
Gas cost that would be saved per year:	$67,600
Total savings:	$189,200
Cost of CHP gas input (8,032 MM Btu/yr natural gas):	$104,416
Estimated net energy savings:	$84,784 first year savings

Note: There may also be a demand savings as well, but this will depend on the unit's anticipated and non-anticipated downtime, the existence of ratchet provisions, and so on. As such, it will not be included in this analysis.)

Estimated total project cost:	77 kW × $4,300/kW = $331,100
Simple payback:	$331,100/$84,784 = 3.9 years

The lifecycle cost would be a better indicator of how attractive the investment is. Assuming a 20-year CHP unit life, a 3% average annual energy cost escalation, and a 5% discount rate:

Year	Savings	PV of Savings
0	Initial Investment: $331,100	
1	$84,784	$80,747
2	$87,328	$79,209
3	$89,947	$77,700
4	$92,646	$76,220
5	$95,425	$74,768
6	$98,288	$73,344
7	$101,237	$71,947
8	$104,274	$70,576
9	$107,402	$69,232
10	$110,624	$67,913
11	$113,943	$66,620
12	$117,361	$65,351
13	$120,882	$64,106
14	$124,508	$62,885
15	$128,243	$61,687
16	$132,091	$60,512
17	$136,053	$59,360
18	$140,135	$58,229
19	$144,339	$57,120
20	$148,669	$56,032
Total PV Savings:		$1,353,558

Savings-to-investment ratio (SIR) = 4.1. The SIR ratio would make this appear to be a very attractive investment (refer to Chapter 16). However, a more thorough analysis would include the cost of annual maintenance, the cost of overhauls when necessary, and any utility demand charges that might be incurred when the unit is not operating and electricity must be taken from the grid. A full service maintenance contract on a reciprocating engine CHP unit (where the equipment supplier is completely responsible for the unit) might run on the order of $1.60–$1.70 per operating hour.

Example Problem 11–3

A YMCA facility installed a CHP diesel engine generator. Management wants to know if the unit is achieving its minimum guaranteed 70% overall efficiency. Data is collected for a 1-hour test period during which 4 gallons of fuel oil is consumed

(140,000 Btu/gallon) and 50 kWh of electricity is produced. The heat exchanger recovered 60% of the wasted heat during the trial period.

Heat Input = 4 gal x (140,000 Btu/gal) x (1/3,412 Btu/kWh)

= 164 kWh

Output = 50 kWh

Electrical Efficiency = output/input = 50/164 = 30.5%

Wasted heat = 164 kW − 50 kW = 114 kW or 389,000 Btu/hr

Recovered heat = 0.6 x 389,000 Btu/hr = 233,000 Btu/hr or 68.2 kW

Recovered energy = electricity + recovered waste heat

= 50 kW + 68.2 kW

= 118.2 kW

Overall efficiency = 118.2 kW / 164 kW = 72%

The minimum guaranteed overall efficiency was met and exceeded during this 1-hour test period.

11.3 LIMITATIONS

There are limitations to the use of reciprocating engine, fuel cell, and microturbine CHP systems in commercial buildings compared to the use of conventional systems.

1. Systems are generally more economically viable in larger buildings that have larger amounts of electricity and natural gas consumption.
2. Most commercial building owners and building maintenance personnel are not familiar with CHP systems.
3. A year-round, reliable use for the waste heat produced is needed.
4. CHP projects in commercial buildings can be expensive. As such, projects need to be carefully evaluated with respect to fundamental economic and technical criteria. Such evaluations are most often performed by consulting firms who must conduct a comprehensive energy audit of the building and investigate special state and utility permitting requirements. This can further add to the upfront cost.

5. Unlike electricity delivered to a building, CHP systems typically require specialized maintenance that is performed by the manufacturer. Such maintenance contracts can be expensive.
6. Grid-delivered electricity is available on demand at all times. CHP system availability may range from 95% to 98%. CHP systems may go offline for whatever reason, such as annual scheduled maintenance, unscheduled maintenance, system component replacement, or system overhauls. When a system is down for whatever reason, a backup electricity source is required, which is typically the local utility, and this may incur a demand charge by the utility in addition to the generation charge. Moreover, the impact on potential demand cost savings can be dramatic. If a unit trips out or shuts down for maintenance (even for an hour), ratchet provisions, if applicable, may kick in and continue for the next 12 months even though there would be significant demand reduction while the unit is operating.
7. CHP systems that rely on fossil fuels, typically natural gas, may be problematic in the future with the market's move toward electrification/decarbonization. The use of hydrogen and biofuels may be a more prudent but expensive direction in which to move.

REFERENCES

1. U.S. EPA, Combined heat and power partnership, What is CHP, www.epa.gov/chp/what-chp
2. U.S. DOE, Combined heat and power technology, fact sheet series, overview of CHP technologies, www.energy.gov/chp.

12 Solar PV Renewable Energy Systems

Photovoltaic (PV) technologies, more commonly known as solar panels, are often being used as a complementary energy improvement measure for commercial buildings, particularly in support of the switch to renewable energy sources and decarbonization projects.

12.1 EQUIPMENT DESCRIPTION

PV materials and devices convert sunlight into electrical energy. A single PV device is known as a cell. An individual PV cell is usually small, typically producing about 1 or 2 watts of power. These cells are made of different semiconductor materials and are often less than the thickness of four human hairs. To withstand the outdoors for many years, cells are sandwiched between protective materials in a combination of glass and/or plastics.

To boost the power output of PV cells, they are connected together in chains to form larger units known as modules or panels. Modules can be used individually, or several can be connected to form arrays. One or more arrays is then connected to the electrical grid as part of a complete PV system (see Figure 12.1). Because of this modular structure, PV systems can be built to meet almost any electric power need, small or large.

PV modules and arrays are just one part of a PV system. Systems also include mounting structures that point panels toward the sun, along with the components that take the direct current (DC) electricity produced by modules and convert it to the alternating current (AC) electricity to meet the power requirements of the building.

PV arrays must be mounted on a stable, durable structure that can support the array and withstand wind, rain, hail, and corrosion over decades. These structures tilt the PV array at a fixed angle determined by the local latitude, orientation of the structure, and electrical load requirements. To obtain the highest annual energy output, modules in the Northern Hemisphere are pointed due south and inclined at an angle equal to the local latitude. Rack mounting is currently the most common method because it is robust, versatile, easy to construct and install.

For PV arrays mounted on the ground with solar tracking, the tracking mechanisms automatically move panels to follow the sun across the sky, which provides more energy and higher returns on investment. One-axis trackers are typically designed to track the sun from east to west. Two-axis trackers allow for modules to remain pointed directly at the sun throughout the day. Tracking involves more upfront costs and requires more maintenance.

Inverters are used to convert the DC electricity generated by solar PV modules into AC electricity, which is used for local transmission of electricity. PV systems

DOI: 10.1201/9781032692777-12

FIGURE 12.1 112 kW roof-top solar PV installation.

either have one inverter (string inverter) that converts the electricity generated by all the modules, or microinverters that are attached to each individual module. A string inverter is generally less expensive and can be more easily cooled and serviced when needed. Its typical lifetime is approximately 11 years, which means it has to be replaced mid-lifetime of the PV solar system. This will add to the cost. Microinverters, on the other hand, allow for independent operation of each panel, which is useful, for example, if some modules might be shaded or dirty. Microinverters have an expected lifetime more compatible with the EUL of a PV solar system, which is 20–25 years.

Batteries allow for the storage of solar photovoltaic energy so that it can be used at night or when weather elements keep sunlight from reaching PV panels.

12.2 ELECTRICITY PRODUCTION

There are many issues that must be addressed to determine the feasibility of a solar PV installation. These include:

- Understanding the client's objectives.
- Knowing the desired location for the system (e.g., roof-top, canopy, ground) and the potential for shading.
- Gaining familiarity with the building structure (for roof-top units), the electrical system, and potential grid-connectivity configuration.

- Carefully analyzing the building's electricity consumption, energy use profile, and electricity cost (rate structure).
- Being aware of any immediate future planned building upgrades that could impact building energy consumption.
- Determining the available area usable for solar panel installation (preferably without shading or other interferents).
- Understanding utility and permitting requirements.

A solar PV contractor will typically conduct a feasibility study and a shading study (if needed). However, a preliminary solar system estimate can be made as part of a building energy improvement upgrade project. Assuming a roof-top installation for an existing commercial building and knowing the building's electricity consumption (kWh/month and kWh/year), it is possible to estimate the size of the solar PV system using the (kWh electricity production per kW) values in Table 12.1. This is

TABLE 12.1
Typical Roof-Top Fixed Array Solar PV System Electricity Production per kW by Major City

State	City	kWh/kW*
Alabama	Montgomery	1,470
Alaska	Anchorage	1,040
Arizona	Phoenix	1,710
Arkansas	Little Rock	1,380
California	Los Angeles	1,660
	San Francisco	1,550
Colorado	Denver	1,600
Connecticut	Hartford	1,270
Delaware	Wilmington	1,360
Florida	Miami	1,590
	Orlando	1,530
Georgia	Atlanta	1,430
Hawaii	Honolulu	1,630
Idaho	Boise	1,450
Illinois	Chicago	1,310
	Springfield	1,360
Indiana	Indianapolis	1,340
Iowa	Des Moines	1,390
Kansas	Topeka	1,440
Kentucky	Frankfurt	1,330
Louisiana	Baton Rouge	1,470
	New Orleans	1,480
Maine	Augusta	1,260
Maryland	Annapolis	1,400
Massachusetts	Boston	1,300
Michigan	Lansing	1,280

TABLE 12.1 (*Continued*)
Typical Roof-Top Fixed Array Solar PV System Electricity Production per kW by Major City

State	City	kWh/kW*
Minnesota	St. Paul	1,340
Mississippi	Jackson	1,420
Missouri	Jefferson City	1,340
Montana	Helena	1,320
Nebraska	Lincoln	1,440
Nevada	Las Vegas	1,720
New Hampshire	Concord	1,290
New Jersey	Trenton	1,330
New Mexico	Santa Fe	1,740
New York	Albany	1,240
	Manhattan	1,310
North Carolina	Raleigh	1,390
North Dakota	Bismarck	1,360
Ohio	Columbus	1,320
Oklahoma	Oklahoma City	1,540
Oregon	Portland	1,130
Pennsylvania	Harrisburg	1,300
Rhode Island	Providence	1,300
South Carolina	Columbia	1,440
South Dakota	Pierre	1,400
Tennessee	Nashville	1,350
Texas	Austin	1,490
	Dallas	1,505
Utah	Salt Lake City	1,470
Vermont	Montpelier	1,210
Virginia	Richmond	1,370
Washington	Seattle	1,100
West Virginia	Charleston	1,260
Wisconsin	Madison	1,310
Wyoming	Cheyenne	1,560

*Estimates from PVWatts; crystalline silicon modules, 15% efficiency, fixed open rack arrays, 14.08% system loss, 20-degree tilt, 180-degree azimuth (south facing); NREL National Solar Radiation Database; TMY conditions for solar radiation; no shading.

accomplished by first estimating the electricity production (kWh) that needs to be provided by the solar system. Remember that this should reflect any energy savings resulting from energy efficiency upgrades accompanying the project; that is, adjusting the electricity consumption values to reflect the electricity savings expected from the installation of energy efficient measures, such as a high efficiency air conditioning system or energy management system. If this is not done, the solar PV system may be oversized and could adversely impact the economics. Once the design kWh is determined, the appropriate kWh/kW value is selected from Table 12.1 and the kW size can be estimated for a fixed array (non-tracking) system. Tracking PV systems often can produce 25%–50% more kWh of electricity per kW. The estimated system size should then be laid out on the roof to be sure it can fit properly. Typical power densities today are 14–19 watts-DC/roof ft^2. Hence, to provide 150,000 kWh/yr in CT, a 120 kW system would be appropriate and need approximately 6,300 ft^2 of roof area. Assuming the roof area is adequate for installation of the panels, a more comprehensive electricity production estimation can be made (e.g., using PVWatts).

12.2.1 PVWatts

PVWatts(1) is a useful tool for obtaining an estimate of energy production from a photovoltaic system for a specific location (address). PVWatts hides much of the complexity of accurately modeling PV systems from the user by making several assumptions about the type, configuration, and operation of the system. Consequently, the results should be interpreted as being a representative estimate for a similar actual system operating in a year with typical (TMY) weather. The error may be as high as ±10% for energy totals in any year and ±30% for energy totals in any single month.

PVWatts requires a minimal set of PV system specifications. The calculator estimates the monthly and annual electricity production using an hour-by-hour simulation over a period of 1 year. To represent the system's physical characteristics, PVWatts requires values for six inputs:

- DC system size (default: 4 kW_{DC}).
- Module type (default: standard poly- or mono-crystalline silicon modules with efficiencies in the range of 14%–17%).
- Array type (default: fixed open rack).
- System losses (default: 14%).
- Array tilt angle (default: site latitude).
- Array azimuth angle (default: 180 degrees in Northern Hemisphere).

PVWatts uses a set of assumptions that are appropriate for flat-plate photovoltaic systems with typical crystalline silicon or thin-film modules. Advanced inputs can include the DC/AC ratio (default is to a ratio of 1.1, i.e., a 4 kWDC system would result in a 3.63 kW AC inverter rating) and the inverter efficiency (default is to 96%).

Example Problem 12–1

An office building in Connecticut currently uses 330,000 kWh/yr (weather-normalized). The building manager will be upgrading the existing air conditioning system, which should save approximately 19,750 kWh/yr. The building owner also wants to add a solar PV system. What size unit would be needed for the solar PV system?

The kWh the solar PV system should produce would be approximately 330,000− 19,750 = 310,250 kWh/yr.

Assume, for estimating purposes to not oversize the system, that the design electricity production of the solar PV system is 90% of the total production rate. This can account for variability in the air conditioning system energy savings, variability in electricity production by the solar PV system, weather variability at the building location, solar modeling uncertainties, and so on.

$$0.9 * 310{,}250 = 280{,}000 \text{ kWh/yr}$$

In Connecticut, the electricity production rate of solar PV systems is approximately 1,270 kWh/kW DC (see Table 12.1):

$$\begin{aligned}&\text{Solar PV system size, kW DC}\\&= \left(280{,}000 \text{ kWh/yr}\right)/\left(1{,}279 \text{ kWh/kW DC}\right)\\&= 219 \text{ kW DC}\end{aligned}$$

Using PVWatts at the building's location (see below), a 219 kW solar PV system would produce 285,586 kWh/yr, although the output may range from 278,876 to 292,784 kWh per year.

PVWatts Output:

Month	Solar Radiation (kWh/m²/day)	AC Energy* (kWh)
January	2.91	16,562
February	4.2	21,319
March	4.87	26,241
April	5.42	27,163
May	5.64	28,298
June	6.23	29,522
July	6.35	30,753
August	5.78	27,982
September	5.11	24,750
October	3.96	20,613
November	3.33	17,562
December	2.64	14,821
Annual	4.7	285,586

*Output may range between 278,876 and 292,784 kWh per year.

12.3 SYSTEM LOSSES

Losses in the system that are not explicitly modeled are provided by the user as a percentage of DC energy. Losses represented by this number include the impacts of soiling, shading, snow cover, mismatches, wiring, connections, light-induced degradation, nameplate rating, system age, and operational availability. PVWatts uses a default total system loss of 14%, which includes a 2% loss for soiling and a 3% loss for shading. It does not include an inverter loss, which would need to be obtained, although a 4% loss would be a reasonable default value. A shading study should be performed if there is any shading on the panels. Details may be found in Reference 2.

12.4 ELECTRICITY DEMAND IMPACT

In many states, the demand charges ($/kW) can be 50%–60% of the entire electric bill. As such, being able to take advantage of any demand reduction resulting from solar PV could have a material impact on the economics of a solar PV system installation. Unfortunately, it is not possible at this time to predict with any degree of certainty how much a PV system will influence a building's peak demand charges. A PV system's impact on peak demand has wide variability (ranging from virtually no impact to as much as a 25% reduction in monthly peak demand). Demand reduction will be a function of the property type and use, daylight hours, and patterns in both energy consumption and PV production. A database does not yet exist to allow this level of analysis. The cost associated with any demand reduction will also be strongly dependent on the applicable utility rate structure. As such, the most conservative route to take in a preliminary analysis is to assume no peak demand reduction savings (assuming no energy storage system has been installed).

REFERENCES

1. https://pvwatts.nrel.gov/
2. Dobos, A.P., PVWatts version 5 manual, NREL technical report NREL/TP-6A20-62641, September 2014, www.nrel.gov/docs/fy14osti/62641.pdf

13 Energy-Saving Water Reduction Fixtures and Appliances

13.1 EQUIPMENT DESCRIPTION

The most common fixtures found in multifamily and select commercial buildings that when upgraded can reduce building energy consumption and at the same time reduce water usage include:

- Faucets (in kitchens and bathrooms)—hot water saved.
- Showerheads (in showers)—hot water saved.
- Washing machines—hot water saved.
- Dishwashers—hot water saved.
- Urinals—pumping energy saved (minimal).
- Cooling towers—pumping energy saved.
- Condensate recovery on steam boilers—fuel reduction.

EPA has prepared guides, U.S. EPA WaterSense Water Efficiency Management Guides, which provide guidance on the energy consumption impact when making upgrades to these systems. The guides may be found at www.epa.gov/watersense/water-score-multifamily-housing.

EPA also provides a water-use breakdown by end-uses for select commercial buildings(1) and best management practices to reduce water usage in commercial and institutional facilities.(2)

13.2 ENERGY (AND WATER) SAVINGS

Energy savings are associated with the reduction in hot water consumption (for those fixtures using hot water). The equations use assumptions identified in EPA WaterSense Water Efficiency Management Guides. Fannie Mae also relies on assumptions which in some cases are different than EPA's. Table 13.1 compares EPA's factors with default factors used by Fannie Mae for multifamily housing.

13.2.1 Kitchen Faucets

Energy savings are based directly on the reduction in energy associated with having to provide less hot water to the faucets.

DOI: 10.1201/9781032692777-13

TABLE 13.1
Default Water Usage (GPD) Assumptions for Common Fixtures

	EPA WaterSense	Fannie Mae Default
Kitchen Faucets	25 min/unit/day	4 min/person/day
Bathroom Faucets	0.6 gal saved/person/day*	2 min/person/day
Showerhead	7.8 min duration/person 0.7 showers/person/day	8 min/person/day
Toilet	5 flushes/person/day	5 flushes/person/day
Clothes Washer	0.3 loads/person/day	0.4 loads/person/day
Dishwasher	0.1 loads/person/day	0.2 cycles/person/day

*Assumes 2.2 gpm faucet replaced with a 1.5 gpm faucet.

Baseline water flow rates for kitchen faucets are:

- Construction date post-1998: 2.2 gpm.
- Construction date 1992–1998: 3 gpm.
- Construction date pre-1992: 2.5–7 gpm.

Kitchen faucets today typically can reduce water flow rates to 1.0–1.5 gpm.

Assuming the replacement faucets at 1.5 gpm, the gallons of water saved annually per kitchen faucet may be estimated by Equation 13–1.

$$\text{Gal. of water saved/yr} = \left[(\text{Existing faucet gpm}) - (1.5\text{ gpm})\right] * (25.7\text{ minutes per faucet/day}) * (\text{no. of days/yr faucet in use}) \qquad (13\text{–}1)$$

For multifamily apartment buildings, assume the number of kitchen faucets equals the number of units (or households).

Since kitchen faucets use hot water, water savings will also save energy. Electricity savings (for electric water heaters) and natural gas savings (for natural gas-fired water heaters) may be estimated as follows:

Electric water heaters (kWh saved per year):

$$\text{kWh saved /yr} = \left[(\text{gal water saved/yr}) * (0.57\text{ fraction hot water}) * (0.183\text{ kWh to heat a gal of water})\right] / \left[\text{fractional efficiency of water heater}\right] \qquad (13\text{–}2)$$

If the electric water heater efficiency is not known, assume an efficiency of 98%.

Natural gas-fired water heaters (thousand cubic feet [kcf] of natural gas saved per year):

$$\text{kcf saved /yr} = \left[(\text{gal water saved/yr}) * (0.57\text{ fraction hot water}) * (0.00061\text{ kcf to heat a gal of water})\right] / \left[\text{fractional efficiency of water heater}\right] \qquad (13\text{–}3)$$

If gas-fired water heater efficiency is not known, assume an efficiency of 80%.

13.2.2 Bathroom Sink Faucets

Replacing faucets flowing at 2 gpm or more with WaterSense models rated at 1.5 gpm or less (but can be as low as 0.5–1 gpm) can reduce water consumption. The gallons of water saved per year may be estimated by Equation 13–4.

$$\text{Gal. water saved/yr} = (\text{gal. water saved/person/day}) * (\text{\# occupants}) * (\text{days/yr in use}) \quad (13\text{–}4)$$

The gallons of water saved per person per day is:

- 0.6 gallons/person/day assuming replacing a standard 2.2 gpm bathroom sink faucet with a 1.5 gpm faucet.
- 1 gallon/person/day replacing a standard 2.2 gpm faucet with a 1 gpm faucet.
- 1.3 gallons/person/day replacing a standard 2.5 gpm faucet with a 1 gpm faucet.

The number of occupants per unit in multifamily housing may be estimated using Equations 13–5 and 13–6.

Market Rate Housing

$$\text{Number of Occupants} = 1.49 + (0.45 * \text{Average No. Bedrooms per unit}) \quad (13\text{–}5)$$

Affordable Housing

$$\text{Number of Occupants} = 1.69 + (0.57 * \text{Average No. Bedrooms per unit}) \quad (13\text{–}6)$$

The average number of bedrooms (BR) per unit in multifamily housing is determined using Equation 13–7.

$$\text{Avg. No. BR / unit} = [(\text{No. units} * 1\ \text{BR/unit}) + (\text{No. units} * 2\ \text{BR/unit}) + (\text{No. units} * 3\ \text{BR/unit}) + \ldots\ldots] / \text{Total No. Units} \quad (13\text{–}7)$$

Since bathroom faucets use hot water, water savings will also save energy. Electricity savings (for electric water heaters) and natural gas savings (for natural gas-fired water heaters) may be estimated using Equations 13–8 and 13–9.

Electric water heaters (kWh saved per year):

$$\text{kWh saved /yr} = [(\text{gal water saved/yr}) * (0.57\ \text{fraction hot water}) * (0.183\ \text{kWh to heat a gal of water})] / [\text{fractional efficiency of water heater}] \quad (13\text{–}8)$$

If the electric water heater efficiency is not known, assume the efficiency 98%.

Natural gas-fired water heaters (thousand cubic feet [kcf] natural gas saved per year):

$$\text{kcf saved /yr} = \left[\left(\text{gal water saved/yr}\right) * \left(0.57 \text{ fraction hot water}\right) * \left(0.00061 \text{ kcf to heat a gal of water}\right)\right] / \left[\text{fractional efficiency of water heater}\right] \quad (13\text{–}9)$$

If gas-fired water heater efficiency is not known, assume an efficiency of 80%.

13.2.3 Showerheads

Older showerheads (pre-1992 showerheads typically were rated at 5 gpm; from 1992 to the present showerheads typically were rated at 2.5 gpm). Today they can be replaced with low flow models rated at 1.5–2.0 gpm. The gallons of water saved per year may be estimated using Equation 13–10.

$$\text{Gal. water saved/yr} = \left[\left(\text{gpm avg flow rate}\right) - \left(\text{gpm low flow rate}\right)\right] * \left[7.8 \text{ minutes/shower}\right] * \left[0.7 \text{ showers/person/day}\right] * \left[\#\text{ occupants showering}\right] * \left[\text{no. of days/yr in use}\right] \quad (13\text{–}10)$$

Since showers use hot water, water savings will also save energy. Electricity savings (for electric water heaters) and natural gas savings (for natural gas-fired water heaters) may be estimated using Equations 13–11 and 13–12.

Electric water heaters (kWh saved per year):

$$\text{kWh saved /yr} = \left[\left(\text{gal water saved/yr}\right) * \left(0.662 \text{ fraction hot water used}\right) * \left(0.183 \text{ kWh to heat a gal of water}\right)\right] / \left[\text{fractional efficiency of water heater}\right] \quad (13\text{–}11)$$

If the electric water heater efficiency is not known, assume the efficiency 98%.

Natural gas-fired water heaters (kcf natural gas saved per year):

$$\text{kcf saved /yr} = \left[\left(\text{gal water saved/yr}\right) * \left(0.662 \text{ fraction hot water}\right) * \left(0.00061 \text{ kcf to heat a gal of water}\right)\right] / \left[\text{fractional efficiency of water heater}\right] \quad (13\text{–}12)$$

If gas-fired water heater efficiency is not known, assume an efficiency of 80%.

Example Problem 13–1

A multifamily property owner is planning a small upgrade to his 20-year-old, all-electric building that consists of 16 one bedroom/one bath units, 43 two bedroom/one bath units, and 6 three bedroom/two bath units. The upgrade is to include replacement of all existing kitchen and bathroom faucets and showerheads. What water and energy savings might be expected?

First, it will be necessary to establish the existing (baseline) and new flow rates of the fixtures. Since the existing fixture flow rates were not provided, they will be estimated based on the building's age.

Fixture	gpm Existing	gpm New	Number
Kitchen Faucet	2.2	1.5	65
Bathroom Faucet	2.2	1.5	71
Showerhead	2.5	2	71

Kitchen Faucets (using Equation 13–1):

$$\text{Gal/yr water saved} = \left[(\text{Existing faucet gpm}) - (1.5\ \text{gpm})\right]$$
$$*25.7\ \text{minutes per faucet/day}$$
$$*\text{no. of days/yr faucet(s) in use}$$

$$= \left[(2.2 - 1.5)\text{gpm}\right] * 25.7\ \text{min/day} * 365\ \text{day/yr}$$

$$= 6{,}566\ \text{gal/yr per faucet}$$

$$= 426{,}790\ \text{gal/yr}$$

Using Equation 13–2 for an electric water heater:

$$\text{kWh/yr saved} = \left[(\text{gal water saved/yr}) * (0.57\ \text{fraction hot water})\right.$$
$$\left. * (0.183\ \text{kWh to heat a gal of water})\right] /$$
$$\left[\text{fractional efficiency of water heater}\right]$$

$$= \left[(426{,}790\ \text{gal/yr}) * 0.57 * 0.183\ \text{kWh/gal}\right] / 0.98$$

$$= 45{,}427\ \text{kWh/yr}$$

Bathroom Faucets (using Equation 13–4):

$$\text{Gal/yr water saved} = (\text{water saved/person/day})$$
$$*(\#\ \text{occupants}) * (\text{days/yr in use})$$

Assuming market rate housing (refer to Equation 13–5):

$$\# \text{ occupants/unit} = 1.49 + (0.45 * \text{Average No. Bedrooms per unit})$$

Use Equation 13–7 to get the average number of bedrooms per unit:

$$\text{Avg. No. Bedrooms/unit} = \left[(\text{No. units} * 1 \text{ BR/unit}) + (\text{No. units} * 2 \text{ BR/unit}) + (\text{No. units} * 3 \text{ BR/unit})\right] / \text{Total No. Units}$$

$$= \left[(16*1) + (43*2) + (6*3)\right] / 65$$

$$= 120/65$$

$$= 1.85$$

The number of occupants per unit would then be:

$$\# \text{ occupants/unit} = 1.49 + (0.45 * 1.85) = 2.3$$

The total number of occupants in the building is:

$$\text{Total No. Occupants} = 2.3 * 65 \text{ units} = 150$$

Since the bathroom faucet upgrade would be going from 2.2 to 1.5 gpm, the water saved per person per day = 0.6 gallons.

From Equation 13–4:

$$\text{Gal/yr water saved} = (0.6 \text{ gal/day/person}) * 150 \text{ occupants} * 365 \text{ days/yr}$$

$$= 32{,}850 \text{ gal/yr}$$

From Equation 13–8:

$$\text{kWh saved/yr} = \left[(\text{gal water saved/yr}) * (0.57 \text{ fraction hot water used}) * (0.183 \text{ kWh to heat a gal of water})\right] / \left[\text{fractional efficiency of water heater}\right]$$

$$= (32{,}850 * 0.57 * 0.183) / 0.98$$

$$= 3{,}497 \text{ kWh/yr}$$

Showerheads (using Equation 13–10):

$$\text{Gal/yr water saved} = \left[(\text{gpm existing flow rate}) - (\text{gpm new flow rate})\right] * \left[7.8 \text{ minutes/shower}\right] * \left[0.7 \text{ showers/person/day}\right] * \left[\# \text{ occupants showering}\right] * \left[\text{no. of days/yr in use}\right]$$

$$= (2.5 - 2) * 7.8 * 0.7 * 150 * 365$$

$$= 149{,}468 \text{ gal/yr}$$

From Equation 13–11:

$$\text{kWh saved/yr} = \left[(\text{gal water saved/yr}) * (0.662 \text{ fraction hot water used}) * (0.183 \text{ kWh to heat a gal of water})\right] / \left[\text{fractional efficiency of water heater}\right]$$

$$= (149{,}468 * 0.662 * 0.183) / 0.98$$

$$= 18{,}477 \text{ kWh/yr saved}$$

Summary

Fixtures	Water Saved gal/yr	Electricity Saved kWh/yr
Kitchen Faucets	426,790	45,427
Bathroom Faucets	32,850	3,497
Showerheads	149,468	18,477
Total	609,108	67,401

13.2.4 Low-Flow Toilets/Urinals

Replacing old toilets/urinals flushing at 1.6 gal/flush (gpf) or more (although pre-1992 toilets may flush as high as 3.5–5 gpf) with WaterSense-labeled models flushing at 1.28 gpf or less (as low as 0.8–1 gpf) can result in significant water savings. Assuming the replacement with a low flow unit at 1 gpf, the gallons per year of water savings may be estimated using Equation 13–13.

$$\text{Gal. water saved / yr} = \left[(\text{gpf}) - (1 \text{ gpf})\right] * (5 \text{ toilet uses / person / day}) * (\# \text{ occupants}) * (\text{no. of days / yr the toilet is in use}) \quad (13\text{–}13)$$

Assuming multifamily housing, Equations 13–5 or 13–6 would be used to estimate the number of occupants. The energy savings that might be realized would be associated with a reduction in water pumping energy, which is negligible.

13.2.5 Washing Machines/Laundries

In a multifamily building complex, the average resident with a washing machine in the unit washes more than five loads per week. Tenants using a common area laundry typically wash more than three loads per week. The estimated water use for various clothes washers is provided below:

	Gallons per Load
Baseline: Inefficient Residential Top-Loading	41
Baseline: Res. Conventional Top-Loading	37.8
ENERGY STAR Certified Res. Unit Top Loading	19.4
ENERGY STAR Certified Res. Unit Standard Front-Loading	14.8
ENERGY STAR Certified Res. Unit Compact Front-Loading	10.5
Conventional Commercial Top-Loading	23.8
Conventional Commercial Front-Loading	15.4
ENERGY STAR Certified Commercial Front Loading	12.6

Assuming an ENERGY STAR commercial front-loading washer, the gallons of water saved per year from washing machine replacement may be estimated by Equation 13–14.

$$\begin{aligned}\text{Gal. water saved/yr} = &\left[\left(\text{existing gal/load}\right) - \left(12.6\ \text{gal/load}\right)\right]\\ &* \left(0.3\ \text{loads/person/day}\right) * \left(\#\ \text{occupants}\right)\\ &* \left(\text{No. days in use/yr}\right)\end{aligned} \tag{13–14}$$

Assuming multifamily housing, Equations 13–5 or 13–6 would be used to estimate the number of occupants.

Since clothes washers also use hot water, water savings will also save energy. Electricity savings (for electric water heaters) and natural gas savings (for natural gas water heaters) may be estimated using Equations 13–15 and 13–16.

Electric water heaters (kWh saved per year):

$$\begin{aligned}\text{kWh saved /yr} = &\left[\left(\text{gal water saved/yr}\right) * \left(0.20\ \text{fraction hot water used}\right)\right.\\ &\left.* \left(0.183\ \text{kWh to heat a gal of water}\right)\right] /\\ &\left[\text{fractional efficiency of water heater}\right]\end{aligned} \tag{13–15}$$

If the electric water heater efficiency is not known, assume an efficiency of 98%.

Natural gas-fired water heaters (kcf natural gas saved per year):

$$\begin{aligned}\text{kcf saved /yr} = &\left[\left(\text{gal water saved/yr}\right) * \left(0.20\ \text{fraction hot water used}\right)\right.\\ &\left.* \left(0.00061\ \text{kcf to heat a gal of water}\right)\right] /\\ &\left[\text{fractional efficiency of water heater}\right]\end{aligned} \tag{13–16}$$

If gas-fired water heater efficiency is not known, assume an efficiency of 80%.

13.2.6 Dishwashers

The average dishwasher in 2017 used approximately 6 gallons per load of dishes, although older models (pre-2017) have used as much as 10 gallons per load. ENERGY STAR-certified standard dishwashers do not exceed 3.5 gallons per load, while compact models common in multifamily housing do not exceed 3.1 gallons per load. Assuming 3.1 gallons per load, the water savings from dishwasher replacement can be estimated using Equation 13–17.

$$\begin{aligned}\text{Gal. water saved/yr} = & \left[(\text{gallons/load}) - (3.1\ \text{gal/load})\right] \\ & * (0.1\ \text{loads/person/day}) * (\#\ \text{occupants}) \\ & * (\text{No. of days in use/yr})\end{aligned} \qquad (13\text{–}17)$$

Assuming multifamily housing, Equations 13–5 or 13–6 would be used to estimate the number of occupants.

Since dishwashers use hot water, water savings will also save energy. Electricity savings (for electric water heaters) and natural gas savings (for natural gas-fired water heaters) may be estimated using Equations 13–18 and 13–19.

Electric water heaters (kWh saved per year):

$$\begin{aligned}\text{kWh saved /yr} = & \left[(\text{gal water saved/yr}) * (1.0\ \text{fraction hot water used})\right. \\ & \left. * (0.183\ \text{kWh to heat a gal of water})\right] / \\ & \left[\text{fractional efficiency of water heater}\right]\end{aligned} \qquad (13\text{–}18)$$

If the electric water heater efficiency is not known, assume an efficiency of 98%.

Natural gas-fired water heaters (kcf natural gas saved per year):

$$\begin{aligned}\text{kcf saved /yr} = & \left[(\text{gal water saved/yr}) * (1.0\ \text{fraction hot water used})\right. \\ & \left. * (0.00061\ \text{kcf to heat a gal of water})\right] / \\ & \left[\text{fractional efficiency of water heater}\right]\end{aligned} \qquad (13\text{–}19)$$

If gas-fired water heater efficiency is not known, assume an efficiency of 80%.

13.2.7 Cooling Towers

Cooling towers can use significant quantities of water. They dissipate heat from circulating water that is used to cool chillers, air conditioning equipment, or other types of process equipment. Water leaving a cooling tower system must be made up. Make-up water consists of water lost to evaporation plus water lost in blowdown plus drift losses plus any water lost due to leaks or overflows.

The quantity of evaporation is not typically targeted for water efficiency, as it is responsible for the cooling effect. Improving energy efficiency within systems using cooling water will reduce the evaporative load on the tower, thus saving water in addition to energy. Regardless of cooling tower operating efficiency, approximately 1.8 gallons of water will be evaporated for every ton-hour of cooling. (Cooling ton-hours is equal to the kWh/yr used by the chiller system and the chiller kW/ton rated efficiency. Hence, if the chiller system uses 500,000 kWh/yr and the chiller is rated at 0.6 kW/ton, the chiller ton-hours equals 500,000/0.6 or 833,333 ton-hours. To get cooling tower ton-hours, multiply the chiller ton-hours by 1.25.)

Blowdown or bleed-off is performed to remove high concentrations of dissolved solids (e.g., calcium, magnesium, chloride, silica) from the cooling tower system. As the water evaporates, the dissolved solids remain behind, and the concentration of total dissolved solids (TDS) in the cooling tower water increases. The concentration of TDS is controlled by the blowdown and replacing it with make-up water.

Drift is the small quantity of water that can be carried from the cooling tower as mist or water droplets. Efficient drift eliminators and effective leak/overflow detection will minimize water losses from drift and leaks. As such, make-up water for a well-managed cooling tower is essentially only based on evaporation and blowdown rates.

If the system uses less water, less water pumping around the cooling tower will be required (e.g., to the tower, away from the tower), which will save some energy, although this generally will be insignificant.

13.2.8 Condensate Recovery on Steam Boilers

Steam boilers generate steam that can be distributed throughout the property. As this occurs, heat is transferred to the ambient environment and, as a result, some steam condenses to water. This condensate is either discharged to the sewer or captured and returned to the boiler for reuse. Installing a condensate recovery system is the most effective way to reduce both water use and energy consumption. Energy is saved because hot condensate returned to the boiler requires much less energy to reheat to produce steam again.

In addition to condensate losses, make-up water is also required to account for blowdown. Blowdown rates typically range from 4% to 8% of the make-up water flow rate. Proper control of blowdown water is critical to ensure efficient boiler operation and minimize make-up water use. Insufficient blowdown can lead to scaling and corrosion, while excessive blowdown wastes water, energy, and chemicals. Switching to an automatic blowdown control system such as a conductivity controller, for example, can reduce a boiler's energy use by 2%–5% and reduce blowdown by as much as 20%.

13.3 WATER USAGE ADJUSTMENT WHEN METERED DATA IS AVAILABLE

EPA WaterSense water usage assumptions are not unreasonable to use when no water meter data is available. However, many commercial and multifamily buildings pay for the water used, which dictates water metering. With water metering, the amount of water used in a building is known. The question then is how can water usage data be broken into end-use consumption (e.g., faucets, showerheads, etc.) for the existing fixtures and for the new upgraded fixture(s).

For residential housing, the Water Research Foundation[3] provides the end-use estimates in Table 13.2. Hence, if water usage in a multifamily apartment complex is known from meter data, end-use consumption can be estimated using the percentages in Table 13.2. I have found it reasonable to divide kitchen and bathroom faucet water consumption into 60% attributed to kitchen faucets and 40% to bathroom faucets. This would mean 11.4% of total water usage attributed to kitchen faucets and 7.6% to bathroom faucets. To get the water savings, the suggested methodology is as follows:

1. Obtain the total gallons of water used per year for the building from metered data (refer to the utility bill).
2. Calculate the existing (baseline) water consumption in gallon/year for each end-use.
3. Calculate the water savings fraction for the end-use (see Table 13.3).
4. Multiply the end-use water savings fraction by the results from step 2.
5. Estimate the energy savings using the appropriate fixture equations presented previously.

Example Problem 13–2

The water utility bills for an apartment complex in the year prior to the fixture upgrade totaled 1,955,200 gal/yr. The building uses natural gas for heating. The fixture upgrade included:

Fixture	gpm Existing	gpm New
Kitchen Faucets	2.2	1.5
Bathroom Faucets	2.2	1.5
Showerheads	2.5	2

Estimate the water and energy savings.

Assuming the end-use distributions in Table 13.2, the assumed faucet breakdown (60/40 Kitchen/Bathroom), and the savings equations in Table 13.3, the following table can be prepared:

Fixture	gpm Existing	gpm New	End-Use %	End-Use gal/yr	% Water Savings	gal Water Saved/yr	kcf Gas Saved/yr
Kitchen Faucets	2.2	1.5	11.4%	222,893	31.8%	70,880	30.81
Bathroom Faucets	2.2	1.5	7.6%	148,595	31.8%	47,253	20.54
Showerheads	2.5	2	20.0%	391,040	20.0%	78,208	39.48
					Total	196,341	90.82

TABLE 13.2
Residential End-Use Water Consumption Estimates

End-use	% of Water Use per Household
Kitchen/Bath Faucets	19%
Shower	20%
Dishwasher	1%
Toilet/Urinal	24%
Clothes Washer	17%
Other	19%

TABLE 13.3
Percent Water Savings for Common Fixtures in Apartment Buildings with Water Meters

End-Use	Equation for Water Savings Percent (Fraction)
Kitchen Faucet	(Existing faucet gpm – Low flow faucet gpm)/Existing faucet gpm
Bathroom Faucet	(Existing faucet gpm – Low flow faucet gpm)/Existing faucet gpm
Showerheads	(Existing shower gpm – Low flow shower gpm)/(Existing shower gpm
Dishwasher	(Existing gal/load – Low flow gal/load)/Existing gal/load
Toilet/Urinal	(Existing gal/flush – Low flow gal/flush)/Existing gal/flush
Clothes Washer	(Existing gal/load – Low flow gal/load)/Existing gal/load

REFERENCES

1. U.S. EPA, WaterSense guide for water use in commercial buildings, www.epa.gov/watersense/types-facilities.
2. U.S. EPA, WaterSense at work: Best management practices for commercial and institutional facilities, EPA 832-F-12-034, October 2012, https://www.epa.gov/sites/default/files/2017-02/documents/watersense-at-work_final_508c3.pdf
3. Water Research Foundation Report: Residential End-Uses of Water, Version 2, April 2016, https://www.circleofblue.org/wp-content/uploads/2016/04/WRF_REU2016.pdf

14 Miscellaneous Energy Efficiency Replacements

14.1 MOTORS

Motor efficiency simply describes how well a motor converts electrical energy into mechanical energy. A highly efficient motor converts a large proportion of the electrical energy into mechanical energy, allowing low operating costs and smooth speed and operation over the required range, with minimum vibration.

There are three basic types of motors used in HVAC applications:

1. Open drip proof (ODP): ODP motors are constructed with openings in the motor casing that provide a cooling air path to the motor windings. ODP motors are designed for dry applications and, if installed outdoors, must be weather protected.
2. Totally enclosed fan-cooled (TEFC): The TEFC motor shaft extends from the motor casing to drive a cooling fan installed within a shroud on the non–drive end of the motor.
3. Totally enclosed air-over (TEAO): The TEAO motor windings are enclosed in a moisture-tight enclosure. The casing is provided with large fins designed to radiate the motor heat to an airstream passing over them.

Using an "energy efficient" motor in lieu of a "standard" motor can typically reduce fan energy consumption by 5%–7%, depending on the motor size. A "premium efficiency" motor can reduce energy consumption by an additional 1%–3%.

Equations 14–1 and 14–2 are the governing equations.

$$\text{Electric Power Input}\left(\text{kW}\right) = \text{Shaft Power Delivered}\left(\text{kW}\right)/\eta_M \tag{14–1}$$

$$\text{Input Power at Full Rated Load, kW} = 0.7457 * \text{hp} * \text{LF}/\eta_M \tag{14–2}$$

where hp = motor horsepower (nameplate rating of the motor)
η_M = motor efficiency at full rated load
LF = load factor (ratio of avg. operating load to peak load, default at 80%)
0.7457 = conversion factor (1 kW = 1.341 hp or 0.7457 kw/hp)

Most electric motors are designed to run at 50% to 100% of rated load. Maximum efficiency is usually near 75% of rated load. A motor's efficiency tends to decrease dramatically below about 50% load. Overloaded motors can overheat and lose efficiency. Many motors are designed with a service factor that allows occasional overloading. Service factor is a multiplier that indicates how much a motor can be overloaded under ideal ambient conditions. For example, a 10-hp motor with a 1.15

DOI: 10.1201/9781032692777-14

TABLE 14.1
Typical Efficiency for 15- to 20-Year-Old Motors

1,800 rpm TEFC Motor HP	Efficiency, %
1	82.5
5	87.5
10	89.5
20	91

service factor can handle an 11.5-hp load for short periods of time without incurring significant damage. Although many motors have service factors of 1.15, running the motor continuously above rated load reduces efficiency and motor life.

The NEMA definition of motor energy efficiency is the ratio of its useful power output to its total power input. Motor efficiency typically varies between 77% and 95%. Assuming a 15- to 20-year-old motor is being replaced (at the end of its useful life), its baseline efficiency would likely have been in the range of efficiencies as shown in Table 14.1. The portion of electric energy that is not converted to mechanical power is converted to heat, which is mostly unusable. For example, assuming that no transmission losses occur, a motor with an efficiency of 80% will draw an electrical power of 1/0.8 = 1.25 kW for each kilowatt of shaft power it delivers. If the motor is 95% efficient, it will draw 1/0.95 = 1.05 kW to deliver 1 kW of shaft work.

If a less energy efficient motor is replaced with a more energy efficient motor, the energy savings at full load would be defined by Equation 14–3.

$$\text{Energy Savings, kWh} = (P_{\text{Baseline}} - P_{\text{New}}) * \text{EFLH}_{\text{M}} \qquad (14\text{–}3)$$

or

$$\text{Energy Savings, kWh} = 0.7457 * \text{hp} * \text{LF} * \text{EFLH}_{\text{M}} * ((1/\eta_{\text{M-Baseline}}) - (1/\eta_{\text{M-New}})) \qquad (14\text{–}4)$$

where P_{Baseline} = power input, kW, at existing motor efficiency
P_{New} = power input, kW, for new high efficiency motor
EFLH_{M} = equivalent annual operating hours at full load (default HVAC fan motor, chiller water pump, and cooling tower pump hours of annual use by property type from the Connecticut TRM are shown in Table 14.2.[1])
$\eta_{\text{M-Baseline}}$ = standard motor efficiency (existing) at full load
$\eta_{\text{M-New}}$ = efficiency of premium (high efficiency) motor at full load

The peak demand savings can be calculated using Equation 14–5.

$$\text{kW peak demand savings} = 0.7457 * \text{hp} * \text{LF} * \left((1/\eta_{\text{M-Baseline}}) - (1/\eta_{\text{M-New}})\right) * \text{CF} \qquad (14\text{–}5)$$

where CF = coincidence factor (may be a summer or winter CF, depending on when savings occur)

TABLE 14.2
Fan and Pump Commercial Hours of Use*

Facility Type	HVAC Fan Motor Hrs of Use	Chilled Water Pump & Cooling Tower Pump Hrs of Use
Banks, Financial Centers	3,748	1,767
Religious Worship/Church	1,955	1,121
College/University		
College—Cafeteria	6,376	2,713
College—Classes/Administrative	2,586	1,348
College—Dormitory	3,066	1,521
Commercial Condos	4,055	1,877
Convenience Stores	6,376	2,713
Convention Center	1,954	1,121
Dining—Cafeteria/Fast Food	6,456	2,742
Dining—Family	4,182	1,923
Entertainment	1,952	1,120
Fast Food Restaurant	6,376	2,713
Food Stores	4,055	1,877
Gym	2,586	1,348
Healthcare	7,666	3,177
Hospital	7,674	3,180
Library	3,748	1,767
Lodging (Hotels/Motels)	3,064	1,521
Medical Offices	3,748	1,767
Multifamily	7,665	3,177
Nursing Homes	5,840	2,520
Office	3,748	1,767
Office/Retail	3,748	1,767
Parking Garage/Parking Lot	4,368	1,990
Performing Arts Theaters	2,586	1,348
Restaurants	4,183	1,923
Retail	4,057	1,878
Schools		
University	2,187	1,205
Pre-School/Elementary	2,187	1,205
Jr/Sr High School	2,187	1,205
Technical/Vocational	2,187	1,205
Warehouse—Non-Refrigerated	2,602	1,354
Warehouse—Refrigerated	2,602	1,354

*The values in the table are default values. When available, actual hours should be used. The values in the table are taken from the Connecticut TRM, Appendix 5.

Equation 14–5 provides the peak demand savings. All things being equal, assuming no ratchet provisions and that billed demand equals read demand, the peak demand savings would likely occur throughout the year. Under these assumptions and assuming the peak demand savings are fairly uniform, the annual peak demand savings would be determined by multiplying the peak demand savings in Equation 14–5 by 12. It is important to understand how the utility charges for demand and whether ratchet provisions exist before estimating demand cost savings. Demand savings would also typically be multiplied by the associated utility-specified coincidence factors (i.e., for summer and for winter).

Example Problem 14–1

What is the energy consumption and peak energy demand savings replacing a 100 HP motor (79% efficient) with a premium high efficiency motor (95% efficient)? Assume 7,400 hours/yr operation at full load and an 80% load factor?

From Equation 14–4:

Energy Savings, kWh

$$=0.7457 * hp * LF * EFLH_M * \left(\left(1/\eta_{M\text{-existing}}\right)-\left(1/\eta_{M\text{-new}}\right)\right)$$

$$=0.7457 * 100 * 0.8 * 7{,}400 * \left[(1/0.79)-(1/0.95)\right]$$

$$=94{,}030 \text{ kWh/yr}$$

From Equation 14–5:

peak kW demand savings

$$=0.7457 * hp * LF * \left(\left(1/\eta_{M\text{-existing}}\right)-\left(1/\eta_{M\text{-new}}\right)\right) * CF$$

$$=0.7457 * 100 * 0.8 * \left[(1/0.79)-(1/0.95)\right] * 0.8$$

$$=10.2 \text{ kW}$$

14.2 VARIABLE FREQUENCY AND VARIABLE SPEED DRIVES

There are two primary approaches, variable frequency drives and variable speed drives, for controlling electrical motors to better match flow rate (and result in lower energy consumption); however, there are subtle differences.

1. *Variable speed drives* (VSDs) control the speed of a motor by changing the input voltage and can be used with both AC and DC motors.
2. *Variable frequency drives* (VFDs) control the speed of a motor by changing both the voltage and frequency and can only be used with AC motors.

VFDs use power electronics to vary the frequency of input power to the motor, thereby controlling motor speed. VSD is a more generic term, frequently used interchangeably with VFD. VSDs are applied to devices that control the speed of either the motor or the equipment driven by the motor (e.g., fan, pump, compressor). VSDs can be either electronic or mechanical.

Both VSDs and VFDs are used to decrease an electric motor's unnecessary power consumption and optimize the efficiency and operation of electric motor systems. VFDs are only connected with AC motors, while VSDs are connected with either AC or DC motors.

The use of VSDs or VFDs for compressors, air handlers, chillers, cooling tower fans, and pumps is now common because they are readily available at reasonable cost and can provide energy savings. These drives are solid-state devices that were at one time add-ons to constant speed drive equipment. They are now frequently part of factory-supplied equipment and save energy whenever motors run at less than full power. Since most HVAC equipment rarely runs at full capacity, substantial energy savings are available using these drives.

The power (affinity) laws governing fans (refer to Chapter 7) also apply to pumps.

$$gpm_2/gpm_1 = rpm_2/rpm_1 \tag{14–6}$$

$$H_2/H_1 = (rpm_2/rpm_1)^2 \tag{14–7}$$

$$P_2/P_1 = (rpm_2/rpm_1)^3 = (gpm_2/gpm_1)^3 \tag{14–8}$$

where gpm_1 = pump flow rate at original operating speed, gal/min
gpm_2 = pump flow rate at new operating speed, gal/min
rpm_1 = original operating speed, rev./min
rpm_2 = new operating speed, rev./min
H_1 = head developed by the pump at original operating condition, feet
H_2 = head developed by the pump at new operating condition, feet
P_1 = pump power at original operating speed, kW
P_2 = pump power at new operating speed, kW

Example Problem 14–2

A cooling water pump is operating at a speed of 1,800 rpm with a flow rate of 400 gpm at a head of 48 ft. The power of the pump is 45 kW. What is the pump flow rate, head, and power requirements if the pump speed is reduced to 1,200 rpm?

From Equation 14-6:

$$\text{Flow Rate: } gpm_2 = gpm_1 * rpm_2/rpm_1$$

$$= 400 * (1200/1800)$$

$$= 267 \text{ gpm}$$

From Equation 14-7:

$$\text{Head: } H_2 = H_1 * (\text{rpm}_2/\text{rpm}_1)^2$$

$$= 48 * (1200/1800)^2$$

$$= 21.3 \text{ ft}$$

From Equation 14-8:

$$\text{Power: } P_2 = P_1 * (\text{rpm}_2/\text{rpm}_1)^3$$

$$= 45 * (1200/1800)^3$$

$$= 13.3 \text{ kW}$$

The pump [hydraulic] power equation is given by Equation 14–9.

$$\text{BHP} = (\text{GPM} * H_{\text{pump}})/(3{,}960 * \eta_{\text{Pump}}) \tag{14–9}$$

where BHP = brake horsepower, hp (*Note*: Brake horsepower is the amount of real horsepower going to the pump, not the horsepower used by the motor.)
GPM = flow rate, gallons/min
H_{pump} = pump head, feet of water
η_{pump} = pump efficiency, fraction
3,960 = constant (water density at 8.34 lb/gal and 33,000 ft-lb/min/hp)

The pump energy equation (Equation 4–10) is the pump power equation (Equation 14–9) multiplied by the hours of operation with the appropriate terms for motor efficiency, load factor and the conversion of horsepower to kilowatts. It allows the calculation of actual electrical power of the pump [pump shaft power].

$$E_{\text{Pump}} = \text{BHP} * 0.7457 * \text{HRS}_{\text{Pump}} * \text{LF}/\eta_{\text{motor}} \tag{14–10}$$

where E_{Pump} = energy use of pump, kWh
HRS_{Pump} = pump operating hours
LF = load factor (ratio of avg. operating load to peak load (nameplate rating of the motor), default at 80%)
η_{motor} = motor efficiency, fraction

The energy savings associated with installing a variable frequency drive on motors encountered in HVAC applications including supply fans, return fans, exhaust fans, chilled water pumps, and boiler feedwater pumps should be evaluated using a customized assessment given the variability of the many operating factors. However, for preliminary estimating purposes, it is not unusual for variable speed motor drives to save anywhere from 800 to 1,700 kWh/yr/hp and 0.07 to 0.25 kW of peak demand savings/hp in the summer months.

14.3 ROOF INSULATION

A roof represents a significant source of heat loss in cold climates and heat gain in warm climates. For roofing systems that have insulation entirely above the deck surface, which is a common roof arrangement in commercial office buildings, it usually is not a major undertaking to add insulation to reduce heat transfer into or out of the building. Adding insulation to the roof will likely require the removal of the existing membrane. Most owners consider this measure when the existing roof needs to be replaced. Also, replacing the membrane with reflective coatings, such as a cool roof, at the time of the insulation installation may also help to decrease overall cooling loads in hot climates. Existing roof penetrations and curbs need to be considered when increasing the insulation to maintain the minimum distance between the top of the membrane and the top of curbs.

Polyisocyanurate is one of the most popular and energy-efficient types of roofing insulation and is used in many commercial installations. Due to its high compatibility with roofing membranes and its high R-value at 6.5 per inch of insulation, it makes an excellent choice for new and renovation projects.

A facility's minimum required insulation R-value is typically dictated by the building code based on the local climate, but the thicker the insulation, the higher the R-value and the more managers can minimize heat gains in the summer and retain heating in winter, both of which can lower energy bills substantially.

But with increased materials and R-value comes increased cost. Two inches of polyisocyanurate board (R-13) per square foot is approximately 50% more costly than a 1-inch board (R-6.5) of the same size, and a 3-inch board (R-20) per square foot is approximately 25% more costly than the 2-inch equivalent.

Since a roof upgrade will save both heating and cooling energy, Equations 14–11 and 14–12 may be used to estimate the savings.

$$\Delta kW_{cooling} = [(1/R_{Baseline} - 1/R_{New}) * A_{Roof} * CDD * 24]/EER / 1,000 \qquad (14\text{–}11)$$

where $\Delta kW_{cooling}$ = cooling energy saved, kWh
$R_{Baseline}$ = R-value of existing roof insulation, hr-ft^2 °F/Btu
R_{New} = R-value of new roof insulation, hr-ft^2 °F/Btu
A_{Roof} = roof area, ft^2
CDD = cooling degree days per yr, °F-day/year
EER = energy efficiency ratio, Btu/watt-hr
24 = conversion factor, hrs/day
1,000 = conversion factor, 1,000 W/kW

$$\Delta kW_{heating} = [(1/R_{Baseline} - 1/R_{New}) * A_{Roof} * HDD * 24]/COP/ 3,412 \qquad (14\text{–}12)$$

where $\Delta kW_{heating}$ = heating energy saved, kWh
HDD = heating degree days/yr, °F-day/year
COP = coefficient of performance, dimensionless (kWh/kWh)
3,412 = conversion factor, 3,412 Btu/kWh

Example Problem 14–3

A two-story hotel in Delaware is replacing an existing 25,000 ft² roof with R-10 insulation with a new roof having R-30 insulation. What are the heating and cooling energy savings? Assume the following.

CDD =	1,121
HDD =	4,786
EER =	11.2
COP =	0.8 (80% efficient boiler)

Using Equations 14–11 and 14–12:

$$\Delta kW_{Cooling} = \left[\left(1/R_{Baseline} - 1/R_{New}\right) * A_{Roof} * CDD * 24\right] / EER / 1{,}000$$

$$= \left[\left(1/10 - 1/30\right) * 25{,}000 * 1121 * 24\right] / 11.2 / 1000$$

$$= 4{,}004 \text{ kWh/yr}$$

$$\Delta kW_{Heating} = \left[\left(1/R_{Baseline} - 1/R_{New}\right) * A_{Roof} * HDD * 24\right] / COP / 3{,}412$$

$$= \left[\left(1/10 - 1/30\right) * 25{,}000 * 4786 * 24\right] / 0.8 / 3{,}412$$

$$= 70{,}138 \text{ kWh/yr}$$

$$= 239 \text{ MM Btu/yr}$$

An alternative approach would be to use the bin method, where you would need to obtain the outdoor air temperature (OAT) and annual hours in each temperature bin. You would also need to know the heating and cooling set-points, the R-value of the insulation, and the roof area (ft²). The heat savings in the reduction of heat loss through the roof by improving the roof insulation would be given by Equation 14–13.

$$Q_s = A_{Roof} * \Delta T * (1/R_{Baseline} - 1/R_{New}) \quad (14\text{–}13)$$

where Qs = heat savings by increasing roof insulation R-value, Btu/hr
A_{Roof} = area of the roof, ft²
ΔT = temperature difference between bin temperature and OAT, °F
$R_{Baseline}$ = R-value of existing roof insulation, hr-ft² °F/Btu
R_{New} = R-value of new roof insulation, hr-ft² °F/Btu

The heat required to make up for the loss in those bins where the OAT requires heating (OATs up to the heating set-point or economizer enabling temperature) is given by Equation 14–14:

$$Q_{BinHeating} = Q_s * H_{Bin} / \eta_B \quad (14\text{–}14)$$

where $Q_{BinHeating}$ = heat savings in heating system for a specific bin due to the reduction in heat loss through the roof due to the insulation upgrade, Btu/yr

H_{Bin} = hours/yr at bin OAT

η_B = boiler efficiency, fraction

The heating required in each bin would be calculated for the existing R-value of the roof insulation ($R_{Baseline}$) and the proposed new upgrade insulation R-value (R_{New}). The difference would be the heating energy savings. Summing the savings over all the bins requiring heating would be the total energy savings on the heating side.

The cooling required to make up for the loss in those bins where the OAT requires cooling (above the cooling set-point) would be:

$$Q_{BinCooling} = (Q_S * H_{Bin} / 12{,}000) * IPLV \quad (14\text{–}15)$$

where $Q_{BinCooling}$ = cooling savings in cooling system for a specific bin due to the reduction in cooling loss through the roof due to the insulation upgrade, kWh/yr

IPLV = integrated part load value of the cooling system, kW input/ton output

12,000 = 12,000 Btu/hr per ton

The cooling required in each bin would be calculated for the existing R-value of the roof insulation ($R_{Baseline}$) and the proposed new upgrade insulation R-value (R_{New}). The difference would be the cooling energy savings. Summing the savings over all the bins requiring cooling would be the total energy savings on the cooling side.

Example Problem 14–4

The outdoor air temperature associated with a particular bin is 10°F for 47 hr/yr. The heating set-point is 68°F. Estimate the bin heating energy savings of upgrading an existing 2,500 ft² roof with an R-value of 11 to an R-value of 30. Assume the boiler efficiency at 80%. For the same building, in another bin the outdoor air temperature is 80°F for 147 hr/yr. The cooling set-point is 76°F and cooling system efficiency is 1.4 kW/ton. Estimate the cooling energy saved in this bin.

For the Bin OAT = 10°F, from Equations 14–13 and 14–14:

$$Q_S = A * \Delta T * (1/R_{Baseline} - 1/R_{New})$$

$$Q_S = 2{,}500\ \text{ft}^2 * (68 - 10)^{\circ}\text{F} * (1/30 - 1/11)\ \text{Btu}/^{\circ}\text{F-ft}^2\text{-hr}$$

$$Q_S = 8{,}352 \text{ Btu/hr}$$

$$Q_{BinHeating} = Q_S * H_{Bin}/\eta_{Boiler}$$

$$= 8{,}352 \text{ btu/hr} * 47 \text{ hr/yr} / 0.8$$

$$= 490{,}680 \text{ Btu/yr}$$

For the Bin OAT = 82°F, from Equations 14–13 and 14–15:

$$Q_s = 2{,}500 \text{ ft}^2 * (80-76) * (1/11-1/30)$$

$$= 576 \text{ Btu/hr}$$

$$Q_{BinCooling} = (Q_S * H_{Bin}/ 12{,}000) * IPLV$$

$$= 576 \text{ Btu/hr} * 147 \text{ hr/yr} / 12{,}000) * 1.4 \text{ kW/ ton}$$

$$= 9.87 \text{ kWh/yr}$$

14.4 REFRIGERATION SYSTEMS

Refrigeration system upgrades often include:

- Replacement of older fluorescent lighting in commercial display refrigerators, coolers, and freezers with LED lighting. Savings are based on the wattage reduction achieved by replacing fluorescent lighting with LED lighting. Interactive refrigeration savings are also achieved due to the reduced heat loads associated with lighting power reduction from more efficient lighting.
- Installation of cooler night covers (retractable covers) for open-type multi-deck refrigerated display cases (covers are deployed during unoccupied times to reduce energy loss). With this measure there is no demand savings because covers will not be in use during the peak period.
- Installation of evaporator fan controls to walk-in coolers and freezers using evaporator fans that run constantly. The savings from this measure are derived from a reduction in fan speed or the number of hours that the evaporator fans are running. If fan motors are also replaced with electronically commutated motors (ECMs) in conjunction with evaporator fan controls, then savings are based on reduced fan motor wattage. Interactive refrigeration savings are also achieved due to reduced fan speed or run-hours.
- Installation of an electronically commutated motor (ECM) in place of an existing integral electric motor connected to evaporator fans in walk-in coolers, freezers, and reach-in display cases. The savings are based on the wattage reduction from replacing an existing permanent split capacitor (PSC) or

shaded pole motor with an ECM motor. Interactive refrigeration savings are also achieved due to reduced heat loads resulting from fan power reduction.
- Installation of a control system to an existing facility where door heaters operate continuously—applicable to walk-in coolers and freezers that have electric heaters on the doors whose purpose is to prevent condensation from forming—the control system shuts off the door heaters when the facility's humidity is too low to allow condensation to occur. The savings result from a reduction in the operating hours of the door heaters.

In view of the large number of variables that can exist in refrigeration systems located in retail buildings such as supermarkets and grocery stores, estimating energy savings should be done on a case-by-case basis. Excellent sources for estimating energy savings can be found in state TRMs.[2–4]

14.5 WINDOWS

Windows provide light, ventilation, and in many cases passive solar heating, but otherwise can be a source of significant heat loss. In the more northern climates, heat loss occurs equally from all windows regardless of which direction they face. Disallowing for wind factors, similar types of glazing lose heat at the same rate. On the other hand, for heat gain calculations, windows facing east and west gain more heat than those facing north and south. This results in larger quantities of air being distributed to rooms with east and west facing windows. This air is necessary for cooling but not for heating. A reasonably insulated wall easily achieves R-21, while most windows do no better than R-3, meaning that a window loses seven times more heat per square foot than a wall does.

Windows are rated according to a standard set by the National Fenestration Rating Council (NFRC). Four values are used to describe the performance of a window:

1. Heat transmission coefficient (U-value)—tells how much heat the window will lose. The U-factor generally refers to the energy efficiency of the complete window assembly, which includes the glazing, window frame, and spacer. Window glaze refers to the glass portion of a window and is often called windowpane. A single pane window is also known as a single glazed window or single glazing. Double pane equals double glaze, triple pane equals triple glaze, and so on. Often, people think glazing refers to a special coating added to the glass pane, which infers some windows are glazed, while others are not. This is not the case. Every window has glazing since every window has glass. The spacer is the component of a window frame that separates the glazing panels, and it often reduces the U-factor at the glazing edges. The performance rating of the glazing alone, independent of the frame, is known as the center-of-glass U-factor, but use of this rating is less common. It is more common to use the U-factor of the entire window assembly. For most energy-efficient windows, the U-factor for the entire window assembly is higher than the U-factor at the center of the glass. When doing energy transfer calculations, be sure to verify that the U-value given is for the entire window assembly. Typical U-values for different window units are presented in Table 14.3.

TABLE 14.3
Typical U-Values for Windows

Window Type	U-Value, Btu/hr-ft² °F
Single pane, clear	1.10
Double-pane, clear	0.49
Double-pane, argon-filled	0.45
Double-pane, Low-E	0.35
Double-pane, Low-E, argon	0.32
Triple-pane, Low-E, argon	0.25
Quadruple-pane, Low-E, krypton	0.17

2. Visible transmission (VT)—represents the percentage of the available visible light that is allowed to pass through the window. Even clear glass is not perfectly transparent, and multiple glazing and Low-E coatings reduce this value.
3. Solar heat gain coefficient (SHGC)—signifies the percentage of the available solar gain that is allowed to pass through the window. As with visible transmission, this value is lower when multiple glazing and Low-E coatings are present.
4. Air leakage (AL)—is the amount of air leaking through the window. This does not include the air leakage around the window unit where it is attached to the wall (hopefully, the window unit is sealed tightly). The air leakage amount for a window that opens is higher than one that does not (due to the greater difficulty in sealing a window that opens). Also, air leakage for double hung or sliding windows is greater than that for casement types.

Double-pane windows have long been the industry standard and represent significant improvement over single-pane windows. Multiple-pane (triple-pane and quad-pane), Low-E, and gas-filled window configurations are even more efficient. Window energy efficiency labels today list the U-factor, solar heat gain coefficient, visible light transmittance, and air leakage rating.

If windows are going to be replaced because they are old and have significant air leakage, then the extra energy savings associated with higher energy-efficient windows will be worth the extra cost.

The heating and cooling energy savings associated with installing more efficient windows may be estimated using Equations 4–16 and 4–17. These savings do not include energy savings that may be due to any reduction in air infiltration.

$$\Delta kW_{cooling} = [(U_{Baseline} - U_{New}) * A_{Window} * CDD * 24]/EER/1{,}000 \qquad (14\text{–}16)$$

where $\Delta kW_{cooling}$ = cooling energy saved, kWh
$U_{Baseline}$ = U-value of existing windows, Btu/hr-ft² °F
U_{New} = U-value of new windows, Btu/hr-ft² °F
A_{Window} = window area, ft²

CDD = cooling degree days/yr, °F-day/year
EER = energy efficiency ratio, Btu/watt-hr
24 = conversion factor, hrs/day
1,000 = conversion factor, 1,000 W/kW

$$\Delta kW_{heating} = [(U_{Baseline} - U_{New}) * A_{Window} * HDD * 24]/COP/ 3,412 \quad (14–17)$$

where $\Delta kW_{heating}$ = heating energy saved, kWh
HDD = heating degree days/yr, °F-day/year
COP = coefficient of performance, dimensionless
3,412 = conversion factor, 3,412 Btu/kWh

This analysis ignores window orientation which would require more comprehensive assessment and preferably building simulation modeling.

Example Problem 14–5

The building with the characteristics given below is getting a window upgrade to LowE windows.What are the energy savings?

EER = 11.2
COP = 0.8 (80% efficient boiler)
CDD = 1,121
HDD = 4,786
127 – 60″ × 60″ windows ($U_{existing}$ = 0.68, U_{new} = 0.43)
124 – 72″ × 60″ windows ($U_{existing}$ = 0.68, U_{new} = 0.33)

Area of Windows = [127 * (60*60/144)] + (124 * (72 * 60/144)] = 6,895 ft²

The energy savings are estimated using Equations 14–16 and 14–17.
Using Equation 14–16:

$$\Delta kW_{Cooling} = [(U_{existing} - U_{new}) * A * CDD * 24] / EER / 1,000$$
$$= [(0.68 - 0.43) * 6,895 * 1,121 * 24] /11.2 /1,000$$
$$= 4,141 \text{ kWh/yr}$$

Using Equation 14-17:

$$\Delta\ kW_{Heating} = [(U_{existing} - U_{new}) * A * HDD * 24] / COP/ 3,412$$
$$= [(0.68 - 0.43) * 6,895 * 4,786 * 24] / 0.8 / 3,412$$
$$= 72,537 \text{ kWh/yr}$$
$$= 248 \text{ MM Btu/yr}$$

14.6 PUMPS

Pumps are used throughout an HVAC system, for example, boiler feedwater pumps, chilled water and hot water pumps, condenser water pumps to cooling tower, and return water pumps from cooling tower.

Centrifugal pumps are variable-torque machines, like centrifugal fans, and they obey the same laws as do centrifugal fans. The term "variable torque" indicates that the horsepower required to turn the pump or fan does not vary directly with their speed. It varies with the cube of the speed. Constant-torque pumps, like most positive displacement pumps, are those in which the horsepower does vary directly with the speed. The laws that define centrifugal pump performance are called the pump affinity laws (analogous to the fan affinity laws). They dictate the changes in pump performance with variations in speed, flow, head, and impeller diameter. The basic affinity laws are given in Equations 4–18 through 4–22. (Note: Some of these equations have previously been presented in Section 14.2.)

$$GPM_1/GPM_2 = rpm_1/rpm_2 \tag{14–18}$$

$$H_1/H_2 = (rpm_1/rpm_2)^2 \tag{14–19}$$

$$BHP_1/BHP_2 = (rpm_1/rpm_2)^3 \tag{14–20}$$

$$GPM_1/GPM_2 = D_1/D_2 \tag{14–21}$$

$$H_1/H_2 = (D_1/D_2)^2 \tag{14–22}$$

where GPM = pump capacity, gal/min
H = pump head, ft
BHP = pump brake horsepower, hp
D = pump impeller diameter, in.
1 = operation (rpm) at condition 1
2 = operation (rpm) at condition 2

The energy required by a pump, as indicated by the affinity laws, depends on its speed and the diameter of its impeller. The energy imparted to the water by a pump is called the water horsepower (W_{hp}). Its equation is:

$$W_{hp} = GPM * H * SG / 3960 \tag{14–23}$$

where GPM = water flow rate, in gal/min
H = head, in ft
SG = specific gravity (most HVAC water systems operate from 32 to 240°F water, where the specific gravity can vary from 1.001 at 32°F to 0.948 at 240°F; the specific gravity is generally assumed to be equal to 1.00)

Example Problem 14–6

If a pump is delivering 1,000 gal/min at a total head of 100 ft, what is the water horsepower?

$$W_{hp} = 1000 * 100 / 3960 = 25.25 \text{ HP}$$

The energy required to operate a pump is determined in brake horsepower (bhp); the difference between water horsepower and pump brake horsepower is the energy lost in the pump. Brake horsepower is the amount of real horsepower going to the pump, not the horsepower used by the motor. The pump brake horsepower equation given by Equation 4–24.

$$BHP = W_{hp} / h_{pump} \qquad (14\text{–}24)$$

where BHP = pump brake horsepower, hp

η_{pump} = pump efficiency, decimal (default to 75% for small pumps, i.e., less than 100 gpm; 80% for pumps up to 500 gpm; 85% for large pumps greater than 500 gpm)

Example Problem 14–7

If the pump in Example Problem 14–6 is operating at an efficiency of 85%, what is the brake horsepower?

$$BHP = 25.25 / 0.85 = 29.7 \text{ hp}$$

The electrical energy (pump kilowatts) for a motor-driven pump must take into consideration the efficiency of the motor on constant-speed pumps and the wire-to-shaft efficiency of the motor and variable speed drive on variable-speed pumps. Again, the difference between the pump brake horsepower and electric power input to the motor or variable-speed drive and motor is the efficiency of these devices. The power must be converted from brake horsepower to kilowatts, so the pump brake horsepower must be multiplied by 0.7457. This is shown in Equation 14–25.

$$\text{Pump kW} = BHP * 0.7457 * LF / \eta_M \qquad (14\text{–}25)$$

where BHP = pump brake horsepower, hp

η_M = efficiency of the electric motor or the motor and variable-speed drive as a decimal

LF = load factor (ratio of avg. operating load to peak load (nameplate rating of the motor), default at 80%)

The pump kW equation becomes:

$$\text{Pump kW} = GPM * H * SG * LF/(5{,}308 * \eta_{pump} * \eta_{Motor}) \qquad (14\text{–}26)$$

The general rule is constant speed pumps for constant-volume systems and variable-speed pumps for variable-volume systems. If a water system has a large variation in its flow, from less than 50% to 100% of design, it is probably a candidate for variable speed. Likewise, if the flow rarely changes and there is little variation in temperature or viscosity, this is a constant-speed pump application. Selection of a variable-speed pump is more complicated than that for a constant-speed pump because the efficiencies of the motor and variable speed drive must be evaluated along with the efficiency of the pump itself.

The increase in cost of energy has thrust the efficiency of operation to the forefront. Highly efficient centrifugal pumps are now available at little added cost over less efficient pumps. The first and foremost evaluation of HVAC pumps should therefore be efficiency.

REFERENCES

1. UI/Eversource C&LM Program Savings Document, 12th Edition, Appendix 5, 2017, https://www.puc.nh.gov/EESE%20Board/EERS_WG/ct_trm.pdf
2. Focus on Energy, Partnering with Wisconsin Utilities, 2017 Technical Resource Manual, prepared by The Cadmus Group, pp. 91–98 and 466–481, 2017, https://assets.focusonenergy.com/production/Focus%20TRM-%20Spring%202017-%20Final%20for%20Posting%20May%202017_0.pdf
3. Arkansas Technical Resource Manual, Version 7.0, Volume 2: Deemed Savings, prepared by Independent Evaluation Monitor, pp. 437–456, August 2017, https://apsc.arkansas.gov/wp-content/uploads/AR_TRM_V9.1_Volume_1_2_and_3_on_8-31-22.pdf
4. New York State Joint Utilities, Technical Resource Manual, New York Standard Approach for Estimating Energy Savings from Energy Efficiency Programs, Version 7, pp. 467–494, April 2019, https://dps.ny.gov/system/files/documents/2022/11/technical-resource-manual-version-7-filed-april-15-2019-effective-january-1-2020.pdf

15 Other Energy Savings Considerations

15.1 CONSIDERATIONS WHEN INSTALLING MULTIPLE EEMS

An important consideration when replacing multiple energy efficiency measures (EEMs), such as an LED lighting upgrade and installing a high efficiency air conditioning system, is the impact one EEM may have on the other EEMs (interactive effects). This is referred to as the "interactivity consideration." As part of the interactivity consideration, it is also important to understand the order in which the EEMs are installed. This is referred to as EEM sequencing. We have already seen how installing even a single EEM (e.g., lighting) can interact with the cooling and space heating end-use energy consumptions (see Chapter 9, Section 9.4). It is important in energy analysis to consider all these interactions.

15.1.1 Interactivity

Interactions between measures need to be accounted for so as to not double count in performing energy savings calculations. One of the most common ECM interactive effects involves an LED lighting upgrade. An LED lighting upgrade will reduce cooling needed in the summer months but will increase heating needed in the winter months (refer to Chapter 9, Section 9.4). Another example would be related to a window upgrade and replacement chiller where the window upgrade measure would reduce the cooling load to be met by the new chiller. Since the cooling end-use energy consumption would be reduced, so would the energy savings of the chiller system.

15.1.2 Sequencing

If building simulation modeling is not being performed, when multiple energy efficiency improvements are made, the sequencing of these improvements may be important because of interactive effects, again, to avoid double counting. The suggested approach is to keep adjusting the impacted end-use energy consumption as each improvement is made; for example, if an LED lighting upgrade and an RTU upgrade are proposed, assume the LED improvement is installed first and the kWh cooling reduction would be subtracted from the baseline space cooling end-use. The high efficiency RTU percent energy savings would then be determined using the adjusted (slightly smaller) baseline space cooling end-use energy consumption.

The suggested sequencing order would be:

1. Energy saving fixtures (multifamily units)—reduces DHW heating end-use energy consumption.

DOI: 10.1201/9781032692777-15

2. Sealing/add insulation (multifamily units)—reduces the space heating and cooling end-use energy consumptions.
3. High performance, Low-E windows—reduces space heating and cooling end-use energy consumptions.
4. LED lighting—reduces the space cooling end-use energy consumption and increases the space heating end-use energy consumption.
5. High efficiency DHW heating systems, high efficiency A/C and chillers, high efficiency heat pumps, and high efficiency boilers/furnaces are next evaluated.
6. The energy savings of the EMS system is next evaluated using the net space heating and cooling end-use energy consumptions after installation of (1) to (5) above.
7. Finally, the solar PV system is sized for the net whole building electrical energy end-use consumption after all EEMs—(1) to (6) above—have been installed.

Example Problem 15–1

The data for a 70,000 ft^2 office building upgrade project is as follows:

Upgrade	Existing	New
RTU	8.0 EER	12.2 EER
Boiler	78% efficient	94% efficient
EMS	No	Yes

Building energy data is as follows:

Whole building existing kWh/yr = 571,680	(27.9 kBtu/SF-yr)
Whole building therms/yr =37,590	(53.7 kBtu/SF-yr)
% weather-normalized cooling end-use energy consumption = 27% of electricity use	
% weather-normalized space heating end-use energy consumption = 95% of natural gas use	

Estimate the energy savings and the building EUI reduction.

$$\text{RTU energy savings} = 571{,}680 \text{ kWh/yr} * 0.27 * \left(1-\left(8.0/12.2\right)\right)$$
$$= 53{,}138 \text{ kWh/yr}$$

$$\text{Boiler energy savings} = 37{,}590 \text{ therms/yr} * 0.95 * \left(1-\left(0.78/0.94\right)\right)$$
$$= 6{,}071 \text{ therms/yr}$$

Assume EMS saves 12% of cooling energy consumption and 18% of space heating energy consumption.

EMS energy savings should be taken AFTER the cooling and space heating energy savings are taken into consideration (impact of sequencing and to avoid double counting).

$$\text{EMS cooling energy consumption savings} \\ = 0.12 * (154{,}354 - 53{,}138)\ \text{kWh/yr} = 12{,}146\ \text{kWh/yr}$$

$$\text{EMS heating energy consumption savings} \\ = 0.18 * (35{,}711 - 6{,}071)\ \text{therms/yr} = 5{,}335\ \text{therms/yr}$$

$$\text{Total cooling energy consumption saved} \\ = 65{,}284\ \text{kWh/yr}\left(\text{or } 3.2\ \text{kBtu/ft}^2\text{-yr}\right)$$

$$\text{Total space heating energy consumption saved} \\ = 11{,}406\ \text{therms /yr}\left(\text{or } 16.3\ \text{kBtu/ft}^2\text{-yr}\right)$$

$$\text{Existing building EUI} = 81.6\ \text{kBtu/SF-yr}$$

$$\text{Upgraded building EUI} = 62.1\ \text{kBtu/SF-yr}$$

$$\text{Total building EUI reduction} = 23.9\%$$

Note: Had sequencing not been considered, the EMS energy savings would have been:

$$\textit{EMS cooling energy consumption savings} \\ = 0.12 * (154{,}354)\ kWh / yr = 18{,}522\ kWh / yr$$

$$\textit{EMS heating energy consumption savings} \\ = 0.18 * (35{,}711)\ therms / yr = 6{,}428\ therms / yr$$

which would have made the total building EUI reduction 26.1%, i.e., greater than it should be because interactivity has not been taken into consideration.

15.2 MIXED-USE BUILDING CONSIDERATIONS

When a building has more than one use (e.g., office and retail), it is referred to as a "mixed-use" building. When a building's primary use occupies at least 75% of the gross floor area (GFA), EPA Portfolio Manager (ENERGY STAR) assumes the entire building's use as this use. As such, if 75% of the building GFA is office space and 25% retail space, Portfolio Manager would assume the entire building's use as office. This assumption may work well when the energy consumption of the secondary building use(s) is close to that of the primary use but can introduce significant error if it is not. To reduce this potential error in mixed-use buildings, it would be best to collect the energy consumption data associated with each use and then depending on the use, estimate end-use energy consumptions for each use to establish an appropriate baseline (refer to Chapter 2).

If energy consumption data for each building use is not available and it is desired to determine end-use energy consumption to establish a baseline and facilitate the estimation of energy savings resulting from energy efficiency upgrades to an individual use, then benchmarking (EUI) data may be helpful (refer to Chapter 2). Once the median EUI of each building use is obtained and the associated (baseline) end-use energy consumptions (e.g., for space heating, cooling, ventilation), the energy savings can be estimated knowing the energy efficiency metrics of the new equipment to be installed to serve that use.

It must be emphasized that this approach is only an estimation since it assumes that each building use is essentially an independent building (i.e., with its own roof and envelope). However, a single mixed-use building (e.g., office on top of retail) may use less energy than the individual stand-alone uses due to lower heating and cooling energy loss because there is less externally exposed building surface area. For example, assuming a single tall building with retail on the first floor and offices on the second floor, the "roof" of the retail space (its ceiling) would be the "floor" of the office space. The net impact would be less total energy consumption for the retail space than if this use was considered an independent (stand-alone) building. The situation becomes even more complicated if each building use is sharing an end-use such as air conditioning or heating.

Example Problem 15–2

A mixed-use, 10,000 ft², all-electric building has a whole building energy consumption of 291,500 kWh/yr, with air conditioning supplied by a single unit. Use 1 occupies 75% of the total building floor area and has an EUI of 64.5 kBtu/ft². Use 2 occupies 25% of the total building floor space and has an EUI of 204.4 kBtu/ft². The cooling end-use energy consumption percentage (of whole building energy consumption) for a Use 1–type building is estimated at 17% and for a Use 2–type building, 24%. If new air conditioning units serving each building use are to be installed, what baseline cooling energy consumptions should be used to estimate the cooling consumption savings?

Whole building energy consumption =	291,500 kWh/yr
Use 1 estimated median EUI from similar buildings =	64.5 kBtu/ft²
Use 2 estimated median EUI from similar buildings =	204.4 kBtu/ft²
Use 1 =	7,500 ft² = 483,750 kBtu
Use 2 =	2,500 ft² = 511,000 kBtu

Use 1: % of total energy consumption = 483,750/(483,750 + 511,000) = 48.6%
Use 2: % of total energy consumption = 511,000/(483,750 + 511,000) = 51.4%
Use 1 estimated energy consumption = 0.486 * 291,500 = 141,669 kWh/yr
Use 2 estimated energy consumption = 0.514 * 291,500 = 149,831 kWh/yr

Use 1 cooling end-use percent of whole building energy use = 17%
Use 2 cooling end-use percent of whole building energy use = 24%
Building Use 1 baseline cooling energy consumption = 24,084 kWh/yr
Building Use 2 baseline cooling energy consumption = 35,959 kWh/yr
Note: If the entire building EUI was equal to the use occupying 75% or more of the GFA (Use 1), the building EUI would have been assumed to have been that of Use 1, which had a much lower EUI than that of Use 2.

Example Problem 15–3

A mixed-use building has a whole building EUI of 82 kBtu/SF-yr. The total building ft² is 100,000, with Use 1 an office occupying 60,000 ft² and Use 2 a retail establishment occupying 30,000 ft². The building has a 10,000 ft² common area. What is the EUI of each building use? The median EUI for an office building in this area is 67.3 kBtu/ft²-yr and for retail is 94.2 kBtu/ft²-yr.

First, distribute the common area ft² to each building end-use:

Office:	60,000/90,000 * 10,000 = 6,666 ft²
Retail:	30,000/90,000 * 10,000 = 3,333 ft²

Calculate energy consumptions of each use:

Office:	67.3 * 66,666 = 4,486,622 kBtu/yr	or 58.8% of total
Retail:	94.2 * 33,333 = 3,139,969 kBtu/yr	or 41.2% of total
	Total = 7,626,591 kBtu/yr	

EUI of each building use:

Office:	0.588 * 82 = 48.2 kBtu/SF-yr
Retail:	0.412 * 82 = 33.8 kBtu/SF-yr

15.3 COMMON MISTAKES MADE IN ENERGY ANALYSIS

In the process of performing technical reviews of energy efficiency upgrade projects proposed in energy audits on thousands of commercial buildings, the author has found a number of common mistakes that resulted in faulty analyses. These included:

- An aggregated electricity cost ($/kWh) was used to estimate energy cost savings (e.g., total kWh consumed divided by the total utility bill invoice). The $/kWh should only reflect costs associated with kWh consumption, and the $/kW should only reflect costs associated with kW demand/delivery.

- When utility bills were collected and reviewed to establish whole building energy consumption, a baseline period was selected that was not representative. The baseline period should be at least a full year. The building should not have had any major renovations that took place in the baseline period. Examples of major renovations might be an increase in building size, addition of energy efficiency measures such as an LED lighting upgrade, or the addition or removal of a data center. Also, building operation and occupancy should be discussed with the building owner to confirm that it was representative in the baseline period.
- Whole building energy use and end-use energy consumption were based on the building's actual energy consumption in a specific 12-month period. Whole building energy use and end-use energy consumption must be weather-normalized when estimating [future] energy savings.
- When performing lifecycle financial analysis of the energy cost savings:
    - An estimated useful life of the energy efficiency measure(s) was used that exceeded industry norms.
    - An average annual energy cost escalator was used but was greater than that experienced historically by the local utility.
    - Equipment performance degradation was not considered in estimating future annual energy consumption savings.
- The applicability of potential utility ratchet provisions was not investigated when conducting the economic analysis.
- When determining the energy consumption savings for multiple energy efficiency measures, interactive effects between these measures were not considered.
- For solar PV systems:
    - When the solar PV system was proposed, only utility electricity bill data was relied upon without any consideration given to other energy efficiency upgrades that also would be installed (such as an LED lighting upgrade). Energy efficiency upgrades need to be considered in solar system design so that the solar system can be right-sized. It is also important in the economic analysis to consider how the utility handles "excess" kWh that may be sent to the grid by an oversized system; for example, the "excess" kWh may be valued at the utility's wholesale rather than retail cost.
    - A shading study was not included in the solar system proposal.
    - For proposed roof-top systems, the remaining useful life of the roof and the building's structural ability to support the solar system under worst-case weather conditions including wind and snow loading were not considered. Ideally, the remaining useful life of the roof should at least equal the estimated useful life of the solar system.
    - The solar proposal included string inverters but the economic analysis failed to include an extended inverter warranty since string inverter life (approximately 11 years) is much less than the estimated useful life of a solar system (20–25 years). This should not be an issue if microinverters are used since they have a much longer life.
    - Performance degradation of electricity production was not considered when solar system lifecycle financial analysis was performed.

15.4 ASTM BUILDING ENERGY PERFORMANCE STANDARD

ASTM, formerly known as the American Society for Testing and Materials, is an international standards organization founded in 1902 that develops and publishes voluntary consensus technical standards for a wide range of materials, products, systems, and services. The ASTM E3224 Standard Guide for Building Energy Performance and Improvement Evaluation in the Assessment of Property Condition[1] is intended to provide a methodology for a user to identify building energy performance compared to peer buildings. If a building is under-performing compared to its peers, the standard provides a methodology to identify potential energy performance improvements and provide an estimate of the cost for such improvements.

There are both business and growing regulatory reasons to consider building energy performance in due diligence undertaken at the time of a commercial real estate transaction.[2,3] Energy efficient buildings are supported by sound business reasons, including:

- Reduced building operating costs due to lower energy costs.
- Increased building valuation and potential sale price premium.
- Potential rental rate premium ("green" buildings are attractive to tenants).
- Less danger of building obsolescence.
- Buildings may be more competitive in the marketplace.

Legislation requiring commercial building energy performance disclosure by owners to prospective purchasers and lenders involved in a real estate transaction and making such data publicly accessible is growing. As of 2023, four states and more than 39 municipalities, including Boston, New York, Philadelphia, the District of Columbia, Montgomery County, Minneapolis, Austin, Seattle, San Diego, Los Angeles, Chula Vista, Chicago, Salt Lake City, Reno, Denver, Kansas City, Orlando, Columbus, St. Louis, Indianapolis, and San Francisco, have already passed such legislation and many more have expressed interest in doing so. One of the goals of legislation requiring evaluation of a building's energy performance and its disclosure is to penalize less energy efficient buildings. Buildings that are energy inefficient compared to peers may be viewed as less competitive in the marketplace and in danger of obsolescence, potentially resulting in a lower market value.

In view of this growing concern about building energy performance and regulatory disclosure requirements, government agencies, including Fannie Mae, Freddie Mac, HUD, and SBA have taken the lead and responsibly begun to develop programs that focus on building energy performance. Fannie Mae, for example, in their Green Rewards Program[4] specifically includes the need to address the situation in property condition assessments and how a property compares with peers. Freddie Mac in their Green Advantage Program[5] has a similar approach, as does HUD in their Green Physical Needs Assessment.[6] SBA's Green 504 Loans Program[7] includes, among other initiatives, loans to acquire more energy efficient property.

As building energy performance disclosure legislation grows, it becomes more and more likely that a prospective purchaser will request that an energy performance assessment be included in the property due diligence process.

15.5 BUILDING CERTIFICATIONS

15.5.1 ENERGY STAR Certification

The ENERGY STAR program was developed by EPA in 1992 as a voluntary measure to allow businesses and industries to find ways to reduce energy consumption and improve energy efficiency without government legislation requiring it.[(8)] To be ENERGY STAR certified, a building must have energy performance among the top 25% of all buildings of its type.

Based upon their energy consumption, commercial buildings being benchmarked are compared to similar buildings nationwide and receive an ENERGY STAR score between 1 and 100. A score of 50 is the median. If the building scores below 50, it means it is performing below 50% of similar buildings nationwide, while a score above 50 means it is performing better than 50% of its peers. A score of 75 or higher means it is a top performer and may be eligible for ENERGY STAR certification. Such certification is given on an annual basis and must be verified by a third-party licensed professional engineer or registered architect. ENERGY STAR certification is good for 1 year and ceases being active if the building has not been re-certified in the prior 12 months.

15.5.2 LEED Certification

LEED, or Leadership in Energy and Environmental Design, is a green building sustainability rating system available for both newly constructed and existing buildings.[(9)] LEED certification is a globally recognized symbol of sustainability achievement. Buildings pursuing LEED certification earn points across several categories, including energy and water use. Based on the number of points achieved, a project can earn one of four LEED rating levels: Certified, Silver, Gold or Platinum. The fact that the building has gone through the comprehensive third-party LEED certification process provides a sustainability value proposition often exceeding that of many peer buildings.

15.5.3 Green Globes Certification

The Green Globes certification program is an ANSI-approved green building certification program that provides guidance in the design, construction, and operation of high-performance interiors and buildings.[(10)] The standard includes language on energy efficiency, resilience, life cycle cost analysis, moisture control analysis, health and effectiveness, and other market advances such as with lighting, plumbing fixtures, and acoustic comfort. New and existing buildings certified through Green Globes typically cost less to operate through lower energy and water bills, and increased recycling. The Green Building Initiative (GBI) delivers its rating system through an ANSI-audited consensus process that verifies each Green Globes certification, ensuring that building owners, facility managers, and property developers meet clearly defined criteria. The rigorous, documented process relies on both prescriptive measures and performance metrics to validate that the building achieved a variety of sustainability requirements.

15.5.4 National Green Building Standard Certification

The National Green Building Standard (NGBS) is an ANSI-approved green building certification program specifically focused on single-family and multi-family residential buildings, remodeling projects, and land developments.(11) In a partnership with the ASHRAE, the International Code Council (ICC), and the National Association of Home Builders (NAHB), the NGBS was developed to provide a uniform national platform for recognizing and advancing green residential construction and development. The NGBS is a point-based system wherein a single-family or multifamily building can attain certification depending on the sustainable and green standards included in design and construction, and in planned operation and maintenance. Projects can qualify for four certification levels (Bronze, Silver, Gold, or Emerald) by earning the required number of points for each level.

15.5.5 The Building Research Establishment Environmental Assessment Method

The Building Research Establishment Environmental Assessment Method (BREEAM) is a sustainability assessment, rating and certification method for many types of real estate.(12) It uses ten different measurement categories that include management, energy, water, waste, pollution, and others, to rate a building's environmental performance. There are five ratings that a building can potentially achieve: Pass, Good, Very Good, Excellent, and Outstanding.

REFERENCES

1. ASTM Standard E3224–19, Building energy performance and improvement evaluation in the assessment of property condition, www.astm.org, 2019.
2. Buonicore, A.J., A new way to measure risk, Scotsman Guide, pp. 56–58, March 2020.
3. Buonicore, A.J., Integrating building energy performance into property condition assessments, *Environmental Bankers Association Journal*, pp. 17–18, June 2018.
4. Fannie Mae Green Rewards Program, https://multifamily.fanniemae.com/financing-options/specialty-financing/green-rewards.
5. Freddie Mac Green Advantage Program, https://mf.freddiemac.com/product/green-advantage.
6. HUD, Green Physical Needs Assessment, www.hud.gov/program_offices/public_indian_housing/programs/ph/capfund/gpnatool.
7. SBA, Green 504 Loan Program, https://cdcloans.com/sba-504-green-loan/.
8. ENERGY STAR, www.energystar.gov
9. LEED, www.usgbc.org.
10. Green Globes, www.greenglobe.com.
11. National Green Building Standard, www.nahb.org.
12. BREEAM Certification, https://smart-cre.com/what-is-breeam-certification-definition-process-and-benefits/#:~:text=What%20is%20BREEAM%20.

16 Energy Savings Financial Metrics

16.1 THE IMPORTANCE OF ENERGY COST SAVINGS

The estimated energy cost savings calculation is critical as it provides the information needed to support the building owner's decision-making process. This is especially important when a contractor is proposing higher efficiency equipment at a premium price versus proposing only code-compliant equipment. Moreover, such cost savings and financial impact analysis can help answer the tough building owner questions, such as:

- "What energy savings can I expect, and how will it impact the building's cash flow?
- "Can you quantify the estimated return on investment difference between high efficiency and code-compliant equipment?"
- "Does the enhanced performance associated with high efficiency equipment justify its price premium?"

16.2 THE FINANCIAL ANALYSIS CHALLENGE

The journey from energy savings to financial impact analysis for an energy efficiency improvement project is not as straightforward as it may seem. In fact, the path often contains a minefield of potential problems that can lead the energy professional astray in ways that may ultimately result in a dissatisfied client. Simply estimating the cash savings from an improvement project is rarely sufficient for today's demanding cash flow–focused building owners who prefer to see savings presented in the financial metrics they are accustomed to in their business.

The list of potential issues that needs to be addressed begins with a careful understanding of the utility rate schedule for the building. It is essential that the proper rate schedule be identified and interpreted correctly. It is important, for example, to understand electricity generation charges, demand-delivery charges, potential ratchet provisions, and time-of-use charges (refer to Chapter 2).

Estimating the savings of these improvements over the equipment's estimated useful lifetime requires several key assumptions to be made, including, for example, predicting the useful life of the equipment, identifying equipment potential performance degradation over time, estimating potential maintenance savings, and forecasting how utility energy costs might increase over the life of the equipment. Making such estimates and assumptions about the future will always be challenging, particularly considering that most energy-consuming equipment in a commercial building has expected useful lifetimes of 15 to 20 years or more.

 DOI: 10.1201/9781032692777-16

Since future energy savings are also dependent on future weather, which is difficult to predict, energy cost savings are based upon typical meteorological year (TMY) weather or long-term average weather for an area (i.e., the savings are weather-normalized). Hence, the projected annual energy consumption savings, while reasonable over the long term, may be off in any particular year. Notwithstanding, since lifecycle financial analysis of energy efficiency upgrades is performed over a long term (refer to equipment EULs), the unpredictability of future weather should not present any problem when conducting such analysis.

16.3 FINANCIAL METRICS

Experience has shown that the most common energy improvement financial metrics needed to make the business case to today's building owners and managers include:

- Simple payback (SPB).
- Net present value (NPV) of lifetime savings.
- Savings-to-investment ratio (SIR).

16.3.1 Simple Payback (SPB)

The easiest and most often used financial analysis method is simple payback. The SPB period refers to the amount of time it takes to recover the cost of an investment, or more simply, the length of time it takes for an investment to reach break-even. For energy efficiency investments, SPB equals the total upfront capital investment divided by the first-year energy cost savings.

Unfortunately, the SPB period does not consider any benefits or costs that occur after the initial investment has been recouped. Moreover, it ignores the time value of money, that is, the idea that money today is worth more than the same amount in the future. It does not account for the costs associated with financing a project over time (interest costs) or other factors affecting future energy cost savings, such as equipment performance degradation, utility cost escalation rates, and so on. Although simple to calculate, there are many other questions and limitations associated with the SPB method that make building owners and managers rarely want to make investment decisions based solely on this metric. A project can initially appear to be unattractive when viewed through the lens of simple payback period, while a more thorough economic analysis may reveal it to be a highly profitable investment.

Many energy professionals rarely go beyond the SPB period calculation which, though relatively easy to use, can be quite misleading. SPB is calculated using Equation 16–1.

$$SPB = C/S \tag{16–1}$$

where SPB = simple payback period, years
C = total project cost, \$
S = project savings (first year), \$/year

Example Problem 16–1

If a lighting retrofit costs $100,000 and saves $15,000 in annual energy costs, what is the simple payback?

Using Equation 16–1:

SPB = C/s = $100,000 / $15,000/year = 6.7 years

16.3.2 Net Present Value (NPV)

Net present value (NPV) offers a more rigorous analysis than simple payback by not only extending the analysis to include all cash flows over the useful life of the project, but also accounting for the time value of money. The project's cash flow analysis includes the first cost, the energy cost savings (which may be assumed to increase with rising energy prices), and all other costs and benefits, such as O&M costs and any salvage value at the end of the analysis term. The calculation of a project's NPV depends on the discount rate selected as well as the length of the analysis term. The discount rate is viewed as the company's cost of capital or expected rate of return for an investment. In other words, it effectively represents the minimum rate of return that a company or investor requires over the life of the investment. Discount rates recently used in energy efficiency project financial analysis typically range between 3% and 7%. However, discount rates will vary based on current or anticipated changes in market interest rates.

In an NPV analysis, the discount rate is used to determine the present value of each cash flow, adjusting all cash outflows and inflows over the life of the project to comparable dollar amounts today. The choice of a discount rate is critical; the chosen rate should reflect the rate of return that could be earned on an investment of similar risk and duration.

The present value of a future cash value in year n is given by Equation 16–2.

$$PV = F * \left(1/(1+d)^n\right) \tag{16–2}$$

where PV = present value of cash in the future
F = cash n years in the future
d = percent discount rate, fraction
n = number of years in the future

The present value of a series of equal annual savings over a period of n years is given by Equation 16–3.

$$PV = A_S * \left(\left[(1+d)^n - 1\right] / \left[d * (1+d)^n\right]\right) \tag{16–3}$$

where PV = present value of constant annual saving A, $
A_S = series of equal annual savings, $
d = percent discount rate, fraction
n = number of years savings realized

The net present value (NPV) is then calculated using Equation 16–4.

$$NPV = PV - C \qquad (16\text{–}4)$$

where NPV = net present value
PV = present value of savings
C = total project cost, $

A positive NPV indicates that the present value of the cash inflows is greater than the present value of the cash outflows over the analysis term. A negative NPV indicates that the investment required is greater than the project's return.

Example Problem 16–2

Using the same lighting retrofit as in Example Problem 16–1, what is the present value of future cash flows, assuming a 6.5% discount rate and a 12-year useful life of the lighting equipment? What is the NPV?

Using Equation 16–3 for the PV:

$$PV = A_S * \left(\left[(1+d)^n - 1\right] / \left[d * (1+d)^n\right]\right)$$

Where A_s = series of equal annual savings ($15,000)
d = discount rate (0.065)
n = period of years over which the savings occur (12 years)

$$PV = \$122{,}728$$

and from Equation 16–4 for NPV:

$$NPV = PV - C$$

$$NPV = \$22{,}728$$

A positive NPV indicates the cash from the energy savings is greater than the capital investment.

Without considering the time value of money, the savings would have been $180,000, resulting in an energy savings much greater than likely to be encountered. (*Note*: You can also use the Financial function in Excel and go to the calculation for PV.)

16.3.3 Savings-to-Investment Ratio (SIR)

A more substantial and representative financial metric for building owners and managers today is the SIR. The SIR equals the ratio of the present value of the project's estimated lifetime energy cost savings to the total installed cost of the energy efficiency improvement project.

An SIR of 1 or greater indicates that the improvement(s) will pay for itself over the project lifecycle. For example, an SIR of 1.4 indicates that for every $1 invested in the project, the investor will receive $1.40 in return.

In calculating the SIR, it is crucial to have the estimated useful life (EUL) for the equipment, a reasonable estimate of performance degradation that might be experienced, a supportable prediction of how energy rates may increase over the equipment's estimated useful life, and an appropriate discount rate.

Appendix IX provides reasonable EULs for common equipment encountered in energy improvement projects. Performance degradation was discussed in Chapter 3 (Section 3.3). To predict how energy rates may escalate in the future is more challenging. One supportable way to do this is to analyze energy cost escalation in the past. For example, the Energy Information Agency (EIA) tracks average electricity cost (annual sales) in \$/kWh by state for residential, commercial and industrial end-use sectors back to 1990; see www.eia.gov/electricity/data/state/.

Calculating the compound annual growth rate over the period in question for the state where the project is located using data from EIA's chart can provide insight into how energy costs have escalated.

Discount rates are also a crucial factor for SIR calculations. Energy efficiency measures typically have relatively high upfront costs, which need to be recovered by savings over longer periods. Discount rates are thus used to attribute a value to future cash flows. The higher the discount rate, the lower the value assigned to future savings. Consequently, high discount rates make energy efficiency measures and supporting policies look less attractive. For energy improvement projects, discount rates typically range between 3% and 7%. In the author's experience, the most common discount rates encountered for energy savings projects are 3%–4%. Reference 1 provides an excellent source for information on lifecycle costing.

The uniform capital recovery factor (UCR) is used to annualize the investment (capital) cost of a project and is calculated by multiplying the project's initial investment cost by the appropriate UCR.

$$\text{UCR} = [(\text{d}*(1+\text{d})^{\text{n}})/((1+\text{d})^{\text{n}} - 1)] \tag{16–5}$$

where UCR = capital recovery factor at a discount rate, d, over the years the system will be operating (life expectancy of the system)

The SIR equals the ratio of the present value of the project's estimated lifetime energy cost savings to the total installed cost of the energy efficiency improvement project.

$$\text{SIR} = \text{PV/C} \tag{16–6}$$

where SIR = savings to investment ratio

PV = present value of net cash flow (energy savings) over useful life

C = total project cost, \$

Example Problem 16–3

An existing air conditioning system uses 50,000 kWh/year at a current electricity cost of \$0.17/kWh, expected to escalate 4% each year. The existing system is expected to last about 5 more years. Replacing the existing system with a new high efficiency air conditioning system will cost $25,000 and should reduce space cooling energy consumption by 25%. The new system is expected to last 20 years. Should the existing system be replaced now or wait until it has no remaining useful life?

General rule of thumb: As long as the annualized investment cost of the new system is less than its expected yearly savings, and the yearly savings are expected to remain constant or increase over time, it is cost-effective to replace the existing system now.

The annualized investment cost of a project is found by multiplying the project's initial investment cost (less the present value of any residual value, assuming there is any residual value) by the appropriate uniform capital recovery (UCR) factor.

The formula for the UCR is given in Equation 16–5:

$$UCR = [(d*(1+d)^{n})/((1+d)^{n}-1)]$$

where d = discount rate (default to 3%)

n = life expectancy of the system

The UCR for an investment with a life of 20 years and a discount rate of 3% is:

$$UCR = 0.067215708\,.$$

Therefore, for a new system with an investment of $25,000, the annualized investment cost is: $25,000 × UCR = $1,680. As long as the annual savings are expected to be at least $1,680, the replacement is economically justified.

$$\text{First Year Annual Savings} = 0.25 * (\$0.17/\text{kWh} * 50{,}000\ \text{kWh}) = \$2{,}125$$

The annual savings of $2,125 is at today's energy prices and does not factor in electricity cost escalation.

The replacement in this example is cost-effective.

Example Problem 16–4

An office building undertakes an energy improvement project at a total cost of $725,000. The project consists of upgrades with an estimated useful life of 15 years and is expected to save 1,205,400 kWh in the first year and reduce demand by 400 kW. The cost savings at the end of the first year (electricity) is estimated to be $110,400. A state energy loan will be used to finance 100% of the project cost at 6% interest. What is the SIR for the project? What is the cash flow expected to look like?

The state loan of $725,000 at 6% for 15 years amounts to a fixed annual cost of $73,416/year.

Assuming electricity costs escalate at an average 3% per year over the 15-year finance term and that the newly installed equipment's energy performance degrades 0.5% per year, the net impact will be an increase each year of 3% – 0.5% = 2.5%, or to be more exact the escalation/degradation factor would be:

$$((1+\text{Escalation}) * (1\text{-Degradation}))\text{^}(\text{Year-1})$$

The following table summarizes the cash flow, assuming a discount rate of 3%:

Year (end)	Escalation/ Degradation Factor	Simplified Project Savings	Constant Loan Payments	PV of Savings	Net Cash Flow
1	1.0000	$110,400	$73,416	$107,184	$33,768
2	1.0249	$113,143	$73,416	$106,649	$33,233
3	1.0503	$115,955	$73,416	$106,115	$32,699
4	1.0764	$118,837	$73,416	$105,585	$32,169
5	1.1032	$121,790	$73,416	$105,057	$31,641
6	1.1306	$124,816	$73,416	$104,532	$31,116
7	1.1587	$127,918	$73,416	$104,009	$30,593
8	1.1875	$131,097	$73,416	$103,489	$30,073
9	1.2170	$134,354	$73,416	$102,971	$29,555
10	1.2472	$137,693	$73,416	$102,457	$29,041
11	1.2782	$141,115	$73,416	$101,944	$28,528
12	1.3100	$144,621	$73,416	$101,435	$28,019
13	1.3425	$148,215	$73,416	$100,927	$27,511
14	1.3759	$151,898	$73,416	$100,423	$27,007
15	1.4101	$155,673	$73,416	$99,921	$26,505
Totals		$1,977,525	$1,101,240	$1,552,696	$451,456
				SIR =	2.14

This amounts to a very attractive project. It should be noted that this analysis does not consider any tax implications such as depreciation or accelerated depreciation.

16.4 TOTAL PROJECT COSTS FOR BUILDING ENERGY IMPROVEMENT MEASURES

Total project cost, C, as we have seen, is a key variable in evaluating energy financial metrics. It consists of the cost of materials (such as the equipment cost) and the cost of labor associated with installing the equipment. Total project cost is also referred to as total installation cost. Estimating a project's total cost will be a function of the specific energy improvement measure(s) being installed, the property type, the location of the property, and the difficulty of installation, among many other things. It is typically available from contractors bidding on the project. "Ballpark" total project costs from the author's experience for select equipment upgrades in commercial buildings to replace existing equipment most often near, at or beyond their estimated useful life with new, high efficiency equipment are presented in Table 16.1.[2] While equipment cost data is readily available from internet research, equipment distributors, manufacturers, experience from prior projects and contractors, installation costs present much more of a challenge. Removal costs of existing equipment, infrastructure code compliance upgrade costs, installation difficulty, unknown and

TABLE 16.1
Total Retrofit Project Cost Estimations for Select High Efficiency Equipment Replacements in Commercial Buildings*

High Efficiency Equipment Replacement	Total Project Cost	Units
Gas-Fired Condensing Boiler	$0.08	per BTUH
Electric Boiler	$0.03	per BTUH
DX Packaged Roof-Top Air Conditioner	$1,900	per ton
DX Packaged Roof-Top Air Conditioner, with Gas Heating	$2,300	per ton
DX Packaged Roof-Top Air Conditioner, with Electric Heating	$2,000	per ton
DX Split System Air Conditioner (AHU with Heating Coils)	$2,500	per ton
Evaporative Pre-Cooler	$640	per ton
Energy Recovery Ventilator	$6	per CFM
Air-Source Heat Pump	$2,900	per ton
Water-Source Heat Pump System	$3,700	per ton
Ground-Source Heat Pump	$5,700	per ton
Variable Refrigerant Flow Heat Pump	$8,300	per ton
Water-Cooled Centrifugal Chiller Plant	$1,330	per ton
Cooling Tower with VFDs	$370	per ton
Air-Cooled Reciprocating Chiller Plant	$1,860	per ton
Air-Cooled Scroll Chiller Plant	$1,520	per ton
Air-Cooled Screw Chiller Plant	$1,640	per ton
Water-Cooled Screw Chiller Plant	$1,550	per ton
Gas-Fired, Condensing 100 gal Storage Water Heater	$0.086	per BTUH
Gas-Fired, Condensing Instantaneous DHW Heater	$0.035	per BTUH
Electric Water Heater with 100 gal Water Storage	$260	per kW
Instantaneous Electric DHW Heater	$395	per kW
Heat Pump DHW Heater, 120 gal Storage	$1,300	per kW
LED Lighting	$2.20	per bldg ft^2
EMS	$2.70	per bldg ft^2
Fan VFD—NEMA 1 Enclosure (>20 HP)	$130–170	per HP
Fan VFD—NEMA 12 Enclosure (>20 HP)	$150–240	per HP
CHP (reciprocating engine package)	$4,400	per kW
Roof-Top Solar PV	$2.95	per watt
Ground Solar PV	$3.50	per watt
Insulated (R-36) Roof (with existing roof removal)—TPO	$10.50	per SF roof area
Insulated (R-36) Roof (with existing roof removal)—EPDM	$12.00	per SF roof area
Insulated (R-36) Roof (with existing roof removal)—PVC	$13.75	per SF roof area

*The cost estimates are 2022 estimates and assume C-PACE projects for a property located in Hartford, Connecticut. These estimations can be impacted by equipment specifics, site-specific conditions, local building codes, the property location, union versus non-union labor, the need for electrical service upgrades, and so forth.

unanticipated findings, and so on can considerably impact the total project cost. An excellent source for estimating project cost is by using RS Means construction cost guides, which are available annually.[3]

It is not unusual that project costs might exist for similar projects that were completed in the past. While these past costs may be viewed as potentially applicable, it would make better sense to update these costs to the present and correct for a different project's location. One way an estimate can be obtained is by looking at the two components of the total project cost: the cost of materials and the cost of labor. The correction to these costs may be estimated using the information in RS Means,[3] such as their Historical Cost Indexes (which can upgrade construction costs for different years in the same city) and the City Cost Indexes (which can correct for different locations).

Example Problem 16–5

A $750,000 building energy improvement project that was installed in 2010 in New Haven, Connecticut, needs to be updated for a similar project to be installed in the same area in 2014. Estimate the cost of this project in 2014.

The RS Means indexes are as follows:

2010 Historical Cost Index:	183.5
2014 Historical Cost Index:	202.7
Cost in 2014 = $750,000 × (202.7/183.5) = $828,474	

Example Problem 16–6

Take the 2014 project cost (in New Haven, Connecticut) in Example Problem 16–5 and correct it for installation in Long Island City, New York.

The RS Means indexes for these cities are as follows:

2014 City Cost Index, New Haven =	110.2 Weighted Average, Materials and Installation
2014 City Cost Index, Long Island City =	128.6 Weighted Average, Materials and Installation
Long Island City Cost in 2014 =	$828,474 × (128.6/110.2) = $966,804

Note: The City Cost Indexes also identify a weighted Material Cost Index and Installation Cost Index, in addition to the weighted average Materials and Installation Cost Index.

Example Problem 16–7

Assume the following is known about an energy efficiency project at an owner-occupied building:

1. A high efficiency air conditioning roof-top unit (RTU) is selected to replace an existing unit beyond its useful life.

2. The total installed cost of the new unit is $600,000 and is paid upfront out of the capital budget.
3. The first-year (end) energy (utility) savings is projected to be $68,000.
4. The RTU is expected to have a 15-year estimated useful life (EUL).
5. A 3% average annual utility cost escalation factor is considered representative over the equipment's EUL.
6. A 0.5% average annual performance degradation is anticipated over the equipment's EUL based on manufacturer's test data.
7. A 3% discount rate can be assumed.

How attractive is this investment?

The SPB period will be $600,000/$68,000 = 8.8 years, which for many building owners might not represent a compelling investment opportunity. A payback of 3 years or less is often cited by building owners as the criteria for investment in energy improvements. However, as previously indicated, the SPB period does not consider the estimated lifetime savings, which may make for a stronger business case.

Expanding the financial analysis to include lifetime savings and incorporate the time value of money would be the next step.

	Year	Cash Flow*	Present Value**
Initial Investment		($600,000)	($600,000)
Annual Savings	1	$68,000	$66,019
	2	$69,690	$65,689
	3	$71,422	$65,361
	4	$73,196	$65,034
	5	$75,015	$64,709
	6	$76,879	$64,385
	7	$78,790	$64,063
	8	$80,748	$63,743
	9	$82,754	$63,424
	10	$84,811	$63,107
	11	$86,918	$62,792
	12	$89,078	$62,478
	13	$91,292	$62,165
	14	$93,561	$61,855
	15	$95,886	$61,545
Total		$1,218,040 Cash Flow	$956,369 PV

*Annual electricity escalation assumed at 3% and performance degradation assumed at 0.5%; and using the escalation/degradation factor.

**Discount rate assumed at 3%.

However, with the time value of money taken into consideration (using discounted cash flow with a 3% discount rate), the lifetime savings over the EUL would be $956,369, less the upfront investment of $600,000. (If the $600,000 investment is not paid upfront out of the capital budget, but rather is 100% financed, the cost of the loan each year would need to be subtracted from the savings in those same years, and there would be no initial investment. In many cases, particularly where long-term, low-interest, fixed-rate government-supported loans for energy efficiency projects exist, the annual savings each year can exceed the annual cost.)

The investment becomes even more appealing when one looks at the SIR. The SIR in this case would be $956,371/$600,000 = 1.6, which suggests an attractive energy efficiency investment, despite the less attractive simple payback period.

16.5 POTENTIAL PROPERTY VALUE APPRECIATION

Sound business considerations, coupled with growing regulatory requirements, are driving commercial real estate owners and investors to better understand how building energy performance can impact building valuation. Energy saving improvements can have a significant impact on a building's net operating income (NOI) and asset value. If the net operating income of a building is increased due to energy saving improvements (and any maintenance cost savings that might be associated with these improvements), it will have a positive impact on building valuation.

One of the most straightforward value propositions supporting energy efficiency improvements to a building is lower utility bills. The resulting energy savings reduce operating expenses and increase net operating income (NOI), which can have a positive impact on the building's value.

Value attributed to these expense savings is often dependent on lease structure. Owners can directly recoup these savings when they are responsible for utility payments, for example, if the building is owner-occupied or if the tenants have a gross lease (defined as a property lease in which the landlord agrees to pay all expenses normally associated with ownership, such as taxes, utilities, and insurance). For these owners, even small energy savings can be magnified during building valuation.

In buildings where tenants have triple-net leases, that is, where they are responsible for utility costs (in addition to taxes and insurance), owners must pay for energy efficiency upgrades and therefore may have little motivation to make such investments because their tenants would be the sole beneficiary of the benefits. Under these types of leases, the owner's NOI would not be reduced by the energy savings. Such situations produce a misaligned (or split) incentive.

While tenants are always interested in paying lower utility bills, making an investment in a building they do not own can be unrealistic expectation. Fortunately, a relatively new type of lease, the "green lease," is growing in popularity that better aligns tenant and owner interests for investments in energy efficiency and sustainability. An important clause included in green leases involves cost sharing. Cost sharing clauses allow owners to pass all or a portion of the cost of an energy efficiency investment on to their tenants. Owners benefit from an investment that increases the desirability and value of their asset, and tenants benefit from reduced monthly operating costs.

Operational cost savings on maintenance can also convey value. Building owners stand to lower maintenance costs by installing longer-lived or more durable components such as LED lighting.

In a growing number of markets, rental premiums are emerging for energy efficient buildings as forward-thinking tenants, often with corporate sustainability policies to reside in sustainable space, are increasingly willing to pay a premium. For these tenants, leasing energy efficient space is an opportunity to demonstrate a commitment to sustainability, attract the best employees and improve productivity.

Occupancy premiums can also support energy efficiency investment. If it can be determined that a more energy efficient building will result in higher occupancy than an otherwise similar but less energy efficient building, a powerful argument can be made for a higher value. Moreover, savings may be experienced from tenant retention and the corresponding reduction in lost rents, reduced retrofit costs upon releasing of space, lower vacancy at turn-over, and improved lease terms.

Finally, energy efficient building value may also show up in the risk-mitigating protections these assets offer to owners and banks. In the appraisal and underwriting process, energy efficient buildings can offer hedges against changing consumer preferences as well as new laws and increasing energy prices.

Valuation of a building is designed to represent its current marketable value. One method of valuation frequently used in commercial real estate investing is the net income approach. The net income approach often relies on determining the annual capitalization rate (most often referred to simply as the cap rate) for an investment. It typically is derived from the sale of properties with similar risk and return profiles—to reflect investor requirements for a particular property type. It can be calculated by dividing the building's projected annual net income by the current value of the property. So, if an office building is currently valued at \$5,200,000 and the annual income from rentals is \$504,400, the cap rate can be calculated as follows:

$$\$504{,}400 \,/\, \$5{,}200{,}000 = 0.097 \text{ or } 9.7\%$$

Cap rates are generally available for most market and submarket locales for each major commercial building type, and local commercial brokers closely follow them.[4,5] Average national cap rates for different property types in 2022 are presented in Table 16.2. The NOI of a property is its rent less its operating expenses (excluding finance costs) with energy cost generally a major component of these expenses (assuming owners are responsible for energy costs). For example, energy costs can be the single largest operating expense in commercial office buildings, often representing approximately one-third of the operating cost budget.

Hence, if the NOI of a building is increased because of energy saving improvements (and any maintenance cost savings that might be associated with these improvements), the impact on valuation can be estimated.

$$\text{Value} = (\text{NOI}_{\text{After}} - \text{NOI}_{\text{Before}})/\text{Cap Rate} \tag{16–7}$$

TABLE 16.2
National Cap Rates for Select Property Types in 2022

Property Type	National Cap Rate
Office	8.36%
Retail	8.07%
Multifamily	5.32%
Lodging	7.76%
Industrial	5.94%

Assuming all other operating expenses remain the same except for the reduction in utility costs:

$$(NOI_{After} - NOI_{Before}) = \text{Energy Cost Savings} \quad (16\text{–}8)$$

Therefore, the energy cost savings associated with energy efficiency or renewable energy improvements have a direct and proportional impact on the asset value.

Example Problem 16–8

A leading brokerage firm's report indicates that Class A office buildings located in the central business district (CBD) of Wilmington, North Carolina, currently reflect cap rates in the range of 6.80%. An energy audit was conducted on a Class A office building in the CBD of Wilmington where tenants had gross leases.

The audit report recommended five energy efficiency improvements that would result in an annual energy consumption savings of 480,500 kWh with an annual energy cost savings of $40,850. What is the potential positive impact on building valuation, after consideration of the costs to implement the five energy efficiency improvement measures, and assuming all expenses other than utility costs remain the same?

The potential positive impact can be estimated as follows:

$$\text{Valuation Impact} = (\text{Energy Savings})/\text{Cap Rate}$$

$$= \$40{,}850/0.068$$

$$= \$600{,}735$$

Hence, the investment in energy efficiency improvements might result in an estimated asset value increase of slightly more than $600,000.

REFERENCES

1. Kneifel, J. and Webb, D., NIST handbook 135: Life cycle costing manual for the federal energy management program, prepared for U.S. DOE, May 2022, https://doi.org/10.6028/NIST.HB.135e2022-upd1.
2. Buonicore, A.J., Rule-of-thumb total project cost estimates, derived from C-PACE energy efficiency retrofit projects on existing commercial buildings from 2016–2022, Private Files.
3. RS Means, Building construction cost data or green building cost data, see www.rsmeans.com.
4. Integra Realty Resources, Annual commercial real estate trends report (2023 Released Report), www.irr.com/news/just-released-viewpoint-2023-22883.
5. CBRE North America Cap Rate Surveys, Second half of 2022 cap rate survey, www.cbre-ea.com/publications/deconstructing-cre-post/deconstructing-cre/2023/03/29/data-available-for-h2-2022-cap-rate-survey.

17 Estimating Greenhouse Gas Emission Savings

The methodology presented in this chapter can be used to estimate carbon emissions from a building. Carbon emissions presented as the amount of carbon dioxide equivalent (CO_2eq) represent emissions of combustion-related greenhouse gases (i.e., carbon dioxide, methane and nitrous oxide) emitted annually to the atmosphere from the combustion of fossil fuels to produce heat and power for the building. Total CO_2eq is determined for both direct and indirect emissions.

17.1 DIRECT GHG EMISSION SAVINGS

Direct carbon dioxide equivalent (CO_2eq) emissions result from the on-site combustion of fuels. To determine the carbon emission savings resulting from on-site fuel consumption reduction, the MMBtu/yr (million Btu/yr) in fuel savings is multiplied by the lb CO_2eq/MMBtu in Table 17.1.[(1)] CO_2eq includes CH_4 and N_2O emissions using CO_2 equivalency factors of 21 for CH_4 and 310 for N_2O. Dividing lb CO_2eq GHG emissions by 2000 will result in tons of GHG emissions. However, many sources prefer to use metric tons (mt) for GHG emissions. In that case, divide lb CO_2eq GHG emissions by 2205.

17.2 INDIRECT GHG EMISSION SAVINGS

Indirect emissions result from off-site combustion of fuels at power plants that, for example, deliver electricity to the building. To determine the carbon emission savings resulting from electricity delivered to the building, convert the annual kWh energy savings to annual MMBtu savings using the conversion factor of 0.0034144 MMBtu/kWh. Then select the greenhouse gas emission factor (lb CO_2eq/MMBtu), including transmission and distribution losses, from the last column in Table 17.2 for the state where the property is located (the estimation to account for line losses assumes the emission factor is multiplied by 1.045 which is the U.S. average from eGrid).

TABLE 17.1
Direct GHG Emission Factors for Stationary Combustion

Fuel Source	lb CO_2eq/MM Btu
Natural Gas	117
Propane	138.636
Fuel Oil #2	163.089
Fuel Oil #6	165.603

DOI: 10.1201/9781032692777-17

TABLE 17.2
Indirect GHG Emission Factors for Purchased Electricity[(2)]

Data Year	State Abbreviation	State Annual CO_2 Equivalent Non-Baseload Output Emission Rate (lb/MWh)	CO_2eq Emission Rate (lb/MM Btu)	CO_2eq Emission Rate with Losses (lb/MM Btu)
YEAR				
2021	AK	1,285.808	376.8	393.8
2021	AL	1,109.458	325.2	339.8
2021	AR	1,543.595	452.4	472.7
2021	AZ	1,222.965	358.4	374.5
2021	CA	989.106	289.9	302.9
2021	CO	1,617.894	474.2	495.5
2021	CT	919.560	269.5	281.6
2021	DC	395.983	116.1	121.3
2021	DE	991.889	290.7	303.8
2021	FL	1,024.598	300.3	313.8
2021	GA	1,681.243	492.7	514.9
2021	HI	1,760.783	516.0	539.3
2021	IA	1,687.520	494.6	516.8
2021	ID	927.795	271.9	284.1
2021	IL	1,748.413	512.4	535.5
2021	IN	1,983.366	581.3	607.4
2021	KS	2,150.371	630.2	658.6
2021	KY	1,974.846	578.8	604.8
2021	LA	1,175.034	344.4	359.9
2021	MA	932.920	273.4	285.7
2021	MD	1,622.543	475.5	496.9
2021	ME	724.873	212.4	222.0
2021	MI	1,701.798	498.7	521.2
2021	MN	1,590.170	466.0	487.0
2021	MO	1,795.359	526.2	549.8
2021	MS	1,032.304	302.5	316.2
2021	MT	2,039.541	597.7	624.6
2021	NC	1,383.003	405.3	423.6
2021	ND	2,114.881	619.8	647.7
2021	NE	2,098.311	615.0	642.6
2021	NH	966.018	283.1	295.9
2021	NJ	975.577	285.9	298.8
2021	NM	1,666.799	488.5	510.5
2021	NV	1,035.690	303.5	317.2
2021	NY	1,009.099	295.7	309.0
2021	OH	1,646.232	482.5	504.2
2021	OK	1,411.873	413.8	432.4
2021	OR	906.971	265.8	277.8

TABLE 17.2 (*Continued*)
Indirect GHG Emission Factors for Purchased Electricity[2]

Data Year	State Abbreviation	State Annual CO_2 Equivalent Non-Baseload Output Emission Rate (lb/MWh)	CO_2eq Emission Rate (lb/MM Btu)	CO_2eq Emission Rate with Losses (lb/MM Btu)
YEAR				
2021	PA	1,503.642	440.7	460.5
2021	PR	1,622.693	475.6	497.0
2021	RI	925.014	271.1	283.3
2021	SC	1,552.021	454.9	475.3
2021	SD	1,648.014	483.0	504.7
2021	TN	1,714.971	502.6	525.2
2021	TX	1,238.213	362.9	379.2
2021	UT	1,602.786	469.7	490.9
2021	VA	1,068.841	313.2	327.3
2021	VT	359.804	105.4	110.2
2021	WA	1,147.888	336.4	351.6
2021	WI	1,666.630	488.4	510.4
2021	WV	2,071.650	607.1	634.5
2021	WY	2,385.187	699.0	730.5

The approximate pounds of greenhouse gas emission savings in CO_2eq representing the reduction in annual electricity consumption due to installation of ECMs will then be the emission factor multiplied by the building's projected electricity savings in MMBtu/yr.

17.3 THE CHALLENGE OF ACHIEVING GHG EMISSION GOALS

Responding to the ever-growing health and economic threats posed by GHG emissions, the Biden administration established an ambitious new goal to reduce the nation's GHG emissions by 50% by 2030.[3] Moreover, 33 states and the District of Columbia have established GHG emission reduction goals.[4] The goals of select states are presented in Table 17.3.[5] Fossil fuel combustion attributed to commercial and residential buildings accounts for approximately 40% of U.S. total energy consumption and 29% of total U.S. GHG emissions.[4,6]

In New York City, for example, buildings account for approximately two-thirds of the city's GHG emissions. City officials have pledged to address these emissions as part of a plan to make the city carbon neutral by 2050. In April 2019, the city's Climate Mobilization Act included one of the most ambitious plans in the nation for reducing GHG emissions—Local Law 97 (LL-97), which is designed to achieve a 40% reduction in GHG emissions from buildings by 2030. Under LL-97, most buildings over 25,000 square feet today are required to meet new energy efficiency and

TABLE 17.3
Select State GHG Emissions Reduction Goals

State	% Reduction	Baseline Level	By
CA	40	1990	2030
CO	50	2005	2030
CT	45	2001	2030
MA	80	1990	2050
MD	40	2006	2030
ME	45	1990	2030
MN	30	2005	2025
NJ	80	2006	2050
NV	45	2005	2030
NY	40	1990	2030
OR	75	1990	2050
RI	45	1990	2035
VT	40	1990	2030
WA	45	1990	2030

GHG emission limits (effectively a cap on GHG emissions), with stricter limits coming into effect in 2030. If a building's reported GHG emissions exceed the allowable cap (viewed as a non-compliance situation), the building's owner will be fined $268 per ton for emissions in excess of the limit. LL-97 has become a model for other cities throughout the country; for example, Boston, Denver, and Seattle also recently passed performance standards for existing buildings.(7)

Given the commercial building sector's significant GHG emissions and the industry's increasing urgency to reduce its environmental and climate impact, today's forward-thinking building owners are planning and implementing GHG emissions reduction measures. A key component of these building GHG emission reduction plans is the installation of more energy efficient equipment.

Each year it is estimated that an average of 5% of the existing building stock will have energy consuming equipment reaching or exceeding its useful life.(8) Typically, in the small to medium-sized building (SMB) sector, replacement equipment is selected by the building owner or their contractor to comply with minimum local building energy code requirements, which often do not reflect the state of the art in energy efficiency performance. Installing the highest efficiency equipment therefore represents a significant opportunity for building owners to further reduce energy costs and GHG emissions. While such equipment is often more expensive (typically a 20%–30% premium) than minimally code-compliant equipment, the energy savings (reduced energy consumption and energy demand) can often recover the extra cost in a few years, making the return on investment attractive.

Institutional owners of large buildings are increasingly more open to the cost premium associated with higher energy efficiency equipment that is needed to meet their sustainability and environmental, social, and governance (ESG) program goals. Moreover, the greater lifetime energy cost savings, cash flow benefits, GHG emissions reductions, and the increase in property value can make for an attractive investment.

Unfortunately, this is not typically the case with SMB owners, who often focus solely on first cost, which is usually the only cost provided in contractor proposals as contractors want to be competitive and win the project. That is also understandable as the typical installation contractor is unable to provide an estimate of lifetime energy cost savings, impact on cash flow, and GHG emissions reduction for different equipment energy efficiencies. This is unfortunate as such knowledge could influence the building owner to choose a more expensive but more efficient system with a more attractive cash flow potential.

It is indeed unfortunate that the sale of HVAC equipment having the highest energy efficiency practically achievable, particularly in the SMB market, is often not a top priority for competitive reasons. To change this would require a shift from a "lowest price proposal strategy" designed solely to meet minimum local energy code requirements to a "return on investment (ROI) proposal strategy." The ROI strategy would focus not only on first cost, but also on lifecycle energy cost savings, GHG emissions reduction, and the impact on cash flow to make the business case for the building owner to invest in high efficiency equipment.

"Everybody wins" with an ROI strategy:

- Building owners win as they have the information needed to make a confident investment decision, particularly when more expensive, but more energy efficient equipment is purchased.
- Equipment manufacturers win as they accelerate sales of higher efficiency equipment.
- Contractors win as their increased sales of higher efficiency equipment drives ticket size and profit margin and enables the inclusion of utility incentives that can reduce project cost for their building owner clients.
- State and local governments win as they can take credit for GHG emissions reduction captured via these projects and can more quickly reach their ambitious GHG emission goals.

17.4 ELECTRIFICATION-DECARBONIZATION

Building electrification is the movement away from polluting fossil fuel combustion equipment, such as gas and oil-fired boilers and domestic hot water heaters, toward all-electric equipment such as high efficiency heat pumps that provide reliable clean electric heating and cooling to buildings. Electrification, with the move toward improving building energy performance, will play a major role in helping commercial buildings become more energy efficient and reduce their carbon emissions (decarbonization).

Many factors are driving the push towards electrification. In the U.S., state and local climate policies and codes are being enacted to require all-electric construction in major real estate hubs like New York City. Washington became the first state to pass legislation to require building electrification in all commercial buildings. Facility managers have been hearing about the importance of building electrification as a greenhouse gas emissions reductions strategy for years. But Washington's legislation—joining several cities like New York, Denver, and Eugene, Oregon—is a sure sign that building electrification strategy is reaching critical mass in commercial facilities.

On the demand side, businesses and shareholders looking to achieve ambitious ESG and carbon reduction goals are realizing the role electrifying buildings can play in future-proofing assets from volatile fossil fuel prices while improving occupant comfort, health, and safety.

17.4.1 Barriers to Electrification

For the commercial building sector, the ultimate barriers to electrification are economic, not technical.[9] Fuel prices and differences in the capital costs of equipment are the chief determinants of the relative economics of electric compared to non-electric technologies. Electric alternatives exist for all major building energy end-uses. Space heating with electric heat pumps is economically viable in a wide variety of buildings today. Electrification of end-uses generally is relatively cost-effective in new (versus existing) buildings, in settings where a single heat pump can replace the need for the capital cost of an air conditioning unit as well as a space heating unit, and in areas with milder winters.

The potential cost barriers to electrification include:

- The cost of electricity (versus fossil fuel pricing) and their anticipated escalation rates.
- If the fuel switch occurs before the end of the useful life of the existing fossil fuel-fired heating equipment, this effectively raises the cost of the replacement equipment. (A common scenario is that the building's air conditioning system needs to be replaced and the proposal under consideration is to replace it with a more expensive high efficiency heat pump that can also produce space heating electrically and very efficiently. However, this may be replacing an existing gas-fired boiler with remaining useful life.)
- Converting existing fossil fuel equipment to electric may require an expensive upgrade to a building's electric service feed to power the new equipment; also, the building's utility/electric room may be too small for the new switch gear.
- The building's electric demand (kW) may increase with the switch to 100% electric heating and increase the building's electric bill. The net effect will depend upon the increase in building kW demand charges resulting from the switch to electric heating versus the decrease associated with lower energy consumption of the new, presumably higher efficiency equipment.
- Electrification increases the load on the utility's electric delivery infrastructure and the street service may not have enough spare capacity. (While incremental changes in specific buildings are unlikely to have a significant impact, extensive changes by many buildings in the same area could require distribution system upgrades with its own attendant problems, which may ultimately have an impact on building owners.)

In the final analysis, efficient energy consumption in buildings is one of the most affordable ways to reduce building operating expenses and decrease GHG emissions.

Example Problem 17–1

A building owner is planning to convert to 100% electric as part of an electrification project. The small office building currently has a 25-ton packaged air conditioning unit with natural gas for space heating at the end of its useful life. The plan is to convert to a 25-ton air source heat pump for both air conditioning and space heating. The building owner has asked that you take a quick look to see what the financial impact might be in the first year. The cost of electricity is \$0.19/kWh and \$15/kW demand. The cost of natural gas is \$22.28 per thousand cubic feet (with no demand/delivery charge). The following data is also available:

Existing RTU with Natural Gas	ASHP
EER = 8.5	EER = 10.5
80% efficiency, natural gas-firing	Average COP = 2.6
Space heating end-use energy consumption = 250 MM Btu/yr	

The EFLH for cooling from the state TRM is 790 hr/yr.

To annualize peak demand savings, assume billed demand equals read demand and that there are no ratchet provisions. Also assume air conditioning for 4 months of the year and space heating for 5 months.

Existing System:
Cooling Cost:

$$\text{Electricity output} = (25\ \text{ton}) * (12{,}000\ \text{Btu/ton}) * (790\ \text{hr})$$

$$= 237\ \text{MM Btu/yr output}$$

$$\text{Electricity input} = \text{Electricity Output} / (\text{EER} * 1000)$$

$$= 27{,}882\ \text{kWh/yr}$$

$$\text{Cost of electricity} = (27{,}882\ \text{kWh/yr}) * \$0.19/\text{kWh} = \$5{,}298/\text{yr}$$

$$\text{Peak demand estimate: } (27{,}882\ \text{kWh/yr})/790\ \text{hr} = 35\ \text{kW}$$

$$\text{Cooling in 4 summer months, kW/yr} * 4 = 140\ \text{kW/yr}$$

$$\text{Cooling Demand cost} = (140\ \text{kW/yr}) * \$15/\text{kW} = \$2{,}100/\text{yr}$$

$$\text{Total current cooling electricity cost} = \$7{,}398\ /\text{yr}$$

Heating Cost:

Cost of natural gas = (250 MM Btu/yr) * $22.28 / MM Btu = $5,570/yr

Total cost for cooling and heating = $12,968/yr

New ASHP System:
Cooling Cost:

Cooling electricity output = 237 MM Btu/yr output = 69,441 kWh/yr output

Electricity input = (237,000,000 Btu/yr) / (10.5 * 1,000)
= 22,571 kWh/yr

Peak demand estimate: (22,571 kWh/yr)/790 hr/yr = 28.6 kW

Cooling demand cost/season = 28.6 kW * 4 = 114.4 kW/yr

Cooling Cost = 22,571 kWh/yr * $0.19/kWh = $4,288/yr

Demand cost = 114.4 kW/yr * $15/kW = $1,716/yr

Total cooling cost = $6,004/yr

Heating Cost:

Natural gas energy consumption output
= Input energy consumption * efficiency

= 250 MM Btu/yr * 0.8 = 200 MM Btu/yr

or 58,600 kWh/yr output

Electricity input kWh to achieve output
= (58,600 kWh/yr output)/(2.6 kWh output/kWh input)

= 22,538 kWh/yr input

Peak demand input: (22,538 kWh/yr) /680 hr/yr = 33.1 kW

For 5 month heating season: kW * 5 = 165.5 kW

Heating Cost:

22,538 kWh/yr * $0.19 = $4,282/yr

165.5 kW * $15/kW = $2,483/yr

Total for heating = $6,765/yr

Total cost for cooling and heating = $6,004 + $6,765 = $12,769 /yr

Conversion to an ASHP will be approximately break-even in the first year (saves about $200 in first-year energy costs). However, the energy savings become more attractive when looking at lifecycle savings.

The following should be noted in this example.

1. The electrification project will increase utility demand cost due to the addition of electric heating. The electricity demand cost penalty adversely impacts ASHP savings.
2. The analysis assumes that the existing building will be able to accommodate the increased electricity load associated with the ASHP.
3. The analysis assumes the cost of an ASHP and RTU with natural gas is approximately the same.
4. The analysis is very sensitive to natural gas and electricity costs. The ASHP benefits from high natural gas cost and low electricity cost. In this example, we have a high natural gas cost and a high electricity cost which effectively are working against each other.
5. The analysis assumed that there is no demand/delivery cost associated with natural gas.
6. The analysis would be more favorable if the new ASHP had a higher cooling and heating efficiency (as would likely be available today).

Example Problem 17–2

In Example Problem 17–1, how would the greenhouse gas emissions be impacted by the electrification project? Assume the building is located in New York City.

Existing system:

Electricity input = 27,882 kWh/yr = 95.2 MM Btu/yr
Natural gas input = 250 MM Btu/yr

ASHP:

Electricity input = 45,109 kWh/yr = 154 MM Btu/yr

GHG Emission Factors:

Direct emissions = 117 lb CO_2eq/MM Btu (for natural gas)
Indirect emissions = 309 lb CO_2eq/MM Btu (for NY)

GHG Emissions	lb CO_2eq RTU w/NG System	lb CO_2eq ASHP
Direct	29,250	
Indirect	29,417	47,586
Total	58,667	47,586
tons CO_2eq/yr	29.3	23.8

This electrification project would save almost 6 tons per year of GHGs (or 5 mt/yr).

REFERENCES

1. U.S. EPA Center for Corporate Climate Leadership, Direct emissions from stationary combustion sources, Table A-3, December 2020, see www.epa.gov/sites/default/files/2020-12/documents/stationaryemissions.pdf
2. The Emissions & Generation Resource Integrated Database, eGrid 2021, EPA Office of Atmospheric Programs, Clean Air Markets Division, January 30, 2023, https://www.epa.gov/egrid/historical-egrid-data
3. White House Release, April 22, 2021 (50% reduction: 2030 vs. 2005), https://www.whitehouse.gov/briefing-room/press-briefings/2021/04/22/press-briefing-by-press-secretary-jen-psaki-special-presidential-envoy-for-climate-john-kerry-and-national-climate-advisor-gina-mccarthy-april-22-2021/
4. Center for Climate and Energy Solutions, March 2021, https://www.c2es.org/document/climate-action-plans/
5. National Conference of State Legislatures, September 22, 2021, https://www.ncsl.org/
6. Alliance to Save Energy, https://www.ase.org/ and https://energyefficiencyimpact.org/introduction/
7. Building Energy Reporting and Disclosure Ordinance, Boston, MA, October 2021, https://www.boston.gov/departments/environment/building-emissions-reduction-and-disclosure. Building Energy Performance Regulations, City and County of Denver, CO, December 2021, https://www.denvergov.org/files/assets/public/v/1/climate-action/documents/energize-denver-hub/rules-regulations-governing-energize-denver-building-energy-performance-requirements-11-18-2022.pdf. Executive Order 2021–09: Driving Accelerated Climate Action, Seattle, WA, November 2021, https://www.seattleclimate.org/strategies/governments/seattle/cop26-executive-order
8. Assumed 20 year weighted average useful life of multiple energy conservation measures, e.g., high efficiency air conditioning and heating systems, LED lighting, etc.
9. Deason, J., et al., Electrification of buildings and industry in the United States, report prepared by the Lawrence Berkeley National Laboratory for the U.S. Department of Energy, March 2018, https://eta-publications.lbl.gov/sites/default/files/electrification_of_buildings_and_industry_final_0.pdf

Appendix I
Helpful Conversion Factors

- 1 ton = 12,000 Btu/hr
- 1 Btu/hr = 0.293 W
- 1 kW = 1,000 W
- 1 kWh = 3,412 Btu = 0.03412 therms
- 1 kW = 3,412 Btu/hr
- 1 W = 3.412 Btu/hr
- 1 MM Btu = 1,000,000 Btu = 1,000 kBtu = 293 kWh
- 1 kBtu = 1,000 Btu
- EER = 3.412 × COP
- EER = 12/(kW/ton)
- EER ~ (1.12 × IEER) – (0.02 × $IEER^2$)
- IEER ~ [1.12 – (1.2544–0.08 × $EER)^{0.5}$]/0.04
- kW/ton = 12/EER
- COP = EER/3.412
- kW/ton = 12/(COP × 3.412)
- HSPF = COP × 3.412
- 1 therm = 29.3 kWh
- 1 hp = 0.7457 kW
- 1 boiler-horsepower (BHP) = 33,745 Btu/hr = 9.89 kW
- CF natural gas (HHV) = 1,029 Btu
- CCF natural gas = 102,900 Btu
- 1 therm = 100,000 Btu
- 1 decatherm (Dtherm) = 10 therms = 10 CCF = 1 MM Btu
- 1 gallon of fuel oil (#2) = 138,690 Btu

Appendix II
Building Types

In the Commercial Buildings Energy Consumption Survey (CBECS), buildings are classified according to principal building activity, which is the primary business, commerce, or function carried on within each building. Buildings used for more than one of the activities described below are assigned to the activity occupying the most floorspace. A building assigned to a particular principal activity category may be used for other activities in a portion of its space or at some time during the year.

Building Type	Definition	Example Sub-Categories
Education	Buildings used for academic or technical classroom instruction, such as elementary, middle, or high schools, and classroom buildings on college or university campuses. Buildings on education campuses for which the main use is not classroom are included in the category related to their use. For example, administration buildings are included in the office category, dormitories in the lodging category, and libraries in the public assembly category.	Elementary or Middle School High School College or University Pre-School or Daycare Adult Education Career or Vocational Training Religious Education
Food Sales	Buildings used for retail or wholesale of food.	Grocery Store Food Market Gas Station with Convenience Store
Food Service	Buildings used for preparation and sale of food and beverages for consumption.	Fast Food Restaurant or Cafeteria Bar Catering Service Reception Hall Coffee, Bagel or Doughnut Shop Ice Cream or Frozen Yogurt Shop
Health Care (Inpatient)	Buildings used as diagnostic and treatment facilities for inpatient care.	Hospital Inpatient Rehabilitation
Health Care (Outpatient)	Buildings used as diagnostic and treatment facilities for outpatient care.	Medical Clinic Outpatient Rehabilitation Veterinarian

(Continued)

(Continued)

Building Type	Definition	Example Sub-Categories
Lodging	Buildings used to offer multiple accommodations for short-term or long-term residents.	Motel or Inn Hotel Dormitory, Fraternity, Sorority
Mercantile (Retail other than Mall)	Buildings used for the sale and display of goods other than food.	Retail Store Liquor Store Rental Center Vehicle Dealership
Mercantile (Enclosed and Strip Malls)	Shopping malls comprising multiple connected establishments.	Enclosed Mall Strip Center
Office	Buildings used for general office space, professional office or administrative offices.	Administrative Office Professional Office Bank or Financial Institution Medical Office Sales Office City Hall Religious Office
Public Assembly	Buildings in which people gather for social or recreational activities.	Community Center Lodge Meeting Hall Convention Center Senior Center Gymnasium, Health Club Museum Theater Cinema Sports Arena Casino, Nightclub Library Funeral Home Exhibition Hall
Public Order and Safety	Buildings use for preservation of law and order or public safety.	Police Station Fire Station Penitentiary Courthouse
Religious Worship	Buildings where people gather for religious activities.	Church Houses of Worship
Service	Buildings in which some type of service is provided, other than food service or retail sales of goods.	Vehicle Service/Repair Dry Cleaner or Laundromat Post Office Car Wash Gas Station Beauty Parlor, Barber Shop Copy Center Kennel
Warehouse and Storage	Buildings used to store goods.	Refrigerated Warehouse Non-Refrigerated Warehouse Distribution Center Public Storage, Self-Storage

Appendix III
CDD and HDD Data

Average Heating and Cooling Degree Days in Select U.S. Cities*

State	City	HDD	CDD
Alabama	Birmingham	2,693	1,948
	Mobile	1,662	2,463
Alaska	Anchorage	10,121	5
Arizona	Flagstaff	6,912	108
	Phoenix	941	4,557
	Tucson	1,466	3,180
Arkansas	Little Rock	2,976	2,124
California	Los Angeles	1,284	617
	Oakland	2,584	153
	Sacramento	2,699	1,251
	San Diego	1,117	761
	San Francisco	2,708	142
Colorado	Denver	5,942	777
Connecticut	Bridgeport	5,290	814
	Hartford	5,992	739
Delaware	Wilmington	4,786	1,121
Florida	Jacksonville	1,316	2,624
	Miami	130	4,458
	Orlando	544	3,379
	Tampa	528	3,517
Georgia	Atlanta	2,694	1,841
	Savannah	1,753	2,439
Hawaii	Honolulu	0	4,649
Idaho	Boise	5,658	890
Illinois	Chicago	6,311	842
	Springfield	5,429	1,135
Indiana	Indianapolis	5,322	1,055
Iowa	Des Moines	6,240	1,041
Kansas	Lawrence	4,801	1,466
	Wichita	4,564	1,678
Kentucky	Louisville	4,258	1,390
Louisiana	Baton Rouge	1,610	2,653
	New Orleans	1,318	2,846

(Continued)

(Continued)

Average Heating and Cooling Degree Days in Select U.S. Cities*

State	City	HDD	CDD
Maine	Portland	7,082	365
Maryland	Baltimore	4,567	1,228
Massachusetts	Boston	5,621	750
	Worcester	6,707	453
Michigan	Detroit	6,183	775
	Grand Rapids	6,687	626
	Lansing	6,889	570
Minnesota	Minneapolis–St. Paul	7,565	751
	Duluth	9,425	209
Mississippi	Jackson	2,311	2,265
Missouri	Kansas City	4,593	1,657
	St. Louis	4,504	1,631
Montana	Billings	6,779	636
Nebraska	Omaha	6,127	1,109
Nevada	Las Vegas	2,105	3,348
	Reno	5,201	695
New Hampshire	Concord	7,180	461
New Jersey	Atlantic City	4,950	995
	Newark	4,710	1,242
New Mexico	Albuquerque	4,069	1,348
New York	Albany	6,608	502
	New York	4,828	978
	White Plains	5,563	755
North Carolina	Asheville	4,148	829
	Charlotte	3,081	1,669
	Wilmington	2,388	1,987
North Dakota	Grand Forks	9,167	477
Ohio	Cincinnati	4,754	1,151
	Cleveland	5,904	743
	Columbus	5,322	971
Oklahoma	Tulsa	3,494	2,060
Oregon	Portland	4,222	423
Pennsylvania	Harrisburg	5,228	990
	Philadelphia	4,579	1,273
	Pittsburgh	5,356	875
Rhode Island	Providence	5,591	729
South Carolina	Charleston	1,889	2,328
	Columbia	2,529	2,108
South Dakota	Rapid City	7,027	678
Tennessee	Memphis	2,935	2,214
	Nashville	3,542	1,683
Texas	Austin	1,654	2,989
	Dallas	2,105	2,924
	Houston	1,414	3,001

Average Heating and Cooling Degree Days in Select U.S. Cities*

State	City	HDD	CDD
	San Antonio	1,480	3,115
Utah	Salt Lake City	5,521	1,193
Vermont	Burlington	7,406	496
Virginia	Norfolk	3,244	1,657
	Roanoke	4,098	1,184
Washington	Seattle	4,729	177
	Spokane	6,687	423
West Virginia	Charleston	4,443	1,066
Wisconsin	Green Bay	7,684	470
	Milwaukee	6,774	684
Wyoming	Cheyenne	7,148	324

*Heating and cooling degree days use a 65°F base.

Reference Cities for Five CBECS Climate Zones:

Climate Zone	Reference City
Hot/Very Hot & Humid	Miami, FL
Hot & Dry	Las Vegas, NV
Marine	Seattle, WA
Cool	Chicago, IL
Cold/Very Cold	Duluth, MN

Annual HDD and CDD by Census Region, 2018*

Census Region	CDD	HDD	States
New England	651	6,323	CT, ME, MA, NH, RI, VT
Middle Atlantic	885	5,769	NJ, NY, PA
East North Central	974	6,434	IL, IN, MI, OH, WI
West North Central	1,132	6,974	IA, KS, MN, MO, NE, ND, SD
South Atlantic	2,411	2,638	DE, FL, GA, MD (and DC), NC, SC, VA, WV
East South Central	1,932	3,479	AL, KY, MS, TN
West South Central	2,859	2,252	AK, LA, OK, TX
Mountain	1,584	4,809	AZ, CO, ID, MT, NV, NM, UT, WY
Pacific	1,004	3,172	AK, CA, HI, OR, WA
U.S. Average	**1,579**	**4,291**	

*U.S. EIA, Short Term Energy Outlook Data Browser, U.S. Regional Weather Data Table.
Data for use in support of 2018 CBECS.

Annual HDD and CDD by Census Region, 2020*

Census Region	CDD	HDD	States
New England	645	5,821	CT, ME, MA, NH, RI, VT
Middle Atlantic	848	5,199	NJ, NY, PA
East North Central	831	5,855	IL, IN, MI, OH, WI
West North Central	964	6,326	IA, KS, MN, MO, NE, ND, SD
South Atlantic	2,335	2,263	DE, FL, GA, MD (and DC), NC, SC, VA, WV
East South Central	1,636	3,064	AL, KY, MS, TN
West South Central	2,735	1,812	AK, LA, OK, TX
Mountain	1,682	4,785	AZ, CO, ID, MT, NV, NM, UT, WY
Pacific	1,071	3,219	AK, CA, HI, OR, WA
U.S. Average	**1,519**	**3,914**	

*U.S. EIA, Short Term Energy Outlook Data Browser, U.S. Regional Weather Data Table. Data for use in support of 2020 RECS.

U.S. Average HDD and CDD Data

Year	HDD	CDD
2018	4,291	1,579
2019	4,317	1,496
2020	3,914	1,519
2021	3,934	1,492
2022	4,240	1,556

Appendix IV
Air Change Rates per Hour in Typical Rooms and Commercial Buildings

Building/Room	Air Change Rate
Assembly Halls	4–6
Auditoriums	8–15
Bakeries	20–30
Banks	4–10
Barber Shops	6–10
Bars	20–30
Beauty Shops	6–10
Boiler Rooms	15–20
Bowling Alleys	10–15
Cafeterias	12–15
Churches	8–15
Classrooms	6–20
Club Rooms	12
Clubhouses	20–30
Cocktail Lounges	20–30
Computer rooms	15–20
Courthouses	4–10
Dance Halls	6–9
Dental Offices	8–12
Department Stores	6–10
Dining Halls	12–15
Dining Rooms (Restaurants)	12
Dress Shops	6–10
Drugstores	6–10
Engine Rooms	4–6
Fire Stations	4–10
Garage Repair Shops	20–30
Garage Storage	4–6
Hospital Rooms	4–6
Jewelry Shops	6–10
Kitchens	15–60
Laundries	10–15
Libraries	4
Lunchrooms	12–15

(Continued)

(Continued)

Building/Room	**Air Change Rate**
Luncheonettes	12–15
Nightclubs	20–30
Machine Shops	6–12
Malls	6–10
Medical Centers	8–12
Medical Clinics	8–12
Medical Offices	8–12
Municipal Buildings	4–10
Museums	12–15
Offices, Public	3
Offices, Private	4
Paint shops	10–15
Photo Darkrooms	10–15
Police Stations	4–10
Post Offices	4–10
Residences	1–2
Restaurants	8–12
Retail	6–10
School Classrooms	4–12
Shoe Shops	6–10
Shopping Centers	6–10
Shops, Machine	5
Shops, Paint	15–20
Shops, Woodworking	5
Supermarkets	4–10
Swimming Pools	20–30
Town halls	4–10
Taverns	20–30
Theaters	8–15
Warehouses	2

Source: The Engineering Toolbox Website, www.engineeringtoolbox.com.

Appendix V

CBECS Data

2018 CBECS EUI Data by Census Division

Building Activity	New England	Middle Atlantic	East North Central	West North Central	South Atlantic
Education	67.4	75.1	69.4	59.5	64.1
Food Sales			259.5		
Food Service			339.9		292.6
Health Care	134.7	135.4	165.8	140.6	166.8
Inpatient		194.2	197.5		217.0
Outpatient		82.7	93.1		80.9
Lodging	93.3	92.5	101.3	74.1	84.8
Mercantile		88.9	100.8	78.0	85.4
Retail (not Mall)		59.0	69.0	67.1	62.4
Enclosed/Strip Mall		105.8	134.3		105.1
Office	60.7	76.6	65.0	64.6	66.2
Public Assembly	71.6	95.0	105.4	80.8	71.7
Public Order/Safety					
Religious Worship		51.4	36.0	43.8	32.2
Service	66.4	47.4	49.8	33.9	52.5
Warehouse/Storage	25.1	37.5	35.6	42.3	27.5

Building Activity	East South Central	West South Central	Mountain	Pacific
Education	61.0	55.9	56.7	45.8
Food Sales				
Food Service		244.6	247.6	214.6
Health Care	118.6	147.8	115.5	134.7
Inpatient		183.4		180.6
Outpatient			57.0	85.9
Lodging	55.9	74.3	95.9	77.6
Mercantile	76.9	84.9	87.7	95.5
Retail (not Mall)	64.7	67.5	62.9	58.0
Enclosed/Strip Mall		97.5	112.5	113.5
Office	88.3	57.3	57.1	60.0
Public Assembly		74.4	84.7	55.4
Public Order/Safety				72.0

(Continued)

(Continued)

Building Activity	East South Central	West South Central	Mountain	Pacific
Religious Worship	24.6	33.4		30.0
Service		39.8	65.1	40.0
Warehouse/Storage	30.1	27.1	29.2	20.1

2018 CBECS Select End-Uses as % of Whole Building EUI (kBtu/SF-yr) by Building Type Nationally*

Building Activity	Space Heating	Space Cooling	Ventilation	Water Heating	Lighting	Other
Education	44.0%	11.6%	7.7%	5.9%	8.8%	22.0%
Food Sales	15.8%	3.1%	5.6%	N/A	5.6%	69.9%
Food Service	13.0%	7.0%	6.0%	7.1%	3.9%	63.0%
Health Care	31.0%	6.8%	16.9%	6.4%	8.4%	30.5%
Inpatient	32.4%	7.1%	14.8%	7.6%	6.2%	31.9%
Outpatient	26.6%	6.0%	23.3%	2.7%	15.2%	26.2%
Lodging	18.3%	7.0%	14.6%	19.3%	8.4%	32.4%
Mercantile	25.8%	7.6%	12.2%	6.8%	14.8%	32.8%
Retail (not Mall)	25.7%	10.0%	18.3%	1.4%	18.7%	25.9%
Enclosed/Strip Mall	25.7%	6.4%	9.1%	9.2%	12.7%	36.9%
Office	30.6%	7.8%	19.7%	1.5%	11.9%	28.5%
Public Assembly	51.0%	19.5%	4.9%	1.0%	6.3%	17.3%
Public Order/Safety	35.0%	10.7%	7.1%	12.4%	13.3%	21.5%
Religious Worship	50.4%	11.3%	8.2%	2.3%	9.6%	18.2%
Service	55.1%	8.2%	4.9%	N/A	12.9%	18.9%
Warehouse/Storage	41.3%	10.2%	4.4%	1.3%	13.3%	23.3%

*End-use percentages do not reflect the impact of climate zone. For example, an office building in Miami would have different space heating and space cooling end-use percentages compared to a building in Chicago. The 2018 CBECS reflects 2018 weather. End-use percentages should be adjusted for property location (see Chapter 2).

2003 CBECS Select End-Uses as % of Whole Building EUI (kBtu/SF-yr) by Building Type Nationally*

Building Activity	Space Heating	Space Cooling	Ventilation	Water Heating	Lighting	Other
Education	47.4%	9.6%	10.1%	7.0%	13.8%	12.1%
Food Sales	14.5%	4.9%	3.0%	1.5%	18.4%	57.7%
Food Service	16.7%	6.7%	5.7%	15.6%	10.0%	45.3%
Health Care	37.5%	7.5%	7.1%	16.1%	17.6%	14.2%

Inpatient	36.8%	7.5%	8.0%	19.4%	16.1%	12.2%
Outpatient	40.3%	7.6%	3.5%	2.6%	23.9%	22.1%
Lodging	22.2%	4.9%	2.7%	31.4%	24.3%	14.5%
Mercantile	26.3%	10.8%	6.6%	5.6%	30.1%	20.6%
Retail (not Mall)	33.6%	8.0%	5.0%	1.5%	34.8%	17.1%
Enclosed/Strip Mall	23.6%	12.1%	7.3%	7.5%	28.0%	21.5%
Office	35.3%	9.6%	5.6%	2.2%	24.9%	22.4%
Public Assembly	52.9%	10.2%	16.9%	1.1%	7.5%	11.4%
Public Order/Safety	43.1%	7.7%	8.2%	12.1%	14.2%	14.7%
Religious Worship	60.2%	6.7%	3.2%	1.8%	10.1%	18.0%
Service	46.6%	4.9%	7.8%	1.4%	20.2%	19.1%
Warehouse/Storage	42.7%	2.9%	4.4%	1.3%	29.0%	19.7%

*End-use percentages do not reflect the impact of climate zone. For example, an office building in Miami would have different space heating and space cooling end-use percentages compared to a building in Chicago. The 2003 CBECS reflects 2003 weather. End-use percentages should be adjusted for property location (see Chapter 2).

Appendix VI
RECS 2020 Data

Multifamily (five or more) Units

	Northeast	Mid-West	South	West
Census Divisions	New England Mid-Atlantic	East North Central West North Central	South Atlantic East South Central West South Central	Mountain North Mountain South Pacific
No. Household Units	5,100,000	4,200,000	7,830,000	5,700,000
Avg. SF/Unit	869	912	965	853
EUI, kBtu/SF	41.6	42.8	33.8	34.3
End-use % of EUI				
Space Heating	32.30%	42.60%	15.40%	19.90%
Space Cooling	8.00%	6.40%	18.20%	11.30%
DHW Heating	29.8%	25.60%	27.40%	38.70%
Refrigerators	4.70%	4.60%	6.20%	6.20%
Other	25.20%	20.80%	32.80%	23.90%

All Residential Homes*
2020 CBECS Annual Household Site End-use Energy Consumption Percentages by State*

	End-use Percent of Total Household Energy Consumption				
State	**Space Heating**	**DHW Heating**	**Space Cooling**	**Refrigerators**	**Other**
Alabama	29.80%	17.40%	16.30%	4.80%	31.70%
Alaska	64.80%	19.80%	0.60%	2.10%	12.60%
Arizona	16.40%	17.60%	28.30%	5.20%	32.50%
Arkansas	40.30%	17.40%	12.90%	3.70%	25.60%
California	27.60%	30.40%	10.40%	5.20%	26.40%
Colorado	52.80%	19.10%	5.10%	3.30%	19.70%
Connecticut	56.00%	17.00%	4.50%	2.60%	19.90%
Delaware	41.90%	18.20%	8.50%	3.80%	27.60%
District of Columbia	38.90%	21.90%	9.90%	4.10%	25.20%
Florida	11.30%	17.00%	29.40%	6.30%	36.00%
Georgia	33.40%	18.70%	13.90%	4.30%	29.60%
Hawaii	4.30%	28.70%	32.00%	10.20%	24.80%

(Continued)

(Continued)

All Residential Homes*
2020 CBECS Annual Household Site End-use Energy Consumption Percentages by State*

	End-use Percent of Total Household Energy Consumption				
State	Space Heating	DHW Heating	Space Cooling	Refrigerators	Other
Idaho	47.90%	20.50%	5.00%	3.40%	23.20%
Illinois	59.40%	14.70%	5.10%	2.80%	17.90%
Indiana	53.50%	15.60%	6.20%	3.10%	21.50%
Iowa	57.10%	14.00%	5.10%	3.00%	20.80%
Kansas	48.00%	16.80%	9.70%	3.30%	22.10%
Kentucky	42.00%	16.90%	10.50%	3.90%	26.80%
Louisiana	18.90%	19.30%	22.70%	4.60%	34.50%
Maine	58.30%	16.80%	2.90%	2.60%	19.40%
Maryland	44.50%	17.70%	8.70%	3.90%	25.20%
Massachusetts	54.40%	18.00%	4.80%	2.70%	20.10%
Michigan	62.00%	13.60%	3.90%	2.50%	18.10%
Minnesota	59.10%	15.20%	3.50%	2.70%	19.50%
Mississippi	31.00%	18.10%	17.80%	4.50%	28.60%
Missouri	50.80%	15.10%	8.80%	3.50%	21.80%
Montana	53.50%	20.80%	3.20%	3.00%	19.50%
Nebraska	51.20%	17.10%	6.40%	3.60%	21.60%
Nevada	27.80%	20.70%	16.40%	4.00%	31.00%
New Hampshire	58.20%	16.40%	3.70%	2.90%	18.90%
New Jersey	51.10%	18.60%	5.80%	2.70%	21.80%
New Mexico	45.80%	21.10%	8.10%	3.60%	21.50%
New York	52.20%	19.60%	5.30%	2.80%	20.00%
North Carolina	35.60%	18.70%	13.40%	4.60%	27.80%
North Dakota	59.60%	15.20%	3.10%	3.10%	19.10%
Ohio	56.00%	15.20%	5.60%	3.00%	20.30%
Oklahoma	39.50%	18.40%	12.80%	3.50%	25.80%
Oregon	42.70%	23.90%	5.20%	4.20%	24.00%
Pennsylvania	52.80%	16.20%	5.90%	3.10%	21.90%
Rhode Island	59.40%	17.60%	4.00%	2.60%	16.50%
South Carolina	28.00%	18.50%	16.90%	4.70%	31.90%
South Dakota	55.00%	16.30%	5.20%	3.30%	20.20%
Tennessee	37.00%	17.40%	12.20%	4.00%	29.50%
Texas	24.60%	18.30%	20.50%	4.70%	31.90%
Utah	45.50%	20.30%	6.80%	3.20%	24.20%
Vermont	59.00%	17.60%	2.70%	2.80%	17.80%
Virginia	37.20%	18.60%	10.10%	4.00%	30.10%
Washington	44.70%	23.70%	4.50%	3.90%	23.20%
West Virginia	51.50%	16.30%	7.20%	3.30%	21.70%
Wisconsin	60.10%	13.60%	3.60%	3.10%	19.70%
Wyoming	61.20%	15.90%	3.40%	2.80%	16.60%

*Includes all residential homes: single family, multifamily (two or more units) and mobile homes.

Appendix VII
Select Modeled End-Use Data

1. ***Source: Quantifying the Value of Building Reuse, prepared for the National Trust for Historic Preservation by Quantis International, January 23, 2012 (Modeling Study).***

Large Commercial Office (heating source is natural gas)

	End-Use % of Whole Building EUI in Select Cities			
	Portland	**Phoenix**	**Chicago**	**Atlanta**
EUI, kBty/SF-yr	70	71	93	76
End-Use				
Space Cooling	3%	28%	8%	17%
Space Heating	28%	0%	41%	21%
DHW Heating	2%	1%	1%	1%
Ventilation Fans	17%	20%	12%	14%
Exterior Lighting	7%	7%	5%	6%
Interior Lighting	22%	21%	16%	20%
Miscellaneous	21%	23%	17%	21%

Elementary School (heating source is natural gas)

	End-Use % of Whole Building EUI in Select Cities			
	Portland	**Phoenix**	**Chicago**	**Atlanta**
EUI, kBtu/SF-yr	60	61	80	65
End-Use				
Space Cooling	5%	41%	4%	25%
Space Heating	45%	8%	60%	29%
DHW Heating	10%	10%	8%	9%
Ventilation Fans	10%	11%	6%	9%
Exterior Lighting	2%	2%	1%	2%
Interior Lighting	17%	16%	13%	15%
Miscellaneous	11%	12%	8%	11%

Mid-Rise Multifamily (heating source is electricity)

	End-Use % of Whole Building EUI in Select Cities			
	Portland	Phoenix	Chicago	Atlanta
EUI, kBtu/SF-yr	63	50	64	53
End-Use				
Space Cooling	6%	26%	3%	13%
Space Heating	44%	12%	48%	28%
DHW Heating	22%	28%	22%	26%
Refrigerators	5%	6%	5%	6%
Lighting & Appliances	22%	28%	22%	26%

2. ***Source: New Buildings Institute White Paper, Sensitivity Analysis: Comparing the Impact of Design, Operation, and Tenant Behavior on Building Energy Performance, Appendix C: End-Use by Climate, July 2011 (eQuest Modeling Study, TMY Weather Data, Office Building Constructed in 2007 complying with ASHRAE 90.1–2007).***

Medium Office Building (natural gas heat)

	End-Use % of Whole Building EUI in Select Cities					
	Albuquerque	Atlanta	Baltimore	Boulder	Chicago	Duluth
End-Use						
Space Cooling	10.5%	16.8%	11.5%	6.3%	7.9%	3.3%
Space Heating	18.6%	21.3%	32.7%	31.0%	41.2%	56.3%
DHW Heating	1.5%	1.3%	1.2%	1.4%	0.3%	1.0%
Ventilation Fans	18.5%	13.6%	12.4%	16.4%	11.6%	9.5%
Pumps & Aux.	0.3%	neg.	0.2%	0.3%	0.3%	0.3%
Exterior Lights	6.8%	6.3%	5.6%	6.0%	5.1%	4.0%
Interior Lights	21.4%	19.8%	17.8%	18.8%	16.0%	12.5%
Miscellaneous	22.5%	20.8%	18.5%	19.8%	16.8%	13.1%

	End-Use % of Whole Building EUI in Select Cities				
	Helena	Houston	Las Vegas	Minneapolis	Phoenix
End-Use					
Space Cooling	3.2%	23.0%	22.2%	6.8%	28.0%
Space Heating	44.0%	22.7%	3.0%	48.3%	0.6%
DHW Heating	1.2%	1.0%	1.3%	1.1%	1.1%
Ventilation Fans	13.8%	11.8%	20.6%	10.6%	20.3%
Pumps & Aux.	0.3%	0.1%	0.2%	0.2%	0.1%
Exterior Lights	5.0%	5.5%	7.0%	4.4%	6.7%
Interior Lights	15.9%	17.5%	22.3%	14.0%	21.1%
Miscellaneous	16.7%	18.4%	23.4%	14.7%	22.2%

	End-Use % of Whole Building EUI in Select Cities	
	San Francisco	Seattle
End-Use		
Space Cooling	4.2%	3.3%
Space Heating	11.2%	27.5%
DHW Heating	1.8%	1.5%
Ventilation Fan	19.6%	16.6%
Pumps & Aux.	0.2%	0.3%
Exterior Lights	8.4%	6.8%
Interior Lights	26.6%	21.5%
Miscellaneous	28.0%	22.5%

Note: Seattle represents a mild marine climate, Chicago a cold climate, Phoenix a dry hot climate, and Atlanta a moist hot climate.

3. ***Source: Adopted from 2011 U.S. DOE Buildings Energy Data Book (March 2012), End-use data based on U.S. DOE Commercial Buildings Benchmark Models, Existing Commercial Reference Buildings Constructed In or After 1980, Version 1.3_5.0, November 2010; Buildings comply with ASHRAE 90.1–2004; 2003 CBECS Climate Zones; weather data used in modeling is TMY data – end-use percentages are weather-normalized.***

2011 *U.S. DOE Buildings Energy Data Book*: End Use Energy Consumption Percentages of Whole Building EUI for Select Property Types by Climate Zone (Buildings constructed in 1980 or later)

Existing Large Office Buildings (>100,000 SF)

Typical City	Climate Zone	Heating %	Cooling %	Water Heating %	Lighting %	Other %	Whole Building EUI, kBtu/SF
Miami	1A	0.5	38.0	0.5	27.9	33.1	57.6
Houston	2A	7.4	31.0	0.5	28.2	32.9	57.0
Phoenix	2B	5.5	29.6	0.5	29.4	35.0	54.7
Atlanta	3A	12.3	25.2	0.7	28.7	33.1	56.0
Los Angeles	3B	5.6	24.0	0.8	32.4	37.2	49.6
Las Vegas	3B	9.1	21.4	0.6	31.9	37.0	50.4
San Francisco	3C	11.2	12.5	0.9	35.9	39.6	44.8
Baltimore	4A	17.3	21.2	0.7	28.4	32.4	56.6
Albuquerque	4B	13.8	13.6	0.8	33.6	38.2	47.8
Seattle	4C	20.7	9.2	1.0	32.9	36.1	48.8
Chicago/Hartford	5A	26.1	13.0	0.9	28.3	31.6	56.7

(Continued)

(Continued)

2011 *U.S. DOE Buildings Energy Data Book*: End Use Energy Consumption Percentages of Whole Building EUI for Select Property Types by Climate Zone (Buildings constructed in 1980 or later)

Existing Large Office Buildings (>100,000 SF)

Typical City	Climate Zone	Heating %	Cooling %	Water Heating %	Lighting %	Other %	Whole Building EUI, kBtu/SF
Boulder	5B	19.4	10.0	1.0	32.9	36.7	48.9
Minneapolis	6A	32.2	11.0	0.8	26.4	29.5	60.8
Helena	5B	27.2	7.1	1.0	30.8	34.0	52.2
Duluth	7	38.9	6.1	1.0	25.7	28.4	62.5
Fairbanks	8	55.1	3.2	0.8	19.3	21.5	83.3

Existing Medium Office Buildings (>10,000–100,000 SF)

Typical City	Climate Zone	Heating %	Cooling %	Water Heating %	Lighting %	Other %	Whole Building EUI, kBtu/SF
Miami	1A	1.6	36.1	0.7	27.6	34.0	61.0
Houston	2A	8.0	26.8	0.9	29.1	35.2	57.8
Phoenix	2B	6.7	29.4	0.7	28.3	34.9	59.5
Atlanta	3A	14.0	18.2	1.1	30.3	36.4	55.6
Los Angeles	3B	8.3	16.1	1.0	33.9	40.8	49.7
Las Vegas	3B	9.9	23.3	0.9	29.6	36.4	56.8
San Francisco	3C	12.6	6.3	1.3	36.5	43.3	46.1
Baltimore	4A	21.0	13.9	1.0	29.2	34.9	57.7
Albuquerque	4B	15.2	12.7	1.1	32.0	38.9	52.6
Seattle	4C	22.8	4.8	1.2	32.5	38.7	51.8
Chicago/Hartford	5A	29.1	9.0	1.1	27.5	33.2	61.1
Boulder	5B	21.5	8.2	1.3	31.2	37.8	53.9
Minneapolis	6A	35.7	7.3	1.1	25.4	30.6	66.2
Helena	5B	30.8	4.9	1.2	28.6	34.5	58.8
Duluth	7	41.8	3.5	1.2	24.3	29.3	69.2
Fairbanks	8	57.0	1.7	0.9	18.2	22.2	92.6

Existing Retail Buildings

Typical City	Climate Zone	Heating %	Cooling %	Water Heating %	Lighting %	Other %	Whole Building EUI, kBtu/SF
Miami	1A	0.6	27.8	neg.	45.1	26.5	82.7
Houston	2A	13.3	18.6	neg.	42.7	25.5	87.3
Phoenix	2B	9.8	20.3	neg.	44.1	25.8	84.6
Atlanta	3A	26.5	9.8	neg.	39.6	24.2	94.1
Los Angeles	3B	10.1	4.8	neg.	54.4	30.7	68.5
Las Vegas	3B	18.2	13.7	neg.	44.1	24.0	84.6
San Francisco	3C	28.5	0.9	neg.	47.4	23.2	78.6
Baltimore	4A	40.0	5.8	neg.	34.7	19.5	107.4
Albuquerque	4B	32.1	5.6	neg.	39.6	22.7	94.1
Seattle	4C	40.3	0.9	neg.	39.1	19.7	95.3
Chicago/Hartford	5A	47.9	3.5	neg.	30.0	18.5	124.1
Boulder	5B	40.6	3.0	neg.	35.0	21.4	106.6
Minneapolis	6A	52.6	2.6	neg.	26.0	18.9	143.6
Helena	5B	47.1	1.5	neg.	29.1	22.2	127.9
Duluth	7	58.0	0.7	neg.	23.3	18.0	160.0
Fairbanks	8	68.3	0.2	neg.	16.3	15.2	228.9

Existing Supermarkets

Typical City	Climate Zone	Heating %	Cooling %	Water Heating %	Lighting %	Other %	Whole Building EUI, kBtu/SF
Miami	1A	2.8	15.1	0.5	40.8	40.8	78.1
Houston	2A	21.1	9.5	0.4	31.1	37.9	102.3
Phoenix	2B	21.6	11.3	0.4	32.1	34.6	99.2
Atlanta	3A	34.2	4.5	0.4	26.4	34.6	120.9
Los Angeles	3B	25.2	1.2	0.6	35.6	37.4	89.4
Las Vegas	3B	29.1	7.3	0.4	28.2	35.0	113
San Francisco	3C	42.9	0.3	0.4	27.3	29.1	116.6
Baltimore	4A	45.0	2.6	0.3	22.1	29.9	143.9
Albuquerque	4B	38.8	2.4	0.4	24.4	34.0	130.7
Seattle	4C	47.8	0.3	0.4	23.0	28.5	138.6
Chicago/Hartford	5A	49.6	1.5	0.3	19.4	29.2	164.4
Boulder	5B	43.9	1.3	0.3	21.4	33.0	148.6
Minneapolis	6A	54.0	1.1	0.3	17.2	27.4	185
Helena	5B	50.3	0.6	0.3	18.3	30.4	173.7

(Continued)

(Continued)

Existing Supermarkets

Typical City	Climate Zone	Heating %	Cooling %	Water Heating %	Lighting %	Other %	Whole Building EUI, kBtu/SF
Duluth	7	58.9	0.4	0.3	15.2	25.2	209.6
Fairbanks	8	66.7	0.1	0.2	11.3	21.7	282.1

Existing Large Hotels and Motels

Typical City	Climate Zone	Heating %	Cooling %	Water Heating %	Lighting %	Other %	Whole Building EUI, kBtu/SF
Miami	1A	0.6	65.2	12.7	7.4	14.2	237.7
Houston	2A	3.3	55.5	17.6	8.1	15.5	216.2
Phoenix	2B	2.2	55.9	16.5	8.7	16.7	202.6
Atlanta	3A	6.5	45.4	22.7	8.7	16.7	201.2
Los Angeles	3B	1.8	44.0	25.2	10.0	19.1	176.1
Las Vegas	3B	4.2	44.7	22.1	9.9	19.1	176.6
San Francisco	3C	5.0	30.8	32.0	11.1	21.1	158.6
Baltimore	4A	10.1	40.1	25.1	8.5	16.3	206.5
Albuquerque	4B	8.2	30.7	30.3	10.5	20.3	167.0
Seattle	4C	10.6	27.3	32.1	10.3	19.7	171.0
Chicago/Hartford	5A	14.0	34.1	27.4	8.4	16.1	208.4
Boulder	5B	11.6	27.0	32.2	10.0	19.2	176.2
Minneapolis	6A	17.1	31.0	28.3	8.1	15.4	217.4
Helena	5B	16.2	23.1	33.3	9.4	18.0	187.5
Duluth	7	20.9	23.6	31.9	8.1	15.5	217.2
Fairbanks	8	31.5	13.7	33.1	7.4	14.4	236.6

Existing Small Hotels and Motels

Typical City	Climate Zone	Heating %	Cooling %	Water Heating %	Lighting %	Other %	Whole Building EUI, kBtu/SF
Miami	1A	0.2	31.5	6.9	26.3	35.1	81.7
Houston	2A	3.7	23.2	8.8	28.2	36.2	76.3
Phoenix	2B	2.6	24.5	7.9	28.2	36.9	76.4
Atlanta	3A	7.3	16.2	10.5	29.0	37.0	74.1
Los Angeles	3B	2.5	14.1	11.3	31.9	40.3	67.5
Las Vegas	3B	4.7	18.6	9.3	29.5	37.9	73.0

Existing Small Hotels and Motels

Typical City	Climate Zone	Heating %	Cooling %	Water Heating %	Lighting %	Other %	Whole Building EUI, kBtu/SF
San Francisco	3C	6.6	8.7	12.7	32.3	39.7	66.7
Baltimore	4A	12.1	12.7	11.3	28.4	35.5	75.8
Albuquerque	4B	8.2	12.2	11.7	29.8	38.1	72.1
Seattle	4C	10.9	7.0	13.1	30.9	38.1	69.7
Chicago/Hartford	5A	17.1	9.9	11.9	27.2	34.0	79.1
Boulder	5B	12.3	9.2	12.6	29.1	36.9	74.0
Minneapolis	6A	21.8	8.8	11.9	25.6	31.9	84.0
Helena	5B	18.2	6.5	13.0	27.6	34.6	77.9
Duluth	7	26.2	5.6	12.8	24.8	30.6	86.9
Fairbanks	8	39.3	3.7	11.6	20.3	25.1	105.9

Existing Hospitals

Typical City	Climate Zone	Heating %	Cooling %	Water Heating %	Lighting %	Other %	Whole Building EUI, kBtu/SF
Miami	1A	16.5	42.5	0.9	15.7	24.4	209.2
Houston	2A	19.4	41.2	1.0	15.1	23.3	217.3
Phoenix	2B	20.0	38.9	0.9	15.6	24.5	210.9
Atlanta	3A	21.3	38.9	1.2	15.3	23.3	214.9
Los Angeles	3B	22.1	36.7	1.2	16.0	24.1	205.7
Las Vegas	3B	21.0	35.6	1.1	16.9	25.4	195.0
San Francisco	3C	24.7	33.3	1.4	16.5	24.1	199.5
Baltimore	4A	22.9	37.3	1.3	15.4	23.0	213.7
Albuquerque	4B	20.4	31.6	1.6	18.5	28.0	177.7
Seattle	4C	25.2	32.6	1.5	16.4	24.2	200.3
Chicago/Hartford	5A	25.7	32.9	1.5	16.1	23.9	204.6
Boulder	5B	22.1	29.8	1.7	18.6	27.8	176.7
Minneapolis	6A	27.9	29.9	1.7	16.5	24.2	199.9
Helena	5B	25.5	27.1	1.8	18.4	27.1	178.4
Duluth	7	30.7	26.0	1.8	16.9	24.6	194.8
Fairbanks	8	42.3	16.7	1.9	16.0	23.1	205.6

Existing Outpatient Buildings

Typical City	Climate Zone	Heating %	Cooling %	Water Heating %	Lighting %	Other %	Whole Building EUI, kBtu/SF
Miami	1A	29.1	31.0	0.3	8.2	31.4	224.7
Houston	2A	34.1	25.2	0.4	8.6	31.7	214.5
Phoenix	2B	35.5	24.6	0.3	8.3	31.3	222.7
Atlanta	3A	39.1	19.7	0.4	8.7	32.1	212.3
Los Angeles	3B	41.2	17.6	0.4	8.6	32.2	213
Las Vegas	3B	36.6	21.1	0.4	8.8	33.1	209.1
San Francisco	3C	43.4	12.8	0.5	9.4	33.9	195.7
Baltimore	4A	41.5	16.8	0.4	8.9	32.4	207.1
Albuquerque	4B	38.9	15.5	0.5	9.4	35.7	196.4
Seattle	4C	45.6	11.3	0.5	9.2	33.4	200.9
Chicago/Hartford	5A	44.2	13.6	0.5	8.9	32.8	207.2
Boulder	5B	41.3	12.8	0.6	9.5	35.8	193.5
Minneapolis	6A	46.5	11.9	0.6	8.8	32.2	208.7
Helena	5B	45.1	10.1	0.6	9.4	34.8	196.4
Duluth	7	49.3	8.3	0.6	9.0	32.8	204.1
Fairbanks	8	56.7	5.3	0.7	8.1	29.2	227.9

Existing Primary Schools

Typical City	Climate Zone	Heating %	Cooling %	Water Heating %	Lighting %	Other %	Whole Building EUI, kBtu/SF
Miami	1A	1.0	30.2	2.1	34.7	32.0	68.3
Houston	2A	9.7	20.1	2.6	35.8	31.8	66.3
Phoenix	2B	6.3	21.7	2.5	36.3	33.2	65.3
Atlanta	3A	18.7	11.4	3.0	35.6	31.3	66.7
Los Angeles	3B	7.7	10.7	3.3	41.5	36.8	57.1
Las Vegas	3B	10.4	15.9	2.8	37.3	33.6	63.6
San Francisco	3C	18.2	3.8	3.7	39.7	34.6	59.8
Baltimore	4A	26.4	7.7	3.1	33.7	29.1	70.5
Albuquerque	4B	20.5	7.2	3.4	36.5	32.4	65.0
Seattle	4C	26.3	2.1	3.6	36.7	31.3	64.7
Chicago/Hartford	5A	34.8	5.0	3.1	30.5	26.6	77.7
Boulder	5B	26.9	4.0	3.6	35.1	30.4	57.6
Minneapolis	6A	41.2	3.4	3.1	28.1	24.2	84.5
Helena	5B	36.7	2.1	3.4	31.1	26.7	76.4
Duluth	7	46.8	1.3	3.2	26.2	22.5	90.4
Fairbanks	8	63.5	0.5	2.4	17.9	15.7	132.6

Existing Secondary Schools

Typical City	Climate Zone	Heating %	Cooling %	Water Heating %	Lighting %	Other %	Whole Building EUI, kBtu/SF
Miami	1A	0.9	65.7	1.1	16.4	15.9	112.0
Houston	2A	9.9	51.7	1.6	19.1	17.7	96.1
Phoenix	2B	6.8	55.1	1.3	18.4	18.4	97.7
Atlanta	3A	21.5	36.1	2.0	21.2	19.2	87.0
Los Angeles	3B	8.4	37.1	2.5	27.1	24.9	67.8
Las Vegas	3B	12.8	42.2	1.8	22.4	20.8	82.2
San Francisco	3C	25.0	17.7	2.9	28.6	25.8	64.4
Baltimore	4A	33.7	25.9	2.2	20.2	18.0	91.9
Albuquerque	4B	28.1	20.7	2.6	25.3	23.3	72.8
Seattle	4C	40.7	9.6	2.7	24.9	22.1	74.0
Chicago/Hartford	5A	43.4	18.4	2.2	18.9	17.1	97.5
Boulder	5B	38.4	13.1	2.7	23.9	21.9	77.0
Minneapolis	6A	51.9	13.5	2.1	16.9	15.6	108.7
Helena	5B	50.4	7.4	2.6	20.7	18.9	89.1
Duluth	7	60.4	5.9	2.3	16.3	15.1	112.7
Fairbanks	8	73.7	2.3	1.7	11.3	11.0	162.9

Appendix VIII

CEUS End-Use Data (Weather-Normalized)

PG&E

PG&E Territory
(Major cities: Coastal San Francisco and San Jose)
CA Climate Zones 1–5; CBECS Climate Zone: Marine

Electricity Use

Property Type	Total Bldg. Electricity EUI, kBtu/SF	End-Use Energy Consumption*													
		Heating kBtu/SF	Heating % Total	Cooling kBtu/SF	% Total Cooling	DHW kBtu/SF	% Total DHW	Ventilation kBtu/SF	% Total Ventilation	Int. Lighting kBtu/SF	% Total Int. Lighting	Ext. Lighting** kBtu/SF	% Total Ext. Lighting	Refrigeration kBtu/SF	% Total Refrigeration
Small Office (<30K SF)	46.04	0.82	1.78	8.43	18.31	0.68	1.48	4.88	10.60	12.97	28.17	2.46	5.34	1.67	3.63
Large Office (≥30K SF)	57.24	1.77	3.10	10.58	18.49	0.38	0.66	10.17	17.77	14.47	25.28	1.09	1.91	1.19	2.08
Restaurant	113.04	0.07	0.06	12.05	10.66	1.02	0.91	7.03	6.22	19.49	17.24	4.95	4.38	31.06	27.48
Retail	41.60	0.38	0.90	5.36	12.88	0.38	0.90	5.56	13.37	18.19	43.72	2.12	5.09	3.31	7.96
Food Store	138.36	0.10	0.07	7.44	5.38	0.48	0.35	8.23	5.94	27.65	19.98	2.97	2.15	78.5	56.73
Refrigerated Warehouse	63.14	0.07	0.11	0.78	1.24	0.10	0.16	0.85	1.35	8.50	13.46	1.23	1.95	42.59	67.45
Non-Refrigerated Warehouse	16.62	0.10	0.62	0.89	5.34	0.14	0.82	0.61	3.70	8.53	51.33	1.37	8.21	0.96	5.78
School	23.28	0.38	1.61	2.83	12.17	0.20	0.88	3.14	13.49	9.32	40.03	2.49	10.70	1.64	7.05
College	40.75	1.64	4.02	4.78	11.73	0.55	1.34	9.86	24.20	12.94	31.74	2.63	6.45	1.71	4.20
Health	63.17	2.53	4.00	10.72	16.96	0.20	0.32	13.99	22.15	15.53	24.58	1.71	2.70	2.39	3.78
Lodging	33.38	1.60	4.81	5.19	15.54	0.17	0.51	5.19	15.54	9.80	29.35	1.91	5.73	2.59	7.76

*Energy consumption for major end-uses shown; other (minor) end-uses not shown include cooking, office equipment, air compressors, non-HVAC motors, process equipment, and miscellaneous electrical loads.

**Exterior lighting includes parking area lighting, outside advertising, and building exterior façade.

Natural Gas Use

Property Type	Total Bldg. Natural Gas EUI, kBtu/SF	Heating kBtu/SF	% Total Heating	DHW kBtu/SF	% Total DHW	Cooking kBtu/SF	% Total Cooking
Small Office (<30K SF)	18.02	16.7	92.67	1.2	6.66		
Large Office (≥30K SF)	23.94	20.5	85.63	2.6	10.86		
Restaurant	183.53	7.7	4.20	40.3	21.96	135.3	73.72
Retail	7.31	5.4	73.87	1	13.68		
Food Store	34.46	14.5	42.08	10.4	30.18	9.5	27.57
Refrigerated Warehouse	4.5	0.9	20.00	0.6	13.33	1.9	42.22
Non-Refrigerated Warehouse	4.88	4	81.97	0.7	14.34		
School	21.75	14.9	68.51	5.3	24.37	1.3	5.98
College	37.46	27.2	72.61	8.9	23.76	1.3	3.47
Health	91.65	37.4	40.81	38.4	41.90	2.9	3.16
Lodging	38.14	4.6	12.06	27.3	71.58	4.5	11.80

Total Building EUI and Select End-Uses as a Percent of Total Building EUI

Property Type	Total Bldg. EUI, kBtu/SF	Heating kbtu/SF	Heating %	Cooling kBtu/SF	Cooling %	DHW Heating kBtu/SF	DHW Heating %	Ventilation kBtu/SF	Ventilation %	Int./Ext. Lighting kbtu/SF	Int./Ext. Lighting %
Small Office (<30K SF)	64.1	17.52	27.35	8.43	13.16	1.88	2.94	4.88	7.62	15.43	24.08
Large Office (≥30K SF)	81.2	22.27	27.44	10.58	13.03	2.98	3.67	10.17	12.53	15.56	19.17
Restaurant	296.6	7.77	2.62	12.05	4.06	41.32	13.93	7.03	2.37	24.44	8.24
Retail	48.9	5.78	11.81	5.36	10.95	1.38	2.81	5.56	11.37	20.31	41.52
Food Store	172.8	14.60	8.45	7.44	4.31	10.88	6.29	8.23	4.76	30.61	17.71
Refrigerated Warehouse	67.6	0.97	1.43	0.78	1.16	0.70	1.04	0.85	1.26	9.73	14.38
Non-Refrigerated Warehouse	21.5	4.10	19.08	0.89	4.13	0.84	3.89	0.61	2.86	9.90	46.03
School	45.0	15.28	33.93	2.83	6.29	5.50	12.23	3.14	6.97	11.81	26.23
College	78.2	28.84	36.87	4.78	6.11	9.45	12.08	9.86	12.61	15.56	19.90
Health	154.8	39.93	25.79	10.72	6.92	38.60	24.93	13.99	9.04	17.24	11.13
Lodging	71.5	6.20	8.67	5.19	7.25	27.47	38.41	5.19	7.25	11.71	16.37

SCE

SCE Territory
(Major cities: Coastal San Diego, Interior Los Angeles, San Diego, and Orange County)
CA Climate Zones: 7–10; CBECS Climate Zone: Hot Dry, Mixed Dry

Electricity Use

Property Type	Total Bldg. Electricity EUI, kBtu/SF	End-Use Energy Consumption*													
		Heating kBtu/SF	% Total Heating	Cooling kBtu/SF	% Total Cooling	DHW kBtu/SF	% Total DHW	Ventilation kBtu/SF	% Total Ventilation	Int. Lighting kBtu/SF	% Total Int. Lighting	Ext. Lighting kBtu/SF	% Total Ext. Lighting	Refrigeration kBtu/SF	Refrigeration %
Small Office (<30K SF)	45.2	0.34	0.75	9.93	21.96	0.82	1.81	4.20	9.28	13.21	29.21	4.20	9.28	1.95	4.3
Large Office (≥30K SF)	61.1	0.89	1.45	13.52	22.11	0.38	0.61	10.27	16.81	16.04	26.24	2.39	3.91	1.33	2.18
Restaurant	157.6	0.10	0.06	26.72	16.95	1.54	0.97	13.99	8.88	24.30	15.41	9.35	5.93	35.53	22.54
Retail	52.4	0.14	0.26	9.52	18.16	0.44	0.85	6.59	12.57	22.49	42.90	3.65	6.97	3.41	6.51
Food Store	142.4	0.51	0.36	12.22	8.58	0.51	0.36	9.45	6.64	30.72	21.58	3.52	2.47	74.20	52.12
Refrigerated Warehouse	76.5	0.03	0.04	1.71	2.23	0.03	0.04	0.51	0.67	11.02	14.41	1.02	1.34	51.47	67.29
Non-Refrigerated Warehouse	14.6	0.14	0.93	1.16	7.93	0.17	1.17	1.09	7.46	7.20	49.18	0.72	4.90	0.89	6.06
School	28.1	0.31	1.09	5.39	19.22	0.48	1.70	3.48	12.41	10.51	37.47	2.53	9.00	1.81	6.45
College	46.5	4.03	8.66	7.27	15.64	0.48	1.03	6.79	14.61	15.39	33.11	2.32	4.99	1.06	2.28
Health	69.3	2.59	3.74	14.57	21.03	0.31	0.44	14.33	20.69	16.76	24.19	2.29	3.30	2.49	3.6
Lodging	45.3	1.16	2.56	10.27	22.70	0.00	0.00	5.94	13.12	12.46	27.53	2.39	5.28	3.28	7.24

*Energy consumption for major end-uses shown; other (minor) end-uses not shown include: cooking, office equipment, air compressors, non-HVAC motors, process equipment and miscellaneous electrical loads.

Natural Gas Use

Property Type	Total Bldg. Natural Gas EUI, kBtu/SF	Heating kBtu/SF	% Total Heating	DHW kBtu/SF	% Total DHW	Cooking kBtu/SF	% total Cooking
Small Office (<30K SF)	8	5.4	67.50	2.3	28.75		
Large Office (≥30K SF)	13	9.9	76.15	2	15.38		
Restaurant	249.1	8.4	3.37	60.5	24.29	179.7	72.14
Retail	2.5	1.1	44.00	0.7	28.00		
Food Store	21.8	6.1	27.98	5.9	27.06	9.6	44.04
Refrigerated Warehouse	8.1	0.6	7.41	0.9	11.11		
Non-Refrigerated Warehouse	2.5	2.2	88.00	0.2	8.00		
School	12.1	6.1	50.41	4.9	40.50	1	8.26
College	24	13.2	55.00	6.3	26.25		
Health	68.3	34.8	50.95	24	35.14	3.3	4.83
Lodging	40.9	10.8	26.41	25.5	62.35	3.4	8.31

(Continued)

(Continued)

Total Building EUI and Select End-Uses as a Percent of Total Building EUI

Property Type	Total Bldg. EUI, kBtu/SF	Heating kBtu/SF	Heating %	Cooling kbtu/SF	Cooling %	DHW Heating kBtu/SF	DHW Heating %	Ventilation kBtu/SF	Ventilation %	Int./Ext. Lighting kBtu/SF	Int./Ext. Lighting %
Small Office (<30K SF)	53.2	5.74	10.79	9.93	18.66	3.12	5.86	4.20	7.89	17.41	32.70
Large Office (>=30K SF)	74.1	10.79	14.55	13.52	18.23	2.38	3.20	10.27	13.86	18.43	24.86
Restaurant	406.7	8.50	2.09	26.72	6.57	62.04	15.25	13.99	3.44	33.65	8.27
Retail	54.9	1.24	2.25	9.52	17.34	1.14	2.08	6.59	11.99	26.14	47.60
Food Store	164.2	6.61	4.03	12.22	7.44	6.41	3.91	9.45	5.76	34.23	20.85
Refrigerated Warehouse	84.6	0.63	0.75	1.71	2.02	0.93	1.10	0.51	0.61	12.05	14.24
Non-Refrigerated Warehouse	17.1	2.34	13.63	1.16	6.77	0.37	2.16	1.09	6.37	7.92	46.19
School	40.2	6.41	15.96	5.39	13.43	5.38	13.39	3.48	8.67	13.04	32.47
College	70.5	17.23	24.44	7.27	10.31	6.78	9.62	6.79	9.64	17.71	25.13
Health	137.6	37.39	27.18	14.57	10.59	24.31	17.67	14.33	10.42	19.04	13.84
Lodging	86.2	11.96	13.88	10.27	11.92	25.50	29.60	5.94	6.89	14.85	17.23

SDG&E

SDG&E Territory
(Major Cities: Interior—Fresno)
CA Climate Zone: 13; CBECS Climate Zone: Hot Dry, Mixed Dry

Electricity Use

Property Type	Total Bldg. Electricity EUI, kBtu/SF	Heating kBtu/SF	% Total Heating	Cooling kBtu/SF	% Total Cooling	DHW kBtu/SF	% Total DHW	Ventilation kBtu/SF	% Total Ventilation	Int. Lighting kBtu/SF	% Total Int. Lighting	Ext. Lighting kBtu/SF	% Total Ext. Lighting	Refrigeration kBtu/SF	Refrigeration %
Small Office (<30K SF)	41.40	1.02	2.47	7.71	18.63	1.26	3.05	3.65	8.82	13.45	32.48	2.73	6.60	2.59	6.27
Large Office (≥30K SF)	65.63	1.64	2.50	14.54	22.15	0.58	0.88	11.19	17.06	15.19	23.14	1.95	2.96	2.42	3.69
Restaurant	149.25	0.51	0.34	21.95	14.70	1.77	1.19	14.40	9.65	23.14	15.50	6.62	4.44	35.36	23.69
Retail	52.87	0.31	0.58	8.26	15.62	0.89	1.68	5.53	10.46	22.94	43.38	4.20	7.94	4.85	9.17
Food Store	132.39	0.07	0.05	7.34	5.54	0.27	0.21	7.47	5.65	26.66	20.13	3.41	2.58	77.82	58.78
Refrigerated Warehouse	120.51	0.31	0.25	2.32	1.93	0.55	0.45	6.35	5.27	14.27	11.84	1.57	1.30	77.54	64.34
Non-Refrigerated Warehouse	15.50	0.07	0.44	2.22	14.32	0.31	1.98	1.54	9.91	6.72	43.39	0.82	5.29	1.16	7.49
School	22.83	0.75	3.29	3.38	14.80	0.34	1.49	2.90	12.71	9.08	39.76	2.35	10.31	1.40	6.13
College	38.64	2.70	6.98	8.19	21.20	0.07	0.18	2.63	6.80	11.09	28.71	4.78	12.37	2.01	5.21
Health	64.57	1.54	2.38	14.37	22.25	0.38	0.58	9.86	15.27	16.79	26.00	1.57	2.43	2.08	3.22
Lodging	54.95	1.40	2.55	11.43	20.81	0.00	0.00	9.97	18.14	17.65	32.11	1.98	3.60	3.52	6.4

Natural Gas Use

Property Type	Total Bldg. Natural Gas EUI, kBtu/SF	Heating kBtu/SF	% Total Heating	DHW kBtu/SF	% Total DHW	Cooking kBtu/SF	% Total Cooking
Small Office (<30K SF)	2.1	0.7	33.33	0.9	42.86		
Large Office (>=30K SF)	38.9	25.3	65.04	4.2	10.80		
Restaurant	176.8	3	1.70	35.4	20.02	138.4	78.28
Retail	2.4	0.8	33.33	0.6	25.00		
Food Store	26.5	4.8	18.11	6.3	23.77	15.4	58.11
Refrigerated Warehouse	7.1	2.2	30.99	4.9	69.01		
Non-refrig. Warehouse	2.1	1.7	80.95	0.2	9.52		
School	7	4.5	64.29	2.1	30.00		
College	43	14.9	34.65	11	25.58	3.8	8.84
Health	61	14.1	23.11	39.1	64.10	4.7	7.70
Lodging	61.7	4.8	7.78	45.8	74.23	7.5	12.16

Total Building EUI and Select End-Uses as a Percent of Total Building EUI

Property Type	Total Bldg. EUI, kBtu/SF	Heating kBtu/SF	Heating %	Cooling kBtu/SF	Cooling %	DHW Heating kBtu/SF	DHW Heating %	Ventilation kBtu/SF	Ventilation %	Int. Ext. Lighting kBtu/SF	Int. Ext. Lighting %
Small Office (<30K SF)	43.5	1.72	3.96	7.71	17.73	2.16	4.97	3.65	8.40	16.18	37.19
Large Office (≥30K SF)	104.5	26.94	25.77	14.54	13.91	4.78	4.57	11.19	10.71	17.13	16.39
Restaurant	326.1	3.51	1.08	21.95	6.73	37.17	11.40	14.40	4.42	29.76	9.13
Retail	55.3	1.11	2.00	8.26	14.94	1.49	2.69	5.53	10.00	27.13	49.09
Food Store	158.9	4.87	3.06	7.34	4.62	6.57	4.14	7.47	4.70	30.07	18.92
Refrigerated Warehouse	127.6	2.51	1.96	2.32	1.82	5.45	4.27	6.35	4.97	15.84	12.41
Non-Refrigerated Warehouse	17.6	1.77	10.05	2.22	12.61	0.51	2.88	1.54	8.73	7.54	42.87
School	29.8	5.25	17.60	3.38	11.33	2.44	8.18	2.90	9.72	11.43	38.33
College	81.6	17.60	21.55	8.19	10.03	11.07	13.56	2.63	3.22	15.87	19.44
Health	125.6	15.64	12.45	14.37	11.44	39.48	31.44	9.86	7.85	18.36	14.62
Lodging	116.6	6.20	5.31	11.43	9.80	45.80	39.26	9.97	8.54	19.62	16.82

SMUD

SMUD Territory
Major Cities: Coastal LA and Santa Barbara
CA Climate Zone: 6; CBECS Climate Zone: Marine

Electricity Use

Property Type	Total Bldg. Electricity EUI, kBtu/SF	Heating kBtu/SF	% Total Heating	Cooling kBtu/SF	% Total Cooling	DHW kBtu/SF	% Total DHW	Ventilation kBtu/SF	% Total Ventilation	Int. Lighting kBtu/SF	% Total Int. Lighting	Ext. Lighting kBtu/SF	% Total Ext. Lighting	Refrigeration kBtu/SF	Refrigeration %
Small Office (<30K SF)	42.36	1.30	3.06	7.06	16.68	1.09	2.58	5.53	13.05	11.54	27.24	1.88	4.43	1.84	4.34
Large Office (>=30K SF)	68.09	5.09	7.47	11.47	16.84	0.44	0.65	11.54	16.94	16.14	23.71	1.30	1.90	0.96	1.41
Restaurant	159.76	1.09	0.68	24.13	15.10	0.55	0.34	15.87	9.93	23.48	14.70	3.41	2.14	38.63	24.18
Retail	48.74	0.48	0.98	6.21	12.75	0.55	1.12	7.95	16.32	19.25	39.50	4.10	8.40	3.55	7.28
Food Store	152.87	0.07	0.04	14.54	9.51	0.51	0.33	11.57	7.57	35.73	23.38	2.22	1.45	78.53	51.37
Refrigerated Warehouse	57.51	0.44	0.77	1.02	1.78	0.44	0.77	0.34	0.59	6.86	11.93	1.60	2.79	35.02	60.89
Non-refrig. Warehouse	12.83	0.44	3.46	0.65	5.05	0.10	0.80	0.24	1.86	6.38	49.73	0.38	2.93	1.64	12.78
School	31.26	1.06	3.38	4.47	14.30	0.38	1.20	3.99	12.77	10.58	33.84	3.34	10.70	2.42	7.74
College	37.00	1.47	3.97	7.37	19.93	0.65	1.75	6.72	18.17	9.62	26.01	4.06	10.98	1.64	4.43
Health	78.70	3.07	3.90	14.78	18.78	0.07	0.09	19.56	24.85	18.60	23.63	1.57	1.99	3.17	4.03
Lodging	41.84	2.94	7.01	8.36	19.98	1.30	3.10	5.15	12.32	11.47	27.41	1.26	3.02	4.23	10.11

Natural Gas Use

Property Type	Total Bldg. Natural Gas EUI, kBtu/SF	Heating kBtu/SF	% Total Heating	DHW kBtu/SF	% Total DHW	Cooking kBtu/SF	% Total Cooking
Small Office (<30K SF)	8.1	7	86.42	1.1	13.58		
Large Office (≥30K SF)	19.8	16.3	82.32	2.5	12.63		
Restaurant	183.4	14.1	7.69	51.3	27.97	118	64.34
Retail	6.6	5.2	78.79	0.5	7.58		
Food Store	29	14.8	51.03	4.8	16.55	9.4	32.41
Refrigerated Warehouse	1.6	0.4	25.00	0.6	37.50		
Non-refrig. Warehouse	0.6	0.6	100.00	0	0.00		
School	18.4	13	70.65	4.7	25.54		
College	32.5	26.3	80.92	6	18.46		
Health	74.8	37.9	50.67	27.3	36.50	3.4	4.55
Lodging	41.5	6.7	16.14	29.3	70.60	4.8	11.57

Total Building EUI and Select End-Uses as a Percent of Total Building EUI

Property Type	Total Bldg. EUI, kBtu/SF	Heating kBtu/SF	Heating %	Cooling kBtu/SF	Cooling %	DHW Heating kBtu/SF	DHW Heating %	Ventilation kBtu/SF	Ventilation %	Int./Ext. Lighting kBtu/SF	Int./Ext. Lighting %
Small Office (<30K SF)	50.5	8.30	16.44	7.06	14.00	2.19	4.34	5.53	10.96	13.41	26.58
Large Office (≥30K SF)	87.9	21.39	24.33	11.47	13.05	2.94	3.35	11.54	13.13	17.44	19.84
Restaurant	343.2	15.19	4.43	24.13	7.03	51.85	15.11	15.87	4.62	26.89	7.84
Retail	55.3	5.68	10.26	6.21	11.23	1.05	1.89	7.95	14.37	23.34	42.19
Food Store	181.9	14.87	8.18	14.54	7.99	5.31	2.92	11.57	6.36	37.95	20.87
Refrigerated Warehouse	59.1	0.84	1.43	1.02	1.73	1.04	1.77	0.34	0.58	8.46	14.32
Non-Refrigerated Warehouse	13.4	1.04	7.77	0.65	4.83	0.10	0.76	0.24	1.78	6.76	50.31
School	49.7	14.06	28.31	4.47	9.00	5.08	10.22	3.99	8.04	13.93	28.04
College	69.5	27.77	39.96	7.37	10.61	6.65	9.57	6.72	9.67	13.69	19.69
Health	153.5	40.97	26.69	14.78	9.63	27.37	17.83	19.56	12.74	20.17	13.14
Lodging	83.3	9.64	11.56	8.36	10.03	30.60	36.71	5.15	6.18	12.73	15.27

CEUS Property Type Definitions

Category	Examples
Retail	Lumber
	Garden Supplies
	Nurseries
	Paint Stores
	Hardware Stores
	Department Stores
	Auto Dealers
	Shoe Stores
	Shopping Centers
	Furniture Stores
	Appliance Stores
	Sporting Goods
	Book Stores
	Jewelry Stores
	Fuel Dealers
	Florists
Healthcare	Hospitals (inpatient)
	Outpatient
	Medical/Dental Labs
Food Stores	Grocery stores
	Meat & Fish Markets
	Bakeries
	Liquor stores
Schools	Elementary
	Secondary
	Daycare
Colleges	Colleges
	Universities
	Jr. Colleges
	Vocational

Appendix IX
EUL Data

Equipment	EUL (years)	References
Air Conditioners/Split Systems	18	(1)
Air Handling Units	24	(1)
Boilers	22	(1)
	25	(2)
Building Automation System (BAS, EMS)	10	(3)
Chillers	21	(1)
Reciprocating	20	(2)
Centrifugal	23	(2)
Controls		
Pneumatic	20	(2)
Electronic	15	(2)
Cooling Towers	22	(1)
DHW Water Heaters	14	(1)
Evaporative Condensers	20	(1)
Fans	20	(1)
Furnaces	18	(2)
Heat Pumps	15	(4)
Insulation (walls/roof)		
Molded	20	(2)
Blanket	24	(2)
Lighting (see also Table 9.2)		
Fluorescent	12	(1)
HID	14	(1)
LED	35,000–50,000 hrs	(5)
Motors	18	(1)
Windows (High Performance)	20	(1)
Appliances and Fixtures		
Faucets	10	(6)
Laundry Systems	13	(1)
Showerheads	10	(6)
Toilets	20	(1)
Urinals	10	(1)

REFERENCES

1. Milwaukee energy efficiency property assessed clean energy financing program manual, https://city.milwaukee.gov/ImageLibrary/BBC/Pace/Me2PACEProgramManualv7.pdf
2. ASHRAE life expectancy table, ASHRAE handbook: HVAC applications https://www.buildingsiot.com/blog/improving-the-length-of-your-building-automation-systems-useful-life-bd
3. www.facilitiesnet.com/buildingautomation/tip/What-Is-The-Optimal-Life-Cycle-For-A-BAS—37034
4. Lawrence Berkeley National Laboratory, LBNL-5447-E, July 2012.
5. U.S. DOE Energy Efficiency and Renewable Energy Office, Building technologies program, lifetime of white LEDs, PNNL-SA-50957, September 2009, https://betterbuildingssolutioncenter.energy.gov/sites/default/files/attachments/lifetime_white_leds.pdf
6. CT CL&P/UI 2017 Program savings document (CT technical resource manual), 12th ed., Appendix 4, file:///C:/Users/Anthony%20Buonicore/Desktop/State%20TRMs/2017%20CT%20Program%20Savings%20Document_Final.pdf

NOTE

EUL data may also be available in state utility technical resource manuals, as well as in the following sources:

- https://www4.eere.energy.gov/challenge/sites/default/files/uploaded-files/Me2CommercialPACEEligibleMeasu.pdf
- Database for Energy Efficient Resources (DEER), California Public Utility Commission.
- ASHRAE Equipment Life Expectancies.
- Fannie Mae Estimated Useful Life for Use in Property Condition Assessments.

It is not usual to find variation in the EULs provided by different sources.

Appendix X
ENERGY STAR Portfolio Manager Benchmarking Data

U.S. National Median Site EUI By Property Type*

Property Type	Use	Site EUI (kBtu/SF)
Banking/Financial Service	Bank Branch	88.3
	Financial Office	52.9
Education	Adult Education	52.4
	College/University	84.3
	K–12	48.5
	Pre-School/Daycare	64.8
	Vocational School	52.4
Entertainment/Public Assembly	Convention Center	56.1
	Movie Theater	56.2
	Museum	56.2
	Recreation (Fitness Center, Health Club, Gym, Ice Rink, Roller Rink)	50.8
	Social/Meeting Hall	56.1
	Stadium	56.2
	Zoo, Aquarium	56.2
	Casino	56.2
	Bar/Nightclub	130.7
Food Sales & Service	Convenience Store	350.9
	Fast Food Restaurant	402.7
	Restaurant	325.6
	Supermarket/Grocery Store	196.0
	Wholesale Club	51.4
	Food Sales	231.4
	Food Service	270.3
Healthcare	General Hospital	234.3
	Specialty Hospital	206.7
	Outpatient Rehabilitation/ Physical Therapy	62.0
	Residential Care Facility	99.0
	Senior Living	99.0

(Continued)

(*Continued*)

U.S. National Median Site EUI By Property Type*

Property Type	Use	Site EUI (kBtu/SF)
	Urgent Care/Clinic	64.5
Lodging	Hotel	63.0
	Multifamily Housing	59.6
	Penitentiary	69.9
	Residence Hall/Dormitory	57.9
	Residential Care Facility	99.0
	Senior Living Community	99.0
	Lodging	63.6
Office	Medical Office	51.2
	Office	52.9
	Veterinary Office	64.5
Public Services	Courthouse	101.2
	Fire Station	63.5
	Library	71.6
	Post Office	47.9
	Police Station	63.5
	Transportation Terminal	56.2
Religious Worship	Church/Worship Facility	30.5
Retail	Convenience Store	350.9
	Enclosed Mall	65.7
	Strip Mall	103.5
	Retail Store	51.4
	Vehicle Dealership	71.9
Services	Personal Services (Health/ Beauty, Dry Cleaning, etc.)	47.9
	Repair Services (Vehicle, Shoe, etc.)	47.9
	Other Services	47.9
Warehouse/Storage	Self-Storage	20.2
	Distribution Canter	22.7
	Non-Refrigerated Warehouse	22.7
	Refrigerated Warehouse	84.1

*The National Median Site EUI is the median value of the national population.
Source: U.S. EPA, ENERGY STAR Portfolio Manager, Technical Reference, U.S. Energy Use Intensity by Property Type, August 2023.

Appendix XI
Building Energy Performance Simulation Modeling

BUILDING SIMULATION MODELS

A number of building simulation models exist, the most popular of which include U.S. DOE's EnergyPlus/OpenStudio, DOE's eQUEST, Trane TRACE 700, Carrier HAP, TRNSYS, and Integrated Environmental Solutions IESVE.

U.S. DOE's EnergyPlus Version 23.1.0 (March 23, 2023) is a building energy simulation program that is used to model energy consumption, lighting, plug and process loads, and water use in buildings. The development of the program was funded by the U.S. DOE's Building Technologies Office and managed by the National Renewable Energy Laboratory (NREL). It is open-source software and was collaboratively developed by NREL, various other DOE national laboratories, academic institutions, and private firms.

OpenStudio Version 3.6.0 SDK (software development kit) extends the base capability of EnergyPlus. Open Studio includes graphical interfaces. Graphical applications include the OpenStudio SketchUp Plug-in (that allows users to quickly create the geometry needed for EnergyPlus), OpenStudio Application (a fully featured graphical interface to OpenStudio models including envelope, loads, schedules and HVAC), ResultsViewer (enables browsing, plotting, and comparing simulation output data) and the Parametric Analysis Tool (that enables studying the impact of multiple combinations of OpenStudio measures to a base model).

Trane's TRACE 700, which has been a leading tool for building HVAC simulation, has been replaced by TRACE 3D Plus Version 5.20.117. The software is built on U.S. DOE's EnergyPlus engine enhanced by Trane experience. TRACE 3D Plus can model HVAC systems quickly and factor in many of the intricacies of today's often complex building applications.

Carrier's Hourly Analysis Program (HAP), Version 6, is a powerful tool for designing HVAC systems and sizing system components as well as modeling annual energy performance and energy costs. The tool includes valuable energy analysis capabilities for comparing energy consumption and energy costs of design alternatives.

University of Wisconsin's Transient Systems Simulation Program, TRNSYS 18 Version 18, is based on a component approach with a modular structure. Its library includes a detailed multi-zone building model and components for HVAC systems, renewable energy systems, and so forth. The library also includes component routines to handle the input of weather data or other time-dependent forcing functions and output of simulation results. Its modular nature gives the program considerable flexibility and facilitates the addition to the program of mathematical models not included in the standard library.

Integrated Environmental Solutions' IESVE software is designed to predict building energy consumption, GHG emissions, peak demands, energy cost, and electricity production for renewable energy systems. Whole building energy simulation analysis capabilities include assessment and optimization of energy efficiency upgrades, ventilation, and HVAC performance.

MODEL INPUT DATA

Building simulation modeling requires considerable input on building design and geometry, orientation, construction materials, envelope design, mechanical system design, occupancy, zoning, and control set-points, in addition to local hourly weather data and utility rate data. Review of the building's as-built drawings and specifications can be helpful in this challenging data collection effort.
Typical input data includes:

- Climate data: TMY or average data appropriate for the geographic location.
- Utility rate data applicable for the building.
- Building characteristics:
  - Building location
  - Building use
  - Occupancy
  - Shape and orientation
  - Number of floors and floor-to-floor height
  - Aspect ratio (east-west length of building divided by north-south width)
  - Floor area
  - Interior space layout
  - Roof construction
  - Wall construction
  - Insulation (walls, roof, floors)
  - Vertical fenestration (window [glazing] geometry and construction)
  - Building operational characteristics and schedules
  - Building occupancy
- Equipment characteristics:
  - HVAC systems
    - Zoning
    - Set-points
    - Operating schedules
    - Ventilation
    - Infiltration
    - Space cooling equipment
    - Space heating equipment
  - Service hot water systems
    - Water storage
    - Set-points

- Lighting system
  - Interior
  - Exterior
  - Controls
  - Operation
- Plug and process loads
  - Office equipment
  - Operation
  - Other
- Energy management system and controls
  - HVAC
  - Lighting
  - Other.

GENERAL COMMENTS

For existing buildings, assuming building information and as-built drawings are available with all the information needed, it is not unusual for building simulation modeling by an experienced professional to take at least 2 to 4 weeks between visiting the building, collecting the data, entering the data into the model, running the simulation model, calibrating the model against actual building energy consumption, estimating the energy savings of the planned upgrade, and incorporating the results into a report for the client. The general approach is to develop the model using the weather conditions that existed in the 12-month period for which utility bills have been collected, calibrate the model against the utility data collected, and then run the calibrated model over the same 12-month period only now using the proposed energy efficiency upgrade. This will provide the building owner with an estimate of the energy savings that might have been realized in the 12-month period had the upgrade been installed. The model can then be run using TMY weather data to obtain an estimate of the future energy savings that might be realized under TMY weather conditions (TMY weather conditions would be used since it is not possible to predict future weather, except that it should approach averages over the long term). These savings can then be run over the EUL of the equipment, with adjustments for energy cost escalation and equipment performance degradation.

Assuming the modeling is done by experienced professionals, in my experience calibrated models should be able to predict energy savings for replacement equipment upgrades within approximately 10% of the actual energy savings.

The expense to conduct building simulation modeling may be acceptable if building owner clients were willing to pay for these costs, or if there was complete confidence on the part of a contractor that a project go-ahead would be received (before or after the modeling effort was conducted). However, this rarely is the case, particularly for smaller size projects or in small to medium-sized buildings. As such, this can result in a considerable amount of money and time being wasted for modeling when a project award is not guaranteed (or paid for regardless). In view of the upfront risk, it is rare that energy efficiency consultants and contractors will rely on modeling at the project scoping and proposal stage without a financial commitment from the building owner.

Appendix XII
Acronyms

ACH	Air change per hour
AF/BI	Air foil/backward inclined fan
AHRI	Air conditioning, heating, and refrigeration institute
AHU	Air handling unit
AFUE	Annual fuel utilization efficiency
ASHP	Air source heat pump
ASHRAE	American Society of Heating, Refrigerating and Air Conditioning Engineers
BAS	Building automation system
BHP	Brake horsepower
BI	Backward incline (fan)
BMS	Building management system
BPD	U.S. DOE Building Performance Database
BTU	British thermal unit
CAP	Equipment output capacity (delivered to meet space conditioning needs)
CAV	Constant air volume
CBECS	Commercial Building Energy Consumption Survey
CCF	100 cubic feet (typical measurement for natural gas consumption)
CDD	Cooling degree day
CEUS	California Commercial End-Use Survey
CF	Coincidence factor
CFM	Cubic feet per minute
CFL	Compact fluorescent light bulb
CHP	Combined heat and power
CHWP	Chilled water pump
COP	Coefficient of performance
CV	Constant volume
CWP	Condenser water pump
DHW	Domestic hot water
DC	Direct current
DCV	Demand controlled ventilation
DOAS	Dedicated outdoor air system
DOE	U.S. Department of Energy
DX	Direct expansion
EA	Exhaust air
ECM	Electronically commutated motors
EEM	Energy efficiency measure
EER	Energy efficiency ratio
EF	Energy factor
EFLH	Equivalent full load hours (typically in units of hours/year)

EIA	Energy Information Agency
EMS	Energy management system
EPA	U.S. Environmental Protection Agency
ERV	Energy recovery ventilation
EUI	Energy use intensity
EUL	Effective useful life
FCU	Fan coil unit
FL	Full load
FLH	Annual full load hours
GFA	Building gross floor area
HDD	Heating degree day
HID	High intensity discharge lamp
HP	Horsepower
HRV	Heat recovery ventilator
HVAC	Heating, ventilation, and air conditioning
IECC	International Energy Conservation Code
IEER	Integrated energy efficiency ratio
IPLV	Integrated part load value
LED	Light-emitting diode
LF	Load factor
LPD	Lighting power density
MAU	Makeup air unit
M	Thousand (k may also be used)
MM	Million
NEMA	National Electrical Manufacturers Association
NPLV	Non-standard part load value
NPV	Net present value
NWS	National Weather Service
OA	Outside air
PG&E	Pacific Gas & Electric
PTAC	Packaged terminal air conditioner
PTHP	Packaged terminal heat pump
RA	Return air
RECS	Residential Energy Consumption Survey
RT	Refrigeration-ton (equal to 12,000 Btu/hr)
RTU	Roof-top unit
RUL	Remaining useful life
SA	Supply air
SCE	Southern California Edison
SDG&E	San Diego Gas & Electric
SEER	Seasonal energy efficiency ratio
SF	Square feet (ft^2)
SHGC	Solar heat gain coefficient
SMB	Small to medium-sized buildings (generally less than 50,000–100,000 GFA)
SMUD	Sacramento Municipal Utility District
T	Temperature
TMY	Typical meteorological year

TRM	State Technical Resource Manual or Technical Reference Manual
VAV	Variable air volume
VFD	Variable frequency drive
VRF	Variable refrigerant flow
VRV	Variable refrigerant volume
VSD	Variable speed drive
WSHP	Water source heat pump

Appendix XIII
Equation Nomenclature

$\Delta kW_{cooling}$	Cooling energy saved by roof insulation or window upgrade, kWh/yr
$\Delta kW_{heating}$	Heating energy saved by roof insulation or window upgrade, kWh/yr
ΔSP	Static pressure difference across fan, in. W.C.
ΔT	Temperature difference, °F
η	Efficiency
η_B	Efficiency of the boiler
$\eta_{Baseline}$	AFUE or thermal efficiency of baseline existing unit
η_{CT}	Cooling tower efficiency
η_{Fan}	Fan efficiency
η_{motor} or η_M	Motor (drive) efficiency
η_{new}	AFUE or thermal efficiency of new high efficiency unit
η_{pump}	Pump efficiency
ρ_A	Density of air, lb/ft^3
ρ_W	Density of water, lb/gal
A	Equipment age, years
A_{Roof}	Roof area, ft^2
A_S	Series of equal annual savings, $
A_{Window}	Window area, ft^2
ACH	Air change rate, air changes per hour
AFUE	Average fuel utilization efficiency
BHP	Brake horsepower, bhp
C	Total project installed cost, $
CAP_C	Output (delivered) air conditioning cooling capacity, Btu/hr
CAP_H	Output (delivered) space heating capacity, Btu/hr
$CAP_{H\text{-}In}$	Input space heating capacity, Btu/hr
CDD	Cooling degree days in a year or any 12-month period, degree-days
CDD_{Avg}	Average long-term CDD during the period selected (e.g., 12 months, degree-days)
CDD_i	Cooling degree days in month i, degree-days
CF	Coincidence factor
CF_{summer}	Coincidence factor in summer
cfm or CFM	Air flow rate, ft^3/min
COP	Coefficient of performance, (Btu/hr output/Btu/hr input) or (kWh output/kWh Input)
$COP_{Baseline}$	Coefficient of performance for baseline (existing) equipment
COP_C	Coefficient of performance for cooling, Btu output/Btu input
COP_H	Coefficient of performance for space heating, Btu output/Btu input
COP_{New}	Coefficient of performance for new equipment, kWh output/kWh Input
c_{pA}	Specific heat capacity of air, Btu/lb °F
c_{pW}	Specific heat capacity of water, Btu/lb °F
D	Pump impeller diameter, in.

d	Discount rate, percent or fraction
DS_{kW}	Peak demand savings, kW in 15 (or 30) min. interval in billing period
DSF	Demand savings factor
$E_{In\text{-}C}$	Input (consumed) cooling end-use energy consumption, Btu/yr
$E_{In\text{-}C\text{-}Baseline}$	Baseline input (consumed) cooling end-use energy consumption, Btu/yr
$E_{In\text{-}C\text{-}New}$	Input (consumed) cooling end-use energy consumption of new equipment, Btu/yr
$E_{In\text{-}Fan}$	Energy use of fan, kWh
$E_{In\text{-}H}$	Input (consumed) space heating end-use energy consumption, Btu/yr
$E_{in\text{-}H\text{-}Baseline}$	Baseline input (consumed) space heating end-use energy consumption of existing equipment, Btu/yr
$E_{in\text{-}H\text{-}New}$	Baseline input (consumed) space heating end-use energy consumption of new equipment, Btu/yr
$E_{Out\text{-}C}$	Output (delivered) cooling end-use energy consumption, Btu/yr
$E_{Out\text{-}C\text{-}Baseline}$	Baseline output (delivered) cooling end-use energy consumption, Btu/yr
$E_{Out\text{-}C\text{-}New}$	Output (delivered) cooling end-use energy consumption of new equipment, Btu/yr
$E_{Out\text{-}H}$	Output (delivered) space heating end-use energy consumption, Btu/yr
E_{Pump}	Energy use of pump, kWh
$EEPM_D$	Energy efficiency performance metric factoring in degradation
$EEPM_{NoD}$	Energy efficiency performance metric without factoring in degradation
EER	Energy efficiency ratio, Btu output/watt-hr input
$EER_{Baseline}$	Baseline (existing equipment) energy efficiency ratio, Btu output/watt-hr input
EER_C	Cooling energy efficiency ratio, Btu output/W-hr input
EER_{New}	New equipment energy efficiency ratio, Btu output/watt-hr input
EF	Energy factor (for DHW heaters), Btu output (delivered)/Btu input (consumed)
$EF_{Baseline}$	Energy factor for existing hot water heater
EF_{New}	Energy factor for new (higher efficiency) hot water heater
EFLH	Equivalent full load hours
$EFLH_C$	Equivalent full load hours for space cooling, hr/yr
$EFLH_H$	Equivalent full load hours for space heating, hr/yr
$EFLH_M$	Equivalent annual motor operating hours at full load
$EFLH_{Reduction}$	Annual equivalent full load energy reduction due to an evaporative precooler, hr/yr
$ES_{C\text{-}Btu}$	Space cooling energy savings, Btu/yr
$ES_{C\text{-}kWh}$	Space cooling energy savings, kWh/yr
$ES_{H\text{-}Btu}$	Space heating energy savings, Btu/yr
$ES_{H\text{-}kWh}$	Space heating energy savings, kWh/yr
$ES_{DHW\text{-}Btu}$	DHW heating savings, Btu/yr
ESF	Energy savings factor
EUI	Energy use intensity, kBtu/ft^2-yr
EUI_C	Energy use intensity for space cooling, kBtu/ft^2-yr
EUI_H	Energy use intensity for space heating, kBtu/ft^2-yr
$EUI_{N\text{-}C}$	Normalized energy use intensity for space cooling, kBtu/ft^2-yr
$EUI_{N\text{-}H}$	Normalized energy use intensity for space heating, kBtu/ft^2-yr
F	Fraction of annual kWh energy savings that must be removed by the cooling system
FL	Full load
G_i	Gallons of fuel oil for month (i)

G_T	Total gallons of fuel oil delivered in a 12-month period
GFA	Gross floor area of building in square feet
gpf	Gallons of water used per flush
GPM	Water flow rate, gal/min
GPM-T	Cooling tower water flow rate, gal/min per ton
GPY	Gallons of water per year
H	Head developed by the pump, ft. of water
H_{Bin}	Hrs/yr at bin OAT
H_{Fan}	Fan operating hours
$H_{Lighting}$	Lighting hours of use
H_{Pump}	Pump head, ft. of water
H_R	Total heat of rejection in cooling tower, Btu/hr/ton
HDD	Heating degree days in a year, degree-days
HDD_{Avg}	Average long-term HDD in a year, degree-days
HDD_i	Heating degree days in month (i)
hp	Horsepower
HRS	Annual hours of operation
HRS_M	Annual hrs of motor operation
HRS_P	Annual hrs of pump operation
IEER	Integrated energy efficiency ratio, Btu output/watt-hr input
IPLV	Integrated part load value, kW output/kW input or kW input/ton output or Btu output/W-hr input
$IPLV_{Baseline}$	Baseline integrated part load value, Btu output/W-hr input
$IPLV_{Baseline\text{-}Alt}$	Baseline integrated part load value, kW input/ton output
$IPLV_{New}$	Integrated part load value of new chiller, Btu output/W-hr input
$IPLV_{New\text{-}Alt}$	Integrated part load value of new chiller, kW input/ton output
k	1,000
lb_F	Pounds-force
LF	Load factor
LPD	Lighting power density, watts/ft^2
M	Maintenance factor
MM	Million
N	Number of different lighting fixtures with occupancy sensors
n	Lighting fixture number (or number of years in financial analysis)
NOI	Net operating income
NPV	Net present value
OSF_n	Number of fixtures of type n (with same watts/fixture) that have occupancy sensors
P	Pump power at operating speed, kW
$P_{Baseline}$	Power Input, kW, at existing motor efficiency
P_{New}	Power Input, kW, for new high efficiency motor
PV	Present value of cash in the future
Q_A	Energy in air for specific air ΔT, Btu/hr
$Q_{BinHeating}$	Heat savings in a bin to make up for heat loss through the roof with an insulation upgrade, Btu/yr
$Q_{BinCooling}$	Cooling savings in a bin due to the reduced cooling loss through the roof with an insulation upgrade, kWh/yr

Q_S	Heat savings by increasing roof insulation, Btu/hr
Q_W	Energy in water for a specific water ΔT, Btu/hr
R	R_{Value} of insulation, hr-ft^2 °F/Btu
$R_{Baseline}$	R_{Value} of existing roof insulation, hr-ft^2 °F/Btu
R_{New}	R_{Value} of new roof insulation, hr-ft^2 °F/Btu
rpm	Operating speed, rev./min
S	Project savings (first year), $/year
SF	Square feet, ft^2
SG	Specific gravity
SIR	Savings to investment ratio
SPB	Simple payback period, years
T	Temperature, °F
T_{Design}	Design temperature, °F
T_{WB}	Wet bulb temperature of air, °F
U	Overall heat transfer coefficient, Btu/ft^2-hr °F
UA	Overall heat transfer coefficient for ft^2, Btu/hr °F
UCR	Capital recovery factor
WF_n	Input watts for fixture type n
W_{hp}	Water horsepower
yr	Year

Index

Note: Page numbers in *italic* indicate a figure and page numbers in **bold** indicate a table on the corresponding page.

F

G

H

For Product Safety Concerns and Information please contact our EU representative GPSR@taylorandfrancis.com
Taylor & Francis Verlag GmbH, Kaufingerstraße 24, 80331 München, Germany

www.ingramcontent.com/pod-product-compliance
Ingram Content Group UK Ltd.
Pitfield, Milton Keynes, MK11 3LW, UK
UKHW021504240926
475702UK00035B/354

* 9 7 8 1 0 3 2 6 9 2 7 3 9 *